Thirdspace

For the students who have taught me

THIRDSPACE

*Journeys to Los Angeles and Other
Real-and-Imagined Places*

Edward W. Soja

Blackwell
Publishing

© 1996 by Edward W. Soja

BLACKWELL PUBLISHING
350 Main Street, Malden, MA 02148-5020, USA
9600 Garsington Road, Oxford OX4 2DQ, UK
550 Swanston Street, Carlton, Victoria 3053, Australia

First published 1996

20   2017

Library of Congress Cataloging-in-Publication Data

Soja, Edward W.
    Thirdspace / Edward W. Soja
        p.   cm.
    Includes bibliographical references and index.
    ISBN 978-1-55786-674-5 — ISBN 978-1-55786-675-2 (Pbk)
    1. Space perception. 2. Geographical perception. 3. Sociology,
Urban. 4. Postmodernism—Social aspects. 5. Los Angeles (Calif.)–
–Social conditions. I. Title.
    HM206.S633   1966
    304.2'3—dc20   95–26474
                                                         CIP

A catalogue record for this title is available from the British Library.

Set in 11 on 13 pt Palatino
by Wearset, Boldon, Tyne and Wear

The publisher's policy is to use permanent paper from mills that operate a sustainable
forestry policy, and which has been manufactured from pulp processed using acid-free and
elementary chlorine-free practices. Furthermore, the publisher ensures that the text paper
and cover board used have met acceptable environmental accreditation standards.

For further information on
Blackwell Publishing, visit our website:
www.blackwellpublishing.com

Contents

List of Illustrations

All the illustrations have been designed and produced at Ricos Studios (AntonisR@aol.com) by Antonis Ricos in collaboration with the author. Acknowledgements for the photographs used are presented here rather than in the text. Illustrations are identified in the text only by page number.

Cover artwork   The cover and spine were designed by Ricos–Soja. Background photos adapted from Illustrations 1, 3, and 10; icon of swirling complementary colors represents the Trialectics of Illustrations 2a and 2b.

Acknowledgements

Presenting formal acknowledgements has been made easier by a decision not to seek official permission for every epigraph and passage quoted. But I must express my deepest appreciation for the work of every author whose words I have borrowed to make *Thirdspace* as polyvocal as I know how. I am especially grateful for the poetry that appears: to Gloria Anzaldúa and Spinsters/Aunt Lute Press for the poem extracted from *Borderlands/La Frontera* (1987); to Anna Deveare Smith and Doubleday Anchor Books for the performance poem composed from the words of Homi Bhabha; to Guillermo Gómez-Peña and Graywolf Press for the excerpts from *Warriors for Gringostroika* (1993); and to bell hooks and South End Press for the poetic prose extracted from *Yearning* (1990).

Many of the chapters have drawn on my previously published work. Sections of chapters 3 and 4 originally appeared in Keith and Pile (eds), *Place and the Politics of Identity* (Routledge, 1993) in an essay I co-authored with Barbara Hooper; parts of chapters 5 and 7 are taken from "Heterotopologies" in *Strategies: A Journal of Theory, Culture, and Politics* (1990); the first half of chapter 6 draws heavily from "Postmodern Geographies and the Critique of Historicism" in Jones, Natter, and Schatzki (eds), *Postmodern Contentions* (Guildford Press, 1993); chapter 8 is a much expanded and, I think, improved version of "Inside Exopolis," which first appeared in Michael Sorkin (ed.), *Variations on a Theme Park* (Hill and Wang, 1992); and chapter 9 builds on a 1991 publication of the Center for Metropolitan Research in Amsterdam. My thanks to the publishers, editors, and co-authors.

My greatest appreciation is reserved for the graduate students I have worked with over the past 23 years in what, seen in retrospect because it no longer exists in its erstwhile form, was one of the most intellectually stimulating and innovative interdisciplinary communities of spatial scholars ever to exist at an American university, the Graduate School of Architecture and Urban Planning at UCLA. A professional schools restructuring designed, like corporate restructurings, for leaner and meaner administration (and reduced payrolls) has moved Architecture to the School of the Arts, and Urban Planning to a new School of Public Policy and Social Research. As I write, the new structure is awkwardly sorting itself out, but fortunately the flow of outstanding students continues, adding to the long list of students who have taught me and to whom I dedicate this book. Every university professor not just biding time dreams of that reciprocal moment when teaching turns to being taught. I have been fortunate to have more than my share of such moments, and they have invigorated *Thirdspace* more than I can say.

The list begins with Costis Hadjimichalis and Dina Vaiou, who expanded my knowledge of the writings of Henri Lefebvre in the mid-1970s; and with Margaret FitzSimmons, who has moved with nurturing agility from student to colleague to lifelong friend and intellectual confidant. Dina and Costis also introduced me to Antonis Ricos, then a doctoral student in Critical Studies in the UCLA department of theater, film, and television, and today one of Los Angeles's leading photographers and computer artists. Antonis's creative eye and our warm friendship are responsible for all the visual material that appears in this book. On almost as long an interactive path are Don Parson, Fassil Demissie, Larry Barth, and Marco Cenzatti. I continue to learn from all of them, and from the remarkable cluster of students who received their PhD over the past five years: Susan Ruddick, Beverley Pitman, Clyde Woods, Jane Pollard, Moira Kenney, and Sylvia Sensiper. All have helped me to think differently about space and place, the city and the region.

And now there is the current flow of students, so many of whom have helped me significantly in writing this book. As my research assistant, Olivier Kramsch (at present in Barcelona doing field work for his dissertation) strengthened my understanding of Homi Bhabha, Michel de Certeau, and current debates on regionalism. Mary Pat Brady (currently completing her dissertation in the English Department at UCLA) made me aware of the creative spatial turn taken in the recent Chicana and Chicano literature and effectively enhanced my appreciation for the work of radical women of color. Dora Epstein keeps challenging me with her sensitive writings on the lesbian experience of space and place, and her prose, photos, and insights on the new federal complex in downtown Los Angeles significantly improve chapter 7. My thanks also to Gail Sansbury for her useful essays and comments.

And then, there is Barbara Hooper. An older student and experienced writer who returned to the university after almost two decades of "other" work, Barbara almost instantaneously began by teaching me far more than I was able to teach her. Barbara is the only person who has read and creatively critiqued a complete draft of *Thirdspace*, after responding to earlier drafts as well. She has softened what she calls my rottweiler growls, steadfastly insisted on keeping *Thirdspace* open to constant (and unruly) reinterpretation, fed me informative quotes and ideas, and helped more than any other to make this book better than I could have made it alone.

Finally, there are those that deserve thanks for helping me complete what was beginning to feel like a neverending project. A grant from the Getty Foundation for collaborative research with Janet Abu-Lughod on "The Arts of Citybuilding: New York, Chicago, and Los Angeles" contributed significantly in pushing me to finish *Thirdspace* so that I could move on to new adventures. John Davey at Blackwell was always there, from beginning to end, with his gentle support. And also always there but far from gentle in her pressures to get me out of my smoky garden office-cell, finish the damned book, and do something for a change, was Maureen. She always gets the last word.

Introduction/Itinerary/Overture

My objective in *Thirdspace* can be simply stated. It is to encourage you to think differently about the meanings and significance of space and those related concepts that compose and comprise the inherent *spatiality of human life*: place, location, locality, landscape, environment, home, city, region, territory, and geography. In encouraging you to think differently, I am not suggesting that you discard your old and familiar ways of thinking about space and spatiality, but rather that you question them in new ways that are aimed at opening up and expanding the scope and critical sensibility of your already established spatial or geographical imaginations.

Mobilizing this objective is a belief that the spatial dimension of our lives has never been of greater practical and political relevance than it is today. Whether we are attempting to deal with the increasing intervention of electronic media in our daily routines; seeking ways to act politically to deal with the growing problems of poverty, racism, sexual discrimination, and environmental degradation; or trying to understand the multiplying geopolitical conflicts around the globe, we are becoming increasingly aware that we are, and always have been, intrinsically spatial beings, active participants in the social construction of our embracing spatialities. Perhaps more than ever before, a strategic awareness of this collectively created spatiality and its social consequences has become a vital part of making both theoretical and practical sense of our contemporary lifeworlds at all scales, from the most intimate to the most global.

At the same time as this relevance is rising, however, there is reason to be concerned that the practical and theoretical understanding of space and spatiality is being muddled and misconstrued either by the baggage of tradition, by older definitions that no longer fit the changing contexts of the contemporary moment, or by faddish buzz-words that substitute apparently current relevance for deeper under-standing. It thus becomes more urgent than ever to keep our contemporary consciousness of spatiality – our critical geographical imagination – creatively open to redefinition and expansion in new directions; and to resist any attempt to narrow or confine its scope.

In keeping with these objectives and premises, I use the concept of Thirdspace most broadly to highlight what I consider to be the most interesting new ways of thinking about space and social spatiality, and go about in great detail, but also with some attendant caution, to explain why I have chosen to do so. In its broadest sense, Thirdspace is a purposefully tentative and flexible term that attempts to capture what is actually a constantly shifting and changing milieu of ideas, events, appearances, and meanings. If you would like to invent a dif-ferent term to capture what I am trying to convey, go ahead and do so. I only ask that the radical challenge to think differently, to expand your geographical imagination beyond its current limits, is retained and not recast to pour old wine into new barrels, no matter how tasty the vintage has been in the past.

To help ensure that the magnitude of the challenge being pre-sented is understood, I add a much bolder assertion. In what I am convinced will eventually be considered one of the most important intellectual and political developments of the late 20th century, a growing community of scholars and citizens has, for perhaps the first time, begun to think about the *spatiality* of human life in much the same way that we have persistently approached life's intrinsic and richly revealing historical and social qualities: its *historicality* and *sociality*. For much too long, spatiality has been relatively peripheral to what are now called the human sciences, especially among those who approach knowledge formation from a more critical, politically committed perspective. Whether in writing the biography of a par-ticular individual or interpreting a momentous event or simply deal-ing with our everyday lives, the closely associated historical (or temporal) and social (or sociological) imaginations have always been at the forefront of making practical and informative sense of the sub-ject at hand. Every life, every event, every activity we engage in is usually unquestionably assumed to have a pertinent and revealing historical and social dimension. Although there are significant exceptions, few would deny that understanding the world is, in the most general sense, a simultaneously historical and social project.

Without reducing the significance of these historical and social qualities or dimming the creative and critical imaginations that have developed around their practical and theoretical understanding, a third existential dimension is now provocatively infusing the traditional coupling of historicality–sociality with new modes of thinking and interpretation. As we approach the *fin de siècle*, there is a growing awareness of the simultaneity and interwoven complexity of the social, the historical, and the spatial, their inseparability and interdependence. And this three-sided sensibility of spatiality–historicality–sociality is not only bringing about a profound change in the ways we think about space, it is also beginning to lead to major revisions in how we study history and society. The challenge being raised in *Thirdspace* is therefore transdisciplinary in scope. It cuts across all perspectives and modes of thought, and is not confined solely to geographers, architects, urbanists and others for whom spatial thinking is a primary professional preoccupation.

There is still another attachment I wish to make before beginning to explore the real and imagined worlds of Thirdspace. As I shall argue repeatedly, the most interesting and insightful new ways of thinking about space and spatiality, and hence the most significant expansions of the spatial or geographical imagination, have been coming from what can be described as a radical postmodernist perspective. Given the swirling confusion that fills the current literature both for and against postmodernism, it may help to explain briefly what I mean by radical or critical postmodernism and why, contrary to so many current writings, it does not represent a complete contradiction in terms, a fanciful oxymoron.

To clarify the meaning of radical postmodernism requires a much more substantial look at the flurry of scholarly debates that has come to surround the postmodern critiques of modernism. Later chapters will return to these debates in greater detail, but for the moment let us begin by focusing attention on what I consider to be the most important of these postmodern critiques. I am referring to what postmodern scholars have described as the deconstruction and strategic reconstitution of conventional modernist epistemologies – in other words, the radical restructuring of long-established modes of knowledge formation, of how we assure that the knowledge we obtain of the world can be confidently presumed to be accurate and useful. This epistemological critique has ranged from a formidable attack on the foundations of modern science; to a deep questioning of the established disciplinary canons of the separate social sciences, arts, and humanities; and further, to a reformulation of the basic knowledge structure of scientific socialism or Marxism as well as other fields of radical theory and practice, such as feminism and the struggles against racism and colonialism.

In every one of these targeted arenas, the postmodern epistemological critique of modernism and its tendencies to become locked into "master narratives" and "totalizing discourses" that limit the scope of knowledge formation, has created deep divisions. For some, the power of the critique has been so profound that modernism is abandoned entirely and new, explicitly postmodern ways of thinking take its place in making sense of the contemporary world. For others, the postmodern challenge is either ignored or creatively reconstituted to reaffirm more traditional modes of still avowedly modernist thought and practice. As I shall argue throughout *Thirdspace*, these are not the only choices available. Unfortunately, such categorically postmodernist and modernist responses have dominated and polarized the current literature, leaving little room for alternative views.

The opposing camps are increasingly clearly drawn. On one side are those self-proclaimed postmodernists who interpret the epistemological critique as a license to destroy all vestiges of modernism. They become, as I once called them, the smiling morticians who celebrate the death or, more figuratively, the "end of" practically everything associated with the modern movements of the twentieth century: of the subject and the author, of communism and liberalism, of ideology and history, of the entire enlightenment project of progressive social change. In essence, postmodernism is reduced here to *anti-modernism*, to a strategy of annihilation that derives from modernism's demonstrated epistemological weaknesses and its presumed failures to deal with the pressing problems of the contemporary world. Intentionally or not, this focused form of inflexible and unselective anti-modernism has entered contemporary politics all over the world primarily to support and sustain both premodern fundamentalisms and reactionary and hyperconservative forms of postmodern political practice that today threaten to destroy the most progressive accomplishments of the 20th century.

At the other extreme is a growing cadre of adamant *anti-postmodernists*. Usually marching under the banner of preserving the progressive projects of liberal and radical modernism, these critics see in postmodernism and postmodern politics only a polar opposition to their progressive intentions. Just as reductionist as the anti-modernists, they deflect the power of the epistemological critique of modernism by associating it exclusively with nihilism, with neoconservative empowerment, or with a vacuous anything-goes "new age" philosophy. In this simplistic caricaturing, there is no possibility for a radical postmodernism to exist unless it is self-deluding, really modernism in oxymoronic disguise.

Not only have the debates on modernism and postmodernism polarized around these reductionist stances, a kind of ritual purification has been practiced to rule out any alternative possibilities. If you are a postmodernist, it is proclaimed, then you cannot be a Marxist or be committed to a continuation of the progressive projects of the

European Enlightenment. And vice versa: to be committed to radical social change one must resist the enchantments of postmodern thinking. Simply practicing the methods of deconstruction or expressing sympathy with the writings of Derrida, Lyotard, Foucault, or Baudrillard brands you as either unremittingly neoconservative or deviously apolitical. One particularly misguided purification game, engaged in even by those who appear to reject such simplistic dichotomization, involves searching for traces of modernism in the writings of postmodernists, as if these discoveries were a signal of duplicity, unforgivable inconsistency, or some sort of false consciousness. No mixture or combination is permitted. There is only an either/or choice, especially for those on the political left.

I urge you to begin reading *Thirdspace* with an open mind on these debates. At least temporarily, set aside the demands to make an either/or choice and contemplate instead the possibility of a both/and also logic, one that not only permits but encourages a creative combination of postmodernist and modernist perspectives, even when a specific form of postmodernism is being highlighted. Singling out a radical postmodern perspective for particular attention is not meant to establish its exclusive privilege in exploring and understanding Thirdspace. It is instead an efficient invitation to enter a space of extraordinary openness, a place of critical exchange where the geographical imagination can be expanded to encompass a multiplicity of perspectives that have heretofore been considered by the epistemological referees to be incompatible, uncombinable. It is a space where issues of race, class, and gender can be addressed simultaneously without privileging one over the other; where one can be Marxist and post-Marxist, materialist and idealist, structuralist and humanist, disciplined and transdisciplinary at the same time.

Thirdspace itself, as you will soon discover, is rooted in just such a recombinatorial and radically open perspective. In what I will call a critical strategy of "thirding-as-Othering," I try to open up our spatial imaginaries to ways of thinking and acting politically that respond to all binarisms, to any attempt to confine thought and political action to only two alternatives, by interjecting an-Other set of choices. In this critical thirding, the original binary choice is not dismissed entirely but is subjected to a creative process of *restructuring* that draws selectively and strategically from the two opposing categories to open new alternatives. Two of these critical thirdings have already been introduced. The first revolves around the interjection of a critical spatial imagination into the interpretive dualism that has for the past two centuries confined how we make practical and theoretical sense of the world primarily to the historical and sociological imaginations. The second has shaped the preceding discussion of modernism and postmodernism, creating the possibility for a more open and combinatorial perspective. Still another is implied in this book's title and subtitle. *Thirdspace* too can be described as a creative recombination and extension, one that builds on a Firstspace perspective that is focused on the "real" material world and a Secondspace perspective that interprets this reality through "imagined" representations of spatiality. With this brief and, I hope, helpful and inviting introduction, we are ready to begin our journeys to a multiplicity of *real-and-imagined* places.

Discovering Thirdspace

The six chapters that comprise Part I, "Discovering Thirdspace" are aimed at showing how and why spatiality and the inquisitive spatial imagination have recently entered, as a vital third mode of practical and theoretical understanding, what has heretofore been seen as an essentially two-sided socio-historical project. These chapters collectively establish the points of departure and an itinerary for the journeys Inside and Outside Los Angeles that comprise Part II and will be continued in a companion volume to *Thirdspace* that will be published by Blackwell in early 1997 under the title *Postmetropolis*. As these chapters presume some prior knowledge of the debates and academic discourse that have arisen over the interpretation and theorization of spatiality in recent years, I give more space and time here to assist in comprehending the often complex and perhaps, for some, abstruse arguments they contain.

1   The Extraordinary Voyages of Henri Lefebvre

The intellectual journeys of discovery begin with an appropriately allegorical tour of the life of Henri Lefebvre, a French "metaphiloso-pher" who has been more influential than any other scholar in opening up and exploring the limitless dimensions of our social spatiality; and also in arguing forcefully for linking historicality, sociality, and spatiality in a strategically balanced and transdisciplinary "triple dialectic." I use his term "transdisciplinary" to mean not being the privileged turf of such specialized fields as History, Sociology, and Geography, but spanning all interpretive perspectives. As Lefebvre insistently argued, historicality, sociality, and spatiality are too important to be left only to such narrowed specializations.

There are many such transdisciplinary perspectives, or as Lefebvre described them, "ways to thread through the complexities of the modern world." One might think of literary criticism, psycho-analysis, linguistics, discourse analysis, cultural studies, and critical philosophy, as well as comprehensive and critical interpretations of the historical development and social composition of this modern world. What distinguishes Lefebvre from so many others is that he "chose space" as his primary interpretive thread and, beginning in the 1960s, insistently wove space into all his major writings. How the meaning of what I have described as a "triple dialectic" – Lefebvre called it *une dialectique de triplicité* – relates to his expanding geographical imagination will become clearer as we move on.

I approach Lefebvre's biography in the first chapter as an introductory voyage of discovery, selectively excavating from his adventurous life its most revealing moments of spatial insight. The chapter can thus be seen in part as an attempt to spatialize what we normally think of as biography, to make life-stories as intrinsically and revealingly spatial as they are temporal and social. It is also a more specific historical geography of Lefebvre's triple consciousness of the complex linkages between space, time, and social being, or, as I suspect he would prefer them to be called, the production of space, the making of history, and the composition of social relations or society. En route through his 90 years, this triple consciousness took many different twists and turns, from his early fascination with surrealism and the various mystifications of working-class consciousness; through his Marxist explorations of the spatiality and sociology of everyday life and the equally mystifying "urban condition;" to his later work on the social production of space and what he called "rhythmanalysis." At all times he remained a restless intellectual nomad, a person from the periphery who was able to survive and thrive in the center as well, as a refined barbarian, a Parisian peasant from the Occitanian forelands of the Pyrenees.

In his personal (re)conceptualization of the relation between centers and peripheries comes one of his most important ideas, a deep critique not just of this oppositional dichotomy of power but of all forms of categorical or binary logic. As he always insisted, two terms (and the oppositions and antinomies built around them) are never enough. *Il y a toujours l'Autre*, there is always an-Other term, with *Autre*/Other capitalized to emphasize its critical importance. When faced with a choice confined to the either/or, Lefebvre creatively resisted by choosing instead an-Other alternative, marked by the openness of the both/and also ..., with the "also" reverberating back to disrupt the categorical closures implicit in the either/or logic.

Emanating from his insistent disordering or, to use a more contemporary term, deconstruction of binary logic in thinking about space and other complexities of the modern world are his various recombinations of the center–periphery relation in such concepts as the critique of everyday life, the reproduction of the social relations of production, the bureaucratic society of controlled consumption (the forerunner of what we today call consumer society), the struggle over the right to the city and the right to be different, the urbanization of consciousness and the necessity for an urban revolution, and a more general emphasis on the dynamics of geographically uneven development from the local to the global scales. These conceptualizations and others springing from Lefebvre's creative spatial consciousness infiltrate every chapter of *Thirdspace*.

2   The Trialectics of Spatiality

In chapter 2, I re-engage with Lefebvre's journeys through an alternative reading of *The Production of Space*, arguably the most important book ever written about the social and historical significance of human spatiality and the particular powers of the spatial imagination. *The Production of Space* is a bewildering book, filled with unruly textual practices, bold assertions that seem to get tossed aside as the arguments develop, and perplexing inconsistencies and apparent self-contradictions. Yet its meandering, idiosyncratic, and wholesomely anarchic style and structure are in themselves a creative expression of Lefebvre's expansive spatial imagination. Years ago, when I first read the original French version (*La Production de l'espace*, 1974), I found myself having great difficulty navigating through the chapters that followed the extraordinarily exciting and relatively clearly written introduction, translated in English as "Plan of the Present Work." Lefebvre seemed not to be following his own plan, flying off in lateral directions and posing very different arguments from those presented earlier. There was so much there in the first chapter, however, that I set aside my frustrations with the rest of the text as a product of my own linguistic deficiencies and Lefebvre's complicated writing style.

But I had the same reaction when I read the 1991 English translation. Nearly all that seemed solid and convincing in the "Plan" frustratingly melted into air in the dense and eclectic prose of the subsequent chapters. I dutifully recommended this apparently badly-planned book to my planning students, but told them, quite uncomfortably, to read seriously only the introductory chapter and to browse the rest with a sense of *caveat lector*. It was only when I began writing chapter 2 of *Thirdspace*, after going over dozens of Lefebvre's other writings to prepare chapter 1, that I realized he may not have intended *The Production of Space* to be read as a conventional academic text, with arguments developed in a neat linear sequence from beginning to middle to end. Taking a clue from Jorge Luis Borges, who in his short story, "The Aleph," expressed his despair in writing about the simultaneities of space in such a linear fashion, and from Lefebvre's frequently mentioned love of music, I began to think that perhaps Lefebvre was presenting *The Production of Space* as a musical composition, with a multiplicity of instruments and voices playing together at the same time. More specifically, I found that the text could be read as a polyphonic fugue that assertively introduced its keynote themes early on and then changed them intentionally in contrapuntal variations that took radically different forms and harmonies.

Composing the text as a fugue served multiple purposes. First of all, it was a way of spatializing the text, of breaking out of the conventional temporal flow of introduction–development–conclusion to explore new "rhythms" of argument and (con)textual representation. Similarly, it spatialized the equally temporal, sequential logic of dialectical thinking, always a vital part of Lefebvre's work. Thesis, antithesis, and synthesis are thus made to appear simultaneously, together in every chapter in both contrapuntal harmonies as well as disruptive dissonances. Just as importantly, the fugue formed some protection for Lefebvre against the canonization of his ideas into rigidly authoritative protocols. Although he was frequently vicious and dogmatic in his attacks on the "schools" that developed around the work of other leading scholars, especially his fellow Marxists, Lefebvre always saw his own intellectual project as a series of heuristic "approximations," never as permanent dogma to be defended against all non-believers.

In the first of our all too brief meetings, I almost convinced him to agree that *The Production of Space* was his most pathbreaking work. But he was clearly uneasy. It was to him just another approximation, incomplete, merely a re-elaboration of his earlier approximations as well as those of Marx, Hegel, Nietzsche and others, another temporary stop en route to new discoveries, such as the "rhythmanalysis" he was working on up to his death in 1991. To the end, Lefebvre was a restless, nomadic, unruly thinker, settling down for a while to explore a new terrain, building on his earlier adventures, and then picking up what was most worth keeping and moving on. For him there are no "conclusions" that are not also "openings," as he expressed in the title of the last chapter of *The Production of Space*. Following Lefebvre, I have tried to compose every one of the chapters of *Thirdspace* as a new approximation, a different way of looking at the same subject, a sequence of neverending variations on recurrent spatial themes. Whether or not I have been successful in this effort, I hope the reader will at least keep this intention in mind while plowing through the text.

Given these intentions, what I have done with *The Production of Space* in chapter 2 would probably have discomforted Lefebvre. I have extracted from the introductory "Plan" a central argument and attached to it a specific critical methodology. The central argument I refer to has already been mentioned: the ontological, epistemological, and theoretical rebalancing of spatiality, historicality, and sociality as all-embracing dimensions of human life. This "meta-philosophy," to use Lefebvre's preferred description of his work, builds upon a method that I present as a critical "thirding-as-Othering," with Other capitalized to retain the meaning of Lefebvre's insistent, anti-reductionist phrase *il y a toujours l'Autre*. And for the result of this critical thirding, I have used another term, "trialectics," to describe not just a triple dialectic but also a mode of dialectical reasoning that is more inherently spatial than the conventional temporally-defined dialectics of Hegel or Marx. I then use this method to re-describe and help clarify what I think Lefebvre was writing about in the thematic "Plan" of *The Production of Space* fugue: a trialectics of spatiality, of spatial thinking, of the spatial imagination that echoes from Lefebvre's interweaving incantation of three different kinds of spaces: the *perceived* space of materialized Spatial Practice; the *conceived* space he defined as Representations of Space; and the *lived* Spaces of Representation (translated into English as "Representational Spaces").

It is upon these formulations that I define Thirdspace as an-Other way of understanding and acting to change the spatiality of human life, a distinct mode of critical spatial awareness that is appropriate to the new scope and significance being brought about in the re-balanced trialectics of spatiality–historicality–sociality. This begins a longer story, or journey, that weaves its way through all the chapters. Briefly told, the spatial story opens with the recognition that the mainstream spatial or geographical imagination has, for at least the past century, revolved primarily around a dual mode of thinking about space; one, which I have described as a Firstspace perspective and epistemology, fixed mainly on the concrete materiality of spatial forms, on things that can be empirically mapped; and the second, as Secondspace, conceived in ideas about space, in thoughtful re-pre-sentations of human spatiality in mental or cognitive forms. These coincide more or less with Lefebvre's perceived and conceived spaces, with the first often thought of as "real" and the second as "imagined." What Lefebvre described specifically as lived space was typically seen as a simple combination or mixture of the "real" and the "imagined" in varying doses, although many in the so-called spatial disciplines (Geography, Architecture, Urban and Regional Studies, and City Planning, with capital letters used to signify the formally constituted discipline) as well as scholars in other disciplined fields tended to concentrate almost entirely on only one of these modes of thinking, that is on either Firstspace or Secondspace perspectives.

In the late 1960s, in the midst of an urban or, looking back, a more generally spatial crisis spreading all over the world, an-Other form of spatial awareness began to emerge. I have chosen to call this new awareness Thirdspace and to initiate its evolving definition by describing it as a product of a "thirding" of the spatial imagination, the creation of another mode of thinking about space that draws upon the material and mental spaces of the traditional dualism but extends well beyond them in scope, substance, and meaning. Simultaneously real and imagined and more (both and also . . .), the exploration of Thirdspace can be described and inscribed in journeys to "real-and-imagined" (or perhaps "realandimagined"?) places. Hence the subtitle of this book.

For reasons which I will not attempt to explain here, this new way of thinking about space became most clearly formulated in Paris, in particular in the writings of Lefebvre and his colleagues, but also, much less visibly, in the work of Michel Foucault. For almost 20 years, however, these "Other spaces" (*des espaces autres*, Foucault called them) remained unexplored and often substantially misunderstood by even the greatest admirers of Lefebvre and Foucault. Outside the spatial disciplines, the new importance being given to space and spatiality, when it was noticed at all, was seen primarily as another data set, interpretive language, or collection of modish tropes to be added to the serious business-as-usual of historical and social analysis. Within the spatial disciplines, when noticed, the work of Lefebvre and Foucault was taken as a reconfirming benediction on the long-established scope of conventional spatial or geographical imaginations. What was almost entirely missed by nearly all was the radical critique and disruptive challenge detonated by Lefebvre and Foucault to restructure the most familiar ways of thinking about space across all disciplines and disciplinarities. Rather than accepting the critique and responding to the challenges to think differently about space, the work of Lefebvre and Foucault was obliviously sucked back into unchanged disciplinary cocoons.

In Geography, the field I know best, there continues to be a wholly disciplined absorption or, alternatively, complete rejection of what I describe as the Thirdspace perspectives of Lefebvre and what Foucault called "heterotopology." Given my disciplinary background, I refer relatively infrequently to the work of geographers on the pages of *Thirdspace*. When asked to speak to audiences of geographers about my recent work I tend to emphasize its "bad news" for Geography, especially regarding the formidable rigidity of the Firstspace–Secondspace dualism into which geographers have been so tightly socialized. I do so to compensate for the tendency to use the rising importance being given to the spatial imagination either to reaffirm proudly (and uncritically) the traditional disciplinary project, with Geography crowned as the master discipline of space; or to reject the new approaches completely as not Geography at all, thereby preserving the canonical traditions of the past. Similar reactions occur in Architecture, Urban Planning, and Urban Sociology, alternatively co-opting or rejecting the new modes of thinking I am associating with a Thirdspace perspective. The most significant exceptions seem to arise only among those in the spatial disciplines who have been engaging seriously with the recent literature in the broad new field of critical cultural studies. This moves the spatial story or journey I am recounting to the next chapters.

3 Exploring the Spaces that Difference Makes: Notes on the Margin

Chapter 3 re-opens the voyages of discovery through an excavation of the more contemporary writings of bell hooks, an African-American cultural critic who has been advancing – and reconceptualizing – the frontiers of Thirdspace through creative inquiries into the connected spatialities of race, class, and gender. Although influenced by Lefebvre and Foucault, hooks has not been a spatial theoretician but has instead put into personal and political practice a vivid Thirdspatial imagination, especially in her American Book Award winning *Yearning: Race, Gender, and Cultural Politics* (1990). In essays on "Postmodern Blackness," "Homeplace: A Site of Resistance," and most powerfully "Choosing the Margin as a Space of Radical Openness," hooks recomposes our lived spaces of representation as potentially nurturing places of resistance, real-and-imagined, material-and-metaphorical meeting grounds for struggles over all forms of oppression, wherever they are found. I use hooks (more on this "use" in a moment) to exemplify the contemporary leadership of cultural studies scholars, especially radical women of color, in the creative exploration of Thirdspace and to implant their spatial awareness in the strategic margins of an explicitly but critically postmodern cultural politics, filled with an expanding roster of struggles based not just on race, gender, and class but also on sexuality, age, nation, region, nature, empire, and colony.

In the particular ways she chooses marginality as a space of radical openness, hooks builds upon but also reconstitutes and recontextualizes the Thirdspace insights of Lefebvre and Foucault. Chapter 3 thus serves to initiate another journey of exploration, filling in many of the voids and silences contained in the first two chapters, and I might also add in *Postmodern Geographies: The Reassertion of Space in Critical Social Theory* (1989), my earlier attempt to reconceptualize the geographical imagination. I have chosen to foreground bell hooks in beginning this new exploration for several reasons. First of all, I have found no one better to illustrate the radical openness of Thirdspace, its strategic flexibility in dealing with multiple forms of oppression and inequality, and its direct relevance to contemporary politics, particularly with respect to the journeys that will be taken to Los Angeles and other real-and-imagined places. My own real-and-imagined homeplace for my first 20 years in the Bronx and my academic specialization in African studies for the next 20 years of my life have added other compelling attractions and connections to her work. And just as important, I find hooks's radical openness and chosen marginality a powerful antidote to the narrowed and aggressive centrisms and essentialisms that have deflected most modernist movements based on gender, race, and class into hostile and competitive binary battlegrounds of woman versus man, black versus white, labor versus capital. In a discussion of "the difference postmodernity makes," I elaborate on my own definitions of postmodernity and postmodernism and the comparison between modernist and postmodernist cultural politics.

But there are still a few problems worth mentioning in my "choosing" bell hooks and in her "choosing marginality." Almost impossible to set aside entirely are reactions that here is another example of a powerful, presumably established and affluent, White Western Man liberally attaching himself, in the margins no less, to a radical woman of color, who in her turn is a well-established and presumably affluent scholar. For those who feel compelled to respond in this way, I can only say please continue to read. Perhaps you will find more in my explorations of Thirdspace and hooks's marginality as a space of radical openness than immediately meets your skeptical eye.

4   *Increasing the Openness of Thirdspace*

After exploring the spaces that difference makes with bell hooks in chapter 3 and as a means of preventing the formulations of the first three chapters from solidifying into rigid dogma, chapter 4 charts out additional pathways for increasing the openness of Thirdspace and redefining its meanings. With the uncanny assistance of Barbara Hooper, who has tried with some success to control my impulses to tweak a few of the feminist geographers who seemed to dismiss my admittedly gender-biased *Postmodern Geographies* as masculinist posing *tout court*, I explore and try to learn from the rich spatial feminist literature. While appreciating the pioneering efforts of earlier modernist spatial feminists in developing a rigorous critique of urbanism and the gendering of cityspace (work that I should have recognized more centrally in my 1989 book), and learning a great deal from the most recent work of feminist geographers such as Gillian Rose, I focus my primary attention on the extraordinary discourse that has been developing among those spatially attuned feminists who feel most comfortable, like bell hooks, with being described as radical or critical postmodernists.

What distinguishes this literature for me and what makes it an unusually enlightening terrain for developing new ways to think about Thirdspace has been the active engagement of postmodern spatial feminist writers, poets, artists, film critics, photographers, philosophers, and others in creatively rethinking and retheorizing spatiality not just in conjunction with gender and patriarchy but also in a more polycentric mix of other forms of oppression, exploitation, and subjection. For much too long, radical and progressive politics has been tightly channeled in social movements that have not only remained relatively unaware of the politics of space but have also found it difficult to forge significant and lasting alliances across channels and between movements. The work of postmodern spatial feminists has taken the lead in reconceptualizing the new cultural politics and in making a radical consciousness of the spatiality of human life a foundation and homeground for creating cross-cutting alliances and communities of resistance to contend with those "complexities" of the (post)modern world. I rush through this work much too quickly in the first half of chapter 4, leaving too much unexplored in what may indeed be the richest vein of innovative contemporary writing on what I have conceptualized as Thirdspace.

Moving on, however, I re-emphasize the significance of the postmodern spatial feminist critiques by elaborating further on the "border work" being done by postcolonial feminists such as Gloria Anzaldúa, María Lugones, and Gayatri Chakravorty Spivak. Here, in the overlapping borderlands of feminist and postcolonial cultural criticism is a particulary fertile meeting ground for initiating new pathways for exploring Thirdspace and also for the later journeys to a real-and-imagined Los Angeles. In all too brief sketches, I present first a cluster of imaginative spatial insights emanating from Chicana and Chicano artists and scholars, highlighting Anzaldúa, Lugones, and Guillermo Gómez-Peña. This is followed by the spatial "reworldings" of Spivak and Edward Said, using both to introduce a critical awareness of the space-blinkering effects of historicism that will be built upon in later chapters. Finally, a more lengthy excursion is taken through the writings of Homi Bhabha, who develops his own version of what he called "the Third Space," also a space of radical openness and "hybridity," his term for the spaces of resistance being opened at the margins of the new cultural politics.

5   *Heterotopologies: Foucault and the Geohistory of Otherness*

After this most contemporary re-opening, Thirdspace is reconceptualized again in chapter 5 through a glance backwards in time to review Foucault's original conception of "heterotopology" and "heterotopia." Drawing almost entirely on a short lecture on space he prepared for a group of architects, but never published, I try to show some of the similarities and differences between Foucault and Lefebvre in their coinciding and parallel discoveries of Thirdspace before and after the upheavals of May 1968 in Paris. Like Lefebvre, Foucault begins his explorations with a thirding, a sympathetic critique of the bicameralized spatial imagination that leads us to Other spaces quite similar, yet teasingly different, when compared to Lefebvre's lived spaces of representation. Foucault called these spaces "heterotopias" and described them as "the space in which we live, which draws us out of ourselves, in which the erosion of our lives, our time and our history occurs." Like Lefebvre, but with even greater tenacity and success, he also filled these heterogeneous sites with the trialectics of *space, knowledge, and power*, what Derek Gregory, whose *Geographical Imaginations* (1994) features prominently in this chapter and is returned to in many other places, called Foucault's distinctive "discursive triangle." I go over Foucault's uncharacteristically explicit and didactic discussion of the "principles of heterotopology" and attempt through these formulations to stretch Thirdspace in new and different directions.

But what is most interesting to me about Foucault's pathway to thinking differently about the powers of space and spatiality – and perhaps his most important contribution to the conceptualization of Thirdspace – was the explicitness and insight of his treatment en route of the relations between space and time, between the spatial and the historical imaginations. Amplifying on what I had written earlier in *Postmodern Geographies*, I re-present Foucault's critique of historicism as a vital part of understanding why thinking differently about space and spatiality has been so difficult for at least the past 150 years. In words that have been epigraphically echoed repeatedly in contemporary discussions of space, Foucault asked why is it that time has tended to be treated as "richness, fecundity, life, dialectic" while in contrast space has been typically seen as "the dead, the fixed, the undialectical, the immobile"? He answers his question by referring to a persistent overprivileging of the powers of the historical imagination and the traditions of critical historiography, and the degree to which this privileging of historicality has silenced or subsumed the potentially equivalent powers of critical spatial thought. Breaking down the controlling effects of this particular form of historicism becomes a key step in radically opening up the spatial imagination and in rebalancing the trialectics of historicality–sociality–spatiality; in other words, to exploring both theoretically and practically the lifeworlds, the heterotopias, of Thirdspace.

This critique of historicism is so crucial and so easily misunderstood that it is worthwhile clarifying immediately what it does and does not imply. It is not a rejection of the proven powers of the historical imagination, nor is it a substitution of a spatialism for historicism. It is instead a recognition that historicality and historiography are not enough, and a plea for opening up the historical and tightly interwoven sociological imaginations to a deeper appreciation for the spatiality of human life. This was perhaps not so problematic when the spatial imagination remained tightly encased in its bicameral compartments, either fixed on the forms and patternings of "real" material life or involved with mental and ideational worlds of abstract or "imagined" spaces. Such knowledges could easily be absorbed and subsumed by the free-flowing and infinitely expandable historical imagination, as "things" and "thoughts" that can be best understood by putting them into their historical context, into a narrative, a sequential story. But the times are changing. A new perspective is not only beginning to recompose the spatial or geographical imagination, it is entering disruptively, if still located on the margins, into the ways we think about historicality and sociality, demanding an equivalent empowering voice, no more but no less. We return then to the premises contained in the first paragraphs of this Introduction/Itinerary/Overture.

6   Re-Presenting the Spatial Critique of Historicism

The spatial critique of historicism underpins the middle chapters of *Thirdspace*. It begins in chapter 4 with Gayatri Spivak's reworldings and Edward Said's far-reaching critique of "Orientalism," rooting historicism in Western, Eurocentric, masculinist, modernist, and imperialist intellectual traditions, even when promulgated with revolutionary intentions. It flows through chapter 5 and the efforts of Foucault to construct a different "geohistory of otherness" despite his Eurocentrism; and becomes the central focus in chapter 6. Here I foreground the writings of Hayden White, one of the contemporary world's finest and most open-minded "metahistorians."

White approaches historicality with much the same enthusiasm and critical drive I try to generate with respect to spatiality; and he attempts to open the borders of the historical imagination against a perceived external threat to subsume its power in ways that seem to echo, in reverse, the spatial critique of historicism. In demonstrating "why loving maps is not enough," I take White's history of consciousness and his consciousness of history to task not for completely ignoring spatiality but for unconsciously subordinating and subsuming it under history's *sui generis* "burden." And in "Hayden White meets Henri Lefebvre," based on White's recent revealing review of *The Production of Space*, I show why the failure of this most thoughtful, open-minded, and imaginative theoretician of history to understand or even recognize the need to spatialize his story beyond mere Braudelian references may be the strongest reason for continuing to press the spatial critique of historicism throughout the contemporary human sciences.

Inside and Outside Los Angeles

Part II consists of a cluster of chapters that begin to apply contextually the ideas and theoretical arguments contained in Part I. Here our journeys take a more empirical and visual turn, first into the memorable inner sanctums of the Civic Center of the City of Los Angeles and then to the galactic outer spaces of Orange County, the virtually unbounded extremes of urban centeredness and decenteredness, Endopolis and Exopolis. In these antipodes of Greater Los Angeles core and periphery, city and suburb, seem to be imploding and exploding simultaneously, turning everyday urban life insideout and outside-in at the same time and in the same places. This widening gyre confounds conventional narrative interpretations of urban spatiality, for we are too aware of what is cutting across the storyline laterally, of how the local and the particular are becoming simultaneously global and generalizable. Increasingly unconventional modes of exploring Los Angeles are needed to make practical and theoretical sense of contemporary urban realities – and hyperrealities.

For extrinsic insight, a diverting third tour is taken to the Centrum and regional periphery of Amsterdam, spiraling us into another stimulating itinerary. It is at once a celebration of intimate locality and a re-routing of our journeys into worldwide contexts of urban development and global restructuring. This contemporary comparison of Amsterdam and Los Angeles – themselves the provocative antipodes of urbanism as a way of life – triggers an appreciation for the complex compositeness of difference and similarity, the intricate interweaving of the unique and the general, the local and the global. It also renews our understanding of the dynamics of uneven development over space and time, and especially of what Lefebvre was to describe as the simultaneous tendencies toward homogenization, differentiation, and hierarchical ordering that thread through the specific geographies of the modern world.

Each of these three chapters takes the form of a visual but also
re-envisioning tour of lived spaces. They present very different
perspectives, yet running through them there is much that is the
same.

7   Remembrances: A Heterotopology of the Citadel-LA

The first chapter of Part II continues the spatial critique of historicism through a Foucauldian stroll through an exhibition I helped
organize at UCLA commemorating the bicentennial of the French
Revolution and evoking memories of the synchronic resonances
between Paris and Los Angeles in the period 1789–1989. It both concludes the sequence of chapters dealing specifically with the relations between spatiality and historicality and opens new ways of
looking at and understanding the geohistory of Thirdspace.

Two heterotopological spaces resonate together in this real-and-
imagined journey. The first is a rectangular chunk of downtown Los
Angeles that today contains one of the most formidable agglomerations of the sites of governmental power and surveillance anywhere
in the US. The second space represents the first in a small gallery and
connected places in the UCLA building that housed the exhibition.
Everything is seen as a simultaneously historical–social–spatial
palimpsest, Thirdspace sites in which inextricably intertwined temporal, social, and spatial relations are being constantly reinscribed,
erased, and reinscribed again.

Here, in the Citadel-LA of Southern California, the "little city" of
gigantic powers, I use specific sites and sights as memory aids, geographical *madeleines* for a remembrance of things past and passed:
the historical presence of African-Americans in downtown Los
Angeles from the original siting of the city in 1781 to the violence
and unrest of 1992; the even more impressive presence of a Mexican
city etched into the history of El Pueblo de Nuestra Señora La Reina
de Los Angeles; a recollection of the lifeworlds of Bunker Hill, now a
truncated acropolis of culture waiting to be crowned again in a new
concert hall designed by Frank Gehry for the Walt Disney family; a
look back to the debates that have raged nearby around that citadel
of postmodern cultural studies, the Bonaventure Hotel, perhaps the
first preservation-worthy historical monument of postmodernity;
and finally, a glimpse at the "eye of power" to be seen in the prison-
adjacent *The New World*, a sculptured forum visibly and invisibly
celebrating anarchism and sexual freedom in the middle of an
uptight building complex that serves the US federal government. All
is present within walking distance: the past, the present, the future.

8   Inside Exopolis: Everyday Life in the Postmodern World

Ten scenes from the galaxy of sites that comprise the astral Exopolis, the starry-eyed "city-without-cityness" of Orange County define another tour. Jean Baudrillard rather than Foucault enables this tour to take off and to be defined as a journey to the "hyperreality" of everyday life. For Baudrillard and for many of the inhabitants of Orange County and other exopolises around the world, everyday living has become increasingly embroiled in the "precession of simulacra," in exact copies or representations of everyday reality that somehow substitute for the real itself. We no longer have to pay to enter these worlds of the "real fake" for they are already with us in the normal course of our daily lives, in our homes and workplaces, in how we choose to be informed and entertained, in how we are clothed and erotically aroused, in who and what we vote for, and what pathways we take to survive.

The Exopolis itself is a simulacrum: an exact copy of a city that has never existed. And it is being copied over and over again all over the place. At its best, the Exopolis is infinitely enchanting; at its worst it transforms our cities and our lives into spin-doctored "scamscapes," places where the real and the imagined, fact and fiction, become spectacularly confused, impossible to tell apart. In many ways, Orange County is the paradigmatic Exopolis, a simulated county-city-state of mind that is infused with and diffuses ever-encompassing ideological hyperrealities such as "small government is good government," "the taxpayers' revolt," "the magic of the market," "electronic democracy," "the end of history," "the triumph of capitalism." The tour begins with a quote from *The Wizard of Oz*: "Toto, I've got a feeling we're not in Kansas anymore." But then again, neither is Kansas.

This tour cannot be done on foot. It requires other forms of mobility to experience and comprehend. Despite many amusing diversions, however, the tour must be taken seriously for, whether we like it or not, Orange County offers glimpses into everyday life everywhere in the contemporary world. And brought up to date with the still fulminating financial bankruptcy of this arch-Republican bastion of real-and-imagined fiscal populism, we can begin to see the accumulating signs of an emerging global crisis of postmodernity. Beginning in 1989 with the staged disappearance of the Cold War, exploding most notably in Los Angeles with the so-called Rodney King riots of 1992, and now continuing indefinitely into our futures, a thirty-year cycle of restructuring and postmodernization, detonated by the crises of the 1960s, is beginning to generate its own internal explosions. Much more will be said of this restructuring-generated crisis of postmodernity in *Postmetropolis*, the forthcoming companion volume to *Thirdspace*. For now, our tour inside the Exopolis of Orange County provides only a preview.

9   The Stimulus of a Little Confusion

Part II concludes with a contemporary comparison of Los Angeles and Amsterdam and with, following Henry James, the stimulus of a little confusion. In chapter 9, which in some ways combines the microspatial tour of the citadel of downtown Los Angeles and the macrospatial excursion into the Exopolis of Orange County, I report from my own experiences living for a short period in the Centrum of Amsterdam, the largest and most creatively preserved 17th-century "old town" in Europe. I use my impressions of life on and off Spuistraat, my home street in Amsterdam, to both wrap up our journeys to real-and-imagined places and to open them up again.

Several of my friends who do not live in either Los Angeles or Amsterdam have told me that the original essay upon which this chapter is based was the best interpretive writing I have ever done. Why then do I continue to be uneasy about this compliment? I loved living in Amsterdam and discovering that, in its own secretive way, it was keeping alive the utopian dreams of democratic and humanely scaled urbanism better than any other place I know. Perhaps the excitement of this discovery and my directly personal reflections on it were the explanation.

Yet, I continue to wonder. Was the praise being generated because I was writing about a place other than Los Angeles (but Los Angeles was certainly there too)? Was it because I was speaking personally, without leaning too hard on one or another French philosopher (although Lefebvre was very much with me in this essay)? Or was it because, after working primarily at a macro-geographical scale for so many years, I was becoming more of a micro-geographer, a bit of the *flâneur*, that romantic poet-of-the-streets whose intimate urban insights had become so privileged in much of the current literature, especially over (or is it under) the view from on high that Michel de Certeau, among others, showed to be so limited and misleading? If this last point of view is the source of the essay's perceived quality, then I must add something more to its re-introduction here.

As noted in a postscript to chapter 9, my "contemporary comparison" of Amsterdam and Los Angeles was (and is) intended, in part, to add some stimulating confusion to a growing tendency in postmodern critical urban studies to overprivilege the local – the body, the streetscape, psychogeographies, erotic subjectivities, the microworlds of everyday life and intimate community – at the expense of understanding the city-as-a-whole, or what Lefebvre described as the "urban reality." Macrospatial perspectives are too often labeled taboo by those more attuned to *flânerie*, by critics who see in the view from on high only a dominating masculinist voyeurism, and by what might be called vulgar voluntarists romancing the unconstrainable powers and intentions of human agency against any form of structural analysis or determination. In chapter 9 I try to "third" this debate by exploring an-Other way to approach the micro–macro, local–global, agency–structure oppositions, drawing selectively from both spheres as best I can while pointing toward new directions that transcend any simple additive combination or strict either/or choice. Again, I may not be entirely successful in doing this, but it is useful to make my intentions clear.

Also clarifying my intentions is a second postscript that serves both to conclude this volume and to preview its forthcoming extensions in *Postmetropolis*. Postmetropolis is a composite term I use to describe (a) the new urbanization processes that have reshaped the metropolitan cityscape and everyday urban life over the past thirty years; and (b) the new modes of urban analysis that have been developing in the wake of this profound metropolitan restructuring and postmodernization. The original manuscript for *Thirdspace* contained a lengthy Part III that explored at much greater depth and detail this restructuring and postmodernization of the perceived, conceived, and lived spaces of the exemplary postmetropolis of Los Angeles. Here is a brief outline of its contents.

The first of the three chapters that comprised Part III, "Exploring the Postmetropolis," placed the "conurbation" of Los Angeles – using this almost forgotten term as both a noun and a verb – within a larger geohistorical context that relates the evolution of urban form to two other spatialized timelines: the crisis-filled periodization of capitalist development and the associated "succession of modernities" that have together helped to shape and reshape the perceived (First)spatialities and spatial practices of urbanism over the past three centuries. A panoramic satellite photograph of sprawling "Los Angeles – From Space" initiates the discussion of how this particular conurbation developed over space and time.

The second chapter shifted the interpretive focus to Secondspace, to the conceptual representations of urban spatiality, and more specifically to the "situated" urban imaginary that has consolidated over the past ten years into what some have called a Los Angeles "school" of urban analysis. At the core of this discussion are six discourses on the postmetropolis that represent Los Angeles as: (1) *Flexcity*, a productively postfordist industrial metropolis; (2) *Cosmopolis*, a globalized and "glocalized" world city; (3) *Exopolis*, a cityscape turned inside-out and outside-in through the radical restructuring of urban form; (4) *Polaricity*, a social mosaic of increasing inequalities and polarization; (5) *Carceral City*, a fortressed archipelago where police substitutes for *polis*; and (6) *Simcity*, a hyperreal scamscape of simulations and simulacra. Taking heed of Lefebvre's warning that these Secondspatial representations tend to become hegemonically powerful, I disrupt each of the above discourses with contrapuntal critiques and the "stimulus of a little confusion" to keep them open to continued rethinking and re-evaluation.

The third and erstwhile concluding chapter was infused primarily with a Thirdspace perspective, selectively encompassing the other two spheres of the spatial imagination to open up a distinctive new interpretive realm. Focused on a critical re-envisioning of a singular yet global event, the Los Angeles uprising of April–May 1992, it ends with the same words used to conclude the second postscript to *Thirdspace*: TO BE CONTINUED ...

Before moving on, a few last introductory words are in order. As the reader will soon no doubt realize, the radical openness and limitless scope of what is presented here as a Thirdspace perspective can provide daunting challenges to practical understanding and application. Exploring Thirdspace therefore requires a strategic and flexible way of thinking that is guided by a particular motivating project, a set of clear practical objectives and preferred pathways that will help to keep each individual journey on track while still allowing for lateral excursions to other spaces, times, and social situations. If Firstspace is explored primarily through its readable texts and contexts, and Secondspace through its prevailing representational discourses, then the exploration of Thirdspace must be additionally guided by some form of potentially emancipatory *praxis*, the translation of knowledge into action in a conscious – and consciously spatial – effort to improve the world in some significant way.

The praxis that guides our journeys to Los Angeles and other real-and-imagined places is organized around the search for practical solutions to the problems of race, class, gender, and other, often closely associated, forms of human inequality and oppression, especially those that are arising from, or being aggravated by, the dramatic changes that have become associated with global economic and political restructuring and the related postmodernization of urban life and society. Hovering in the background of all the chapters that will follow is an awareness of the possibility that the contemporary world has entered a new round of turbulent crises that stem not from the many different events that marked the end of the long postwar economic boom and initiated in the late 1960s a still continuing period of restructuring; but from what has seemed to so many to be the most successful examples of economic, political, and urban restructuring and postmodernization in recent times. In other words, our cities and all our lived spaces have been shifting from a period of crisis-generated restructuring to the onset of a new era of restructuring-generated crisis, a crisis deeply imbricated in the post-modernization of the contemporary world. As with all times of crisis, there are both new dangers and new opportunities unleashed by the multiplicity of confusing and often brutal events that have been shaking the world since 1989, from the fall of the Berlin Wall and the collapse of the Soviet Union, to the repercussions of what happened in Tienanmen Square and in the Los Angeles uprising of 1992, to current developments in Bosnia and in the Republican Revolution in the United States. The ultimate goal of *Thirdspace* and the continued journeys to the *Postmetropolis* is to contribute to the progressive reso-lution of at least some of the problems associated with this contem-porary restructuring-generated crisis.

PART I
DISCOVERING THIRDSPACE

1

The Extraordinary Voyages of Henri Lefebvre

The experience of the traveller consisting of a series of moves in space produces a phenomenon of a new order, one by which geography overtakes knowledge. "Our geography invades the planet. This is the second voyage, the reappropriation through knowledge. Geography is nothing else, its birth is there, at the moment at which knowledge becomes universal, in spatial terms and not by virtue of any right." Space makes an inventory of the adventures of knowledge, omitting nothing; knowledge traces a cartography of known lands, omitting nothing. The minute filling in of terrestrial reaches and the exhaustive account of cycles of knowledge are one and the same operation and permit *The Extraordinary Voyages* to establish the difficult relationship between the spatial or geographic model and the model of knowledge as encyclopedia. The (re)emergence of this language of paths, routes, movements, planes, and maps, this spatial language of the writing of the world (geo-graphy), marks the moment of passage toward a new epistemology.... To read and to journey are one and the same act.

(Michel Serres commenting on Jules Verne's *The Extraordinary Voyages* in *Hermes: Literature, Science, Philosophy*, 1982: xxi)

Allegory: Description of a subject under the guise of some other subject of aptly suggestive resemblance; an extended metaphor.

(*Shorter Oxford English Dictionary*)

It is perhaps to the disconcerting quality of postmodernism's continuous transformation of time into space, emptiness into saturation, body into electronics, and absence into presence that one can attribute the premature claim of postmodernism's demise and explain why, like the phoenix, it always emerges stronger from the ashes of such fatuous fires. It is certainly because of that metamorphic quality that I rely in this book on the narrative figure of allegory to describe certain trends in postmodern sensibility. For allegory ... represents a continuous movement toward an unattainable origin, a movement marked by the awareness of a loss that it attempts to compensate with a baroque saturation and the obsessive reiteration of fragmented memories ... like the myriad experiences of a long journey to a mythic homeland.

(Celeste Olalquiaga, *Megalopolis*, 1992: xx–xxi)

In a hospital at Pau, not far from his Occitanian birthplace of Hagetmau and even closer to his long-time summer home in the Pyrenean village of Navarrenx, Henri Lefebvre died some time during the night of June 28–9, 1991, after months of rumors that he had mysteriously "disappeared." Parisians and others were confused by his existential absence in the spring of 1991, for they had become accustomed to his always being there (*être-là*) at that time, ever watchful from his observation post above the Beaubourg (Pompidou Center) in an apartment on the rue Rambuteau that he shared with Catherine Regulier-Lefebvre.

The spring of 1991 was a time of great expectations. Lefebvre's 69th book, *Éléments de rythmanalyse*, written with significant contributions from his wife Catherine, was being prepared for publication,[1] developing an idea that he suggested would "put the finishing touches to the exposition of the production of space" Lefebvre had begun in the early 1970s (1991, 405). Also soon to appear were the long-awaited English translations of *La Production de l'espace* (1974) and *Critique de la vie quotidienne I: Introduction* (1947), reflecting, after decades of remarkable neglect, a rapidly growing interest in Lefebvre's work in the English-speaking world.

There was also a smaller re-awakening of interest in France. A new edition of Lefebvre's impassioned doctoral dissertation, *La Vallée de Campan – Étude de sociologie rurale*, originally published in 1963, was scheduled to appear, as was *Conversation avec Henri Lefebvre*, a small book prepared by Patricia Latour and Francis Combes from their conversations with Lefebvre during the week following New Year's Day at his home in Navarrenx. In the previous year, new editions had been issued of *La Somme et la reste* (1959), Lefebvre's personal discussions of his in-and-out relations with the

---

[1] Henri Lefebvre, *Éléments de rythmanalyse: Introduction à la connaissance des rythmes*, Paris: Éditions Syllepse, 1992. Also contained here is Henri Lefebvre and Catherine Regulier, "*Éléments de rythmanalyse des villes mediterranéennes*," reprinted from *Peuples mediterranéens* 37, 1986. For their first published essay on rhythmanalysis, see "Le Projet rythmanalytique," *Communications* 41, 1985. What was probably Lefebvre's 68th book, *Du Contrat de citoyenneté*, written in collaboration with *le groupe de Navarrenx*, was published in 1991 by Éditions Syllepse in the same collection as *Éléments de rythmanalyse*.

French Communist Party, and *Le Matérialisme dialectique* (1939), his most powerful critique of Stalinist reductionism. And on his desk, awaiting future publication, was a new manuscript for a book he had announced plans for several years earlier, enticingly titled *La Découverte et le secret*.

Seemingly taking his presence for granted, few were aware that his absence from Paris was sparked not just by illness and perhaps a premonitory urge to return home to Navarrenx, but also by another act of displacement and alienation, a subject that had always fascinated, activated, and enraged Lefebvre. He and Catherine had been forced out of their wonderfully ramshackle old dwelling place on the rue Rambuteau by the machinations of the Parisian property market, closing forever the windows through which Lefebvre so brilliantly viewed the rhythms of everyday life.[2] But then again, Lefebvre never stayed very long in the same place. He was always moving on, from moment to moment to moment, even as he entered his tenth decade.

The records state that Lefebvre had turned 90 a few weeks before his death, but he was always playful about his age, frustrating his chroniclers with teasing intimations of a different lived time. During his stay as visiting professor at the University of California in Santa Cruz in 1983–4, organized by Fredric Jameson and the History of Consciousness Program, it was rumored that his passport had a birthdate of 1896; while after one of his visits to the UCLA Graduate School of Architecture and Urban Planning, little paste-ons appeared in the card catalogue of the research library "correcting" his date of birth from 1901 to 1905. He preferred 1901, however, and he is best remembered as entering the world in the first year of the century that both defined his adventurous life and, in many different ways, came to be defined by it. Henri Lefebvre never quite made it to a full hundred years, but I know it would please him if I too played with time a little to suggest that it may have been the 20th century itself, not Henri Lefebvre, that quietly "disappeared" in 1991.[3] His presence/absence will no doubt be felt for many years to come – and will appear again and again in every chapter of this book.

I have chosen to open *Thirdspace* with some brief personal remem-

---

[2] See "Vu de la fenêtre," chapter 3 in *Éléments de rythmanalyse*, 1992. This revealing view has recently been translated as "Seen from the Window," in *Writings on Cities/ Henri Lefebvre*, selected, translated and introduced by Eleonore Kofman and Elizabeth Lebas, Oxford, UK, and Cambridge, MA: Blackwell, 1996. I try to take a similar "view from the window" of everyday life in the Centrum of Amsterdam in chapter 9.

[3] The most comprehensive recent biography of Lefebvre describes his life as "the adventure of the century." See Rémi Hess, *Henri Lefebvre et l'aventure du siècle*, Paris: Éditions A.M. Métailié, 1988. In an intriguing coincidence, Eric Hobsbawm, one of the world's leading radical historians, recently selected 1991 as the endpoint of the "short" 20th century in *The Age of Extremes: A History of the World, 1914–1991*, New York: Pantheon, 1994.

brances of Henri Lefebvre. This overture serves multiple purposes: it is an Acknowledgement, an appreciative Dedication, and an inspiring Travelguide for a long and continuing journey that seeks to make theoretical and practical sense of the steamy historical geographies of the 20th century. Without ever using the specific term, Lefebvre was probably the first to discover, describe, and insightfully explore Thirdspace as a radically different way of looking at, interpreting, and acting to change the embracing spatiality of human life. A few others ventured along similar paths and many today, thanks to these pioneering efforts, are exploring Thirdspace with great insight and political sensibility. But no one has, with such peripatetic persistence, shown the way so clearly to the long hidden lifeworlds of what Lefebvre expansively called *l'espace vécu*, lived space.

As a first approximation of these routes and roots to Thirdspace, I will "travel" through Lefebvre's biography as an explicitly geographical expedition, seeking in this initial journey to real-and-imagined places a way to reconstitute his life as a restless and allegorical voyage of discovery. In this "extended metaphor," I will be purposefully selective and, some might say, manipulative in representing Lefebvre's spatialized biography in too heroic a light, omitting his weaknesses and critical failures, never capturing the essence of his being and becoming. But this is not a traditional "historical" biography; it is instead a voyage, a journey, an exploration, a specifically spatial search. Much will be left out and I may seem to pack into the journey much more than what was actually there. Those who read his life as that of a sociologist or critical philosopher or historical materialist will no doubt be disappointed. But as I hope will become clearer, there is no better place to begin to comprehend the multiple meanings of Thirdspace.

Origins

However vague he may have been about his birthdate and history, Lefebvre was always forthright about his birthplace and geographical origins. *"Je suis Occitan, c'est-à-dire périphérique – et mondial,"* he wrote in his autobiographical *Le Temps des méprises*.[4] "I am Occitanian, that is to say peripheral – and global." He goes into further detail (and I translate freely, with my own emphasis added):

I enjoy my life between the centers and the peripheries; I am at the same time peripheral and central, but *I take sides with the periphery* ... I have known life in peasant communities, among

[4] Henri Lefebvre, *Le Temps des méprises*, Paris: Stock, 1975: 60.

the mountain people, the shephards ... [but] Parisianism, with all its sophistications, is no stranger to me.   (1975: 134)

Paris is my fascination. I have always lived there, yet I am not Parisian. If I count up the years, I have lived much longer in Paris than in the Pyrenees; but in so far as I have roots, a home base, it is in Occitania, where I get my bearings, my sensuality, my appreciation for use values, my nourishment and love....
And then there is the other pole, Paris: detested, admired, fascinating, the abstraction, the elitism, the unbearable elitism of the left, the explosive, the city fermenting with clans and cliques, with a political life that is most odious and most exciting....
These are contradictions, but they are prodigiously stimulating ones.   (1975: 133–4)

Lefebvre always maintained a deeply peripheral consciousness, existentially heretical and contracentric, a spatial consciousness and geographical imagination shaped in the regions of resistance beyond the established centers of power. Yet it was also, simultaneously, a consciousness and imagination peculiarly able to comprehend the innermost workings of the power centers, to know their perils and possibilities, to dwell within them with the critical ambidexterity of the resident alien: the insider who purposefully chooses to remain outside. This simultaneous embrace of centrality and peripheralness made him more evocatively global, worldly, open in his vision, persistently dialectical in his theory and practice – and often irksomely incomprehensible to those positioned more steadfastly at one or the other pole.

The center–periphery relation echoed another dialectic that threads simultaneously through Lefebvre's life and writings. This was the relation between the "conceived" (conçu) and the "lived" (vécu), or as he would later describe it, between the "representations of space" and the "spaces of representation." For Lefebvre, lived spaces were passionate, "hot," and teeming with sensual intimacies. Conceived spaces were intellectual, abstract, "cool," distanciating. They too inflamed passions, but these were centered more in the mind than in the body. He saw these two dialectical pairings (center–periphery, conceived–lived) as homologous, arising from the same sources, and often mapped them directly on one another in the contexts of his own personal life. But even here, although not always successfully, he tried to avoid the rigidity of categorical equivalence (body = hot = periphery, mind = cool = center), for this reeked of what he saw as the deadening of dialectical reasoning in conceptual dualisms, in the construction of compelling binary oppositions that are categorically closed to new, unanticipated possibilities.[5] Two terms are never enough, he would repeatedly write. *Il y a toujours l'Autre.* There is always the Other, a third term that disrupts, disorders, and begins to reconstitute the conventional binary opposition into an-Other that comprehends but is more than just the sum of two parts.[6] I will return to this critical strategy of "thirding-as-Othering" in the next chapter, for it is a keystone to an understanding of both Lefebvre's (meta)philosophy and the conceptualization of Thirdspace that evolves within it.

The "prodigiously stimulating" dialectics of center–periphery and conceived–lived were concretely re-enacted every year for decades in Lefebvre's oscillations between Paris and his home in Navarrenx, where his part-Basque mother was born, a site filled with the passions and memories of radical regional resistance to centralized state power. From these regenerative movements, practically all of Lefebvre's achievements and inspirations can be drawn. And here too, I suggest – and will elaborate further in chapter 2 – can be located other defining qualities of Thirdspace: a knowable and unknowable, real and imagined lifeworld of experiences, emotions, events, and political choices that is existentially shaped by the generative and problematic interplay between centers and peripheries, the abstract and concrete, the impassioned spaces of the conceptual and the lived, marked out materially and metaphorically in *spatial praxis,* the transformation of (spatial) knowledge into (spatial) action in a field of unevenly developed (spatial) power.

Power is ontologically embedded in the center–periphery relation and, hence, also in the ontology of Thirdspace. The presence/ absence of power, however, is not neatly defined or delimited:

> Power, the power to maintain the relations of dependence and exploitation, does not keep to a defined "front" at the strategic level, like a frontier on the map or a line of trenches on the ground. Power is everywhere; it is omnipresent, assigned to Being. It is everywhere *in space.* It is in everyday discourse and commonplace notions, as well as in police batons and armoured cars. It is in *objets d'art* as well as in missiles. It is in the diffuse preponderance of the "visual," as well as in institutions such as school or parliament. It is in things as well as in signs (the signs

---

[5] I say "not always successfully" because, at times, much of Lefebvre's writing seems to revolve around binaries and dualisms that remain uninterrogated or too simply composed. This includes not only mind/body but also man/woman and Western/non-Western, each a focal point for contemporary critiques of Lefebvre, as well as Fredric Jameson, Michel Foucault, and others. I thank Barbara Hooper for her critical remarks on this point and many others in my initial treatment of Lefebvre.

[6] From Lefebvre, *La Présence et l'absence: Contribution à la théorie des représentations,* Paris: Casterman, 1980: 143.

of objects and object-signs). Everywhere, and therefore nowhere.... [P]ower has extended its domain right into the interior of each individual, to the roots of consciousness, to the "topias" hidden in the folds of subjectivity.[7]

Few, including Foucault, ever made this super-charged relationship between space, knowledge, and power so explicit and far-reaching.

Pathways

The complex interplay of centrality and peripheralness defined Lefebvre's distinctive brand of Marxism: constantly open and flexible, always reactive to dogmatic closure, never content with any permanent construct or fixed totalization. His was a restless, passionate, nomadic Marxism that was both at the center of things and yet distinctly marginal: the better to see the entire terrain, to be receptive to new ideas from every potential source, to miss nothing. It was also, for some of the same reasons, both a modern and a postmodern Marxism right from the beginning, another eccentric simultaneity that confounds all those who dwell in neater, more categorical intellectual and political worlds. Lefebvre chose his Marxism as a modernist, always "taking sides" with the periphery and the peripheralized; but he practiced his Marxism as a postmodernist, departing from the tight constraints of the either/or to explore, as a consciously political strategy, the combinatorial openness of the both/and also....

Lefebvre hated what he called *les chapelles*, the "suffocating" (1975: 162) cliques and schools that have always dominated intellectual life in Paris, especially on the Left. Being global and peripheral meant always being open, non-exclusive, available to all those interested in the same project and interested in the potential use value of everything. Some have suggested that the cost of such eclectic flexibility and radical openness was relative isolation from the hegemonic "grand houses" of Western Marxism and Continental Philosophy. But this was an intentional and propitious isolation that protected and enhanced Lefebvre's peripheral and global vision, and at the same time sustained his central position as one of the most original critical philosophers of the 20th century.

For Lefebvre, critical thinking about the world always began by looking at that fundamental dynamic of all social life, the complex

---

[7] From Lefebvre, *The Survival of Capitalism*, tr. F. Bryant, London: Allison and Busby, 1976: 86–7.

and contradiction-filled interplay between the development of the forces of production (knowledge, labor power, technology) and the social relations that are organized around production and, simultaneously, around their own continued reproduction and renewal. The problematic interplay between social production and reproduction is insistently present in all of his works. But this insistence was never more (or less) than a point of departure, an explicitly Marxian home base from which to begin a longer, more meandering and endless journey to still unexplored Other spaces (while never forgetting that there was always a place to return home when necessary, for nourishment, identity, insight, and political choice).[8]

For Lefebvre, the centered peripheral, the "refined barbarian," as he called himself, a nomadic Marxism provided pathways into a *space of radical openness*,[9] a vast territory of infinite possibilities and perils. This space was not located simply "in between" his bi-polar worlds of centers and peripheries, or in some additive combination of them. It lay "beyond," in a (third)world that could be entered and explored through *metaphilosophy*, Lefebvre's summative description of his unsettled and unsettling methodology. Infused with this critical sense of moving beyond, he can perhaps be best described not as a Marxist but as a meta-Marxist. Let us explore further this significant prefixing.

In Greek, *meta-* carries the meaning of both something beyond or after (akin to the Latin *post-*), and also (related to the Latin *trans-*) a change of place or nature, i.e. to transport and/or transcend (as in the roots to the word "metaphor").[10] Lefebvre describes his use of *meta-* in *La Présence et l'absence: Contribution à la théorie des représentations* (1980: 89–90, my translation).

> The term: "meta ... " has many meanings: transgression, a going beyond, an excess, etc. Meta-physics did not imply at first a negation or denial of the physical; it took on this sense only after the long and eventful development of [modern]

---

[8] In my first meeting with Lefebvre in 1978 I clumsily asked him, "Are you an anarchist?" He responded politely,"No. Not now." "Well then," I said, "what are you now?" He smiled. "A Marxist, of course ... so that we can all be anarchists some time in the future."

[9] A phrase I have purposefully borrowed from bell hooks, whose revisioning of Thirdspace will be explored more fully in chapter 3.

[10] Were it not for the Greco-Roman impurity, the term "postmodernity" (and related usages) might better be translated as "metamodernity" to capture the more complex meanings of the Greek versus the Latin prefix: not simply coming after but moving beyond, in the sense of transporting and transcending, moving modernity to a different place, nature, and meaning. It is this wider sense of "post-" that underlies my postmodernism and my approach to a critical understanding of postmodernity and the "postmodern condition" throughout this book.

philosophy. Similarly, meta-language does not signify the abolition of language; it conveys the meaning of a "discourse on discourse".... The term meta-philosophy then is not the abolition of philosophy. To the contrary, it opens up a sphere of reflexion and meditation in which philosophy appears in all its fullness but also with all its limitations.... Meta-philosophy differs from philosophy most notably in its acceptance of the world of representations. It analyzes representations as such, as internal to their world, and from this analysis comes the critique of representations.... The great illusion of philosophy [arises from the belief that it can completely] transcend representations to reach a more concrete and complex Truth.

This commentary on the meanings of *meta-* positions the "critique of representations" at the heart of Lefebvre's metaphilosophy and, as we will soon see, at the source of his conceptualization of the (social) production of (social) space. It also serves not only as a critique of modern philosophy but also as an implicit critique of modern Marxism. Lefebvre's meta-Marxism is an important precursor to what some contemporary critics have begun to call "post-Marxism," a "sphere of reflexion and meditation" in which Marxism appears "in all its fullness but also with all its limitations." This is an important point, especially for those contemporary readers of Lefebvre who are either uncomfortable with and/or dismissive of the Marxism that infuses all his writings; or who, often equally blindly, celebrate his apparent constancy and commitment to modern Marxist traditions and orthodoxies. Neither of these interpretations adequately comprehend Lefebvre's metaphilosophy or, more important for the present argument, his *transgressive* conceptualization of lived space as an-Other world, a meta-space of radical openness where everything can be found, where the possibilities for new discoveries and political strategies are endless, but where one must always be restlessly and self-critically moving on to new sites and insights, never confined by past journeys and accomplishments, always searching for differences, an Otherness, a strategic and heretical space "beyond" what is presently known and taken for granted.

Lefebvre was one of the first to theorize *difference* and *otherness* in explicitly spatial terms and he linked this spatial theorization directly to his meta-Marxist critique of the "representations of power" and the "power of representations." He did so by insisting that difference be contextualized in social and political practices that are linked to *spatio-analyse*, the analysis, or better, the knowledge (*connaissance*) of the (social) production of (social) space.[11] In *Le Manifeste différentialiste* (1971), another of his attempts to reconstitute his metaphilosophical project, Lefebvre opposed the heterogenizing "differential" to the homogenizing "repetitive" and connected this relation to a series of innovative concepts and critiques of what he called everyday life in the modern world, the urban condition and consumer society (*la société bureaucratique de consommation dirigée*), and, especially, the growing power of the State and what he assertively described as a "state mode of production," in part to emphasize his own differences from more orthodox Marxism and also from the prevailing currents of continental philosophy and critical theory in both of which a spatial theorization of the state and its powers was notably absent.

Following up on his idea of *le droit à la ville*, the right to the city, he argued for a need to struggle on a wider terrain for *le droit à la différence*, the right to difference, to be different, against the increasing forces of homogenization, fragmentation, and hierarchically organized power (another of his innovative interpretive triads) that defined the specific geography of capitalism. He located these struggles for the right to be different at many levels, beginning significantly with the body and sexuality and extending through built forms and architectural design to the spatiality of the household and monumental building, the urban neighborhood, the city, the cultural region, and national liberation movements, to more global responses to geographically uneven development and underdevelopment. He embedded these multi-tiered struggles for the right to difference in the contextualized dialectics of centers and peripheries, the conceived and the lived, the material and the metaphorical; and from these concatenated dialectics of uneven development and differentiation he opened up a new domain, a space of collective resistance, a *Thirdspace of political choice* that is also a meeting place for all peripheralized or marginalized "subjects" wherever they may be located. In this politically charged space, a radically new and different form of *citizenship* (*citoyenneté*) can be defined and realized.

Another imbricated mapping is added to the emerging domain of

---

[11] Lefebvre intentionally chose the term *spatio-analyse* for its echoing of an alternative to the *psycho-analyse*, or psychoanalysis, that featured so prominently in French philosophical discourse, especially through the influence of Freud and/or Lacan. As another possible choice, he suggested *spatiology* in obvious reference to sociology, although he was uncomfortable with its intimations of a too narrow "science" of space. Lefebvre did not use either of these terms very often. They appear only when he is searching for new ways to demonstrate to others less spatially inclined the transdisciplinary comprehensiveness and philosophical centrality of his arguments about space and spatiality. What to call this alternative approach to spatiality remains an open question. As will be discussed in chapter 5, Foucault called it "heterotopology." My "postmodern geographies" was another attempt to capture its radical alterity and distinctiveness, an attempt which I am reconstituting here in the notion of Thirdspace.

Thirdspace in Lefebvre's theorization of difference, from the "geography closest in" (to use Adrienne Rich's description of the body) across many other spatial scales to the global divisions of labor and the macrographies of power embedded in a capitalist world of centers, peripheries, and semiperipheries. This nested layering of "regional" scales and differences is itself socially produced and reproduced over time and across space. As with Thirdspace, it is always embroiled in modes of production and reproduction, making Marxism a particularly revealing point of departure – but never the exclusive route to understanding and praxis, even with regard to the specific geography of capitalism.

Lefebvre's radical openness instilled Marxism (and instills Thirdspace) with dimensions that most Marxists (and most spatial analysts) never dreamed were there. He made Marxism (and Thirdspace) intentionally incomplete, endlessly explorable, resistant to closure or easy categorical definition, but persistently faithful in spirit and intent to Marx. He never permitted the formation of a Lefebvrean "school." He was instead one of the most Marx-ist and least Marxist of all 20th-century radical scholars. It is this nomadic and inspired meta-Marxism that opens Thirdspace to an understanding of difference and otherness and so much more as well.

Approximations

Lefebvre's sinuous path to Thirdspace can be traced back to his earliest writings on alienation and the mystification of consciousness, beginning with *La Conscience mystifiée* (1936), written with his close friend, Norbert Guterman.[12] Upset at what he saw among Marxists as an underappreciation for the power of the "conceived" world of ideas and ideology over the "lived" world of material social relations, he persistently sought to transcend, via his inclusive dialectical or, better, *trialectical* materialism, the stubborn bi-polarity and dualism that had developed between Marx's historical materialism and Hegel's philosophical idealism. Stimulated by his attachments to the Surrealist movement and its revisioning of the worlds of representation, Lefebvre set out at once to carve an alternative path, a third way that draws on and encompasses both materialism and idealism, Marx and Hegel, simultaneously, yet remains open to much more than their simple combination. After all, *il y a toujours l'Autre.*

---

[12] Henri Lefebvre and Norbert Guterman, *La Conscience mystifiée*, Paris: Gallimard, 1936 (new edition, Paris: Le Sycomore, 1979).

Introduced to Hegel through his association with André Breton, a key figure in the Surrealist movement, Lefebvre helped to shape the early development of Hegelianized Marxism in France and would become, with such works as *Le Matérialisme dialectique* (1939) and his most popular book, *Le Marxisme* (1948), the mid-century's leading Marxist theoretician.[13] Although authoritatively positioned within the French Communist Party and often criticized for his unbending authoritativeness, Lefebvre was at the same time heretical and original from the start, a precursory meta-Marxist, vigorously resisting any form of dogmatic closure or essentialism, always making sure to keep Marxism radically open to creative renewal.[14]

At the same time, the peripatetic Lefebvre was always ready to move on from even *his* most compelling discoveries. Michel Trebitsch captures this purposeful ambiguity and nomadism well in his Preface to the English translation of Lefebvre's *Critique of Everyday Life*.[15]

Lefebvre has something of the brilliant amateur craftsman about him, unable to cash in on his own inventions; something capricious, like a sower who casts his seeds to the wind without worrying about whether they will germinate. Or is it because of Lefebvre's style, between flexibility and vagueness, where thinking is like strolling, where thinking is *rhapsodic*, as opposed to more permanent constructions, with their monolithic, reinforced, reassuring arguments, painstakingly built upon structures and models? His thought processes are like a

[13] *Le Matérialisme dialectique*, Paris: Alcan, 1939; this original edition was destroyed in 1940, but the book was reprinted many times, beginning with Paris: Presses Universitaires de France, 1947. Translations were published in Spanish (1948), Italian (1949) German (1964), English-UK (1968), Japanese (1971), Dutch (1972), Portuguese (1972), English-US (1973). *Le Marxisme* was published by Presses Universitaires de France in 1948 as part of its *Que sais-je?* series, with translations in Indonesian (1953), Portuguese-São Paulo (1958), Greek (1960), Italian (1960), Spanish (1961), Japanese (1962), Dutch (1969), Swedish (1971), Danish (1973), Arabic (1973), German (1975), Portuguese-Lisbon (1975), Serbo-Croatian (1975), Turkish (1977), Albanian (1977), Korean (1987). An English translation is notably absent. The most useful bibliographic guides to Lefebvre's work are the biography by Rémi Hess (see note 3) and what is described as "The Complete Lefebvre Bibliography" in Rob Shields, "Body, Spirit, and Production: The Geographical Legacies of Henri Lefebvre," paper presented at the annual meeting of the Association of American Geographers, Atlanta, 1993.

[14] To categorize Lefebvre as an "unwavering marxist" or as "standing in the shadows of the totalizing drive of Hegelian Marxism," as Derek Gregory does in his *Geographical Imaginations* (Cambridge, MA, and Oxford, UK: Blackwell, 1994, 353–4), is thus both superficially accurate and a serious misreading of Lefebvre's project.

[15] Henri Lefebvre, *Critique of Everyday Life – Volume I*, tr. J. Moore with a Preface by Michel Trebitsch, London and New York: Verso, 1991 (from *Critique de la vie quotidienne I: Introduction*, Paris: Grasset, 1947; 2nd ed. Paris: L'Arche, 1958).

limestone landscape with underground rivers which only become visible when they burst forth on the surface.

(1991b: ix)

Alienation and the mystification of consciousness[16] are ever-present themes in Lefebvre's writings (and in the conceptualization of Thirdspace), but he considered them only to be the first in a series of what he called "approximations" of his metaphilosophical project. He saw these approximations both as an evolving series and also as always there, immanent in a metaphilosophical and libertarian meta-Marxism, waiting to be discovered. In each approximation, Lefebvre both deconstructs and reconstitutes his neverending project, putting it together in new and different ways without ever breaking entirely from his past journeys.

After *La Conscience mystifiée*, Lefebvre would write the majority of his books to defend Marxism against any form of closure or confinement from within or without, beginning with the rise of Soviet economism and German fascism in the interwar years; moving on in the postwar decades to the reductionism he saw in Sartrean existentialism, Althusserian structuralism, and continental philosophy's often flagrantly apolitical infatuation with language, discourse, and texts; and most recently against the widespread abandonment of Marxism in an era of postmodern global capitalism, with its growing ashpile of "the end ofs."[17] Amidst these defensive efforts, often written with exceptional clarity and directness, he simultaneously moved his own creative project forward along a nomadic and rhapsodic path that, in so many ways, would prefigure the later development of critical postmodernism, spatial feminism, post-Marxism, and much of what has now come to be called critical cultural studies

---

[16] Although there are some similarities between Lefebvre's notion of *la conscience mystifiée* and the powerful arguments about "false consciousness" associated with Lukács' codifications of Marx, there were very significant differences. For Lefebvre, mystified consciousness, especially within the working class, was an integral part of material reality and everyday life, and needed to be treated as such both theoretically and politically. In ways that will become clearer in later chapters, Lefebvre's treatment of *la conscience mystifiée* anticipated the more recent development of the concept of *hyperreality* and its wrappings in simulations and simulacra, especially as defined in the writings of Jean Baudrillard, a former student and colleague of Lefebvre.

[17] For a recent example of Lefebvre's vigilant defense of Marx and (his) Marxism see *Une Pensée devenue monde . . . Faut-il abandonner Marx?* (A Thought Becomes World . . . Must We Now Abandon Marx?), Paris: Fayard (1980). He dedicates this book to Catherine Regulier ("from whom it derives whatever force and ardor it may have") and answers the question posed in the subtitle with a forceful "no." It is useful to see this reaffirmation in conjunction with an earlier book he wrote with Regulier, then a young militant member of the French Communist Party, *La Révolution n'est plus ce qu'elle était* (The Revolution is No Longer What It Used to Be), Paris: Éditions Libres-Hallier (1978).

and the new cultural politics of identity and difference, subjects that will be explored in subsequent chapters.

Lefebvre's postwar approximations can be captured by weaving together the titles of his most pathbreaking books. I present them in diachronological order, but they are best interpreted synchronically, as stopovers on a nomadic journey, each providing a different inflection on the same developing themes.

1947 *Critique de la vie quotidienne I: Introduction*, Paris: Grasset; second edition Paris: L'Arche, 1958. Translations in Japanese (1968), German (1974), Italian (1977), and most recently as *Critique of Everyday Life – Volume I: Introduction* (tr. John Moore, with a Preface by Michel Trebitsch, London and New York: Verso, 1991).

1962 *Critique de la vie quotidienne II: Fondement d'une sociologie de la quotidienneté*, Paris: L'Arche. Translations in Japanese (1972), German (1975), and Italian (1977).

1968 *Le Droit à la ville*, Paris: Anthropos. Translations in Spanish (1968), Portuguese (1968), Japanese (1969), Italian (1970), German (1972), Greek (1976).

1968 *La Vie quotidienne dans le monde moderne*, Paris: Gallimard. Translations in Portuguese (1969), English (1971 – *Everyday Life in the Modern World*, Harmondsworth: Penguin and New York: Harper and Row; tr. Sacha Rabinovich), Spanish (1972), German (1972), Italian (1979), Albanian (1980).

1970 *La Révolution urbaine*, Paris: Gallimard. Translations in German (1972), Spanish (1972), Italian (1973), Serbo-Croatian (1975), Portuguese (Brazil – 1988).

1971 *Le Manifeste différentialiste*, Paris: Gallimard. Translations in Spanish (1972), Italian (1980).

1973 *La Survie du capitalisme: la reproduction des rapports de production*, Paris: Anthropos. Translations in English (1976 – *The Survival of Capitalism: Reproduction of the Relations of Production*, London: Allison and Busby, tr. Frank Bryant), German (1974), Spanish (1974), Portuguese (1974), Greek (1975), Italian (1975), Serbo-Croatian (1975).

1974 *La Production de l'espace*, Paris: Anthropos; third edition 1986. Translations in Italian (1975), Japanese (1975), and at last in English (1991 – *The Production of Space*, Oxford UK and Cambridge US: Blackwell, tr. Donald Nicholson-Smith).

1977 *De l'État III: Le mode de production étatique*, Paris: UGE collection 10/18. Translations in Italian (1978), Spanish (1979), Portuguese (1979), Serbo-Croatian (1982).

1980   *La Présence et l'absence: Contribution à la théorie des représenta-
       tions*, Paris: Casterman. Translations in Spanish (1981) and
       Greek (1982).

1981   *Critique de la vie quotidienne III: De la modernité au modernisme
       (Pour une métaphilosophie du quotidien)*, Paris: L'Arche.

Each of these "bursts" to the surface adds to the complexity and
openness of Lefebvre's eventual conceptualization of Thirdspace. As
he moves from one to the other, there is never a complete abandonment. Instead, the journey is one of creative refocusing and reconstitution, a neverending cumulative adventure of exploring new
horizons, of moving beyond.

Perhaps the best known of Lefebvre's itinerary of approximations
is his *critique of everyday life*, a critical understanding of daily life in
lived space that would occupy Lefebvre throughout all his life.[18] In
the initial volume of his trilogy on everyday life (1947, 1962, 1981),
written in the postwar flush of French liberation, Lefebvre optimisti-
cally saw the critique of everyday life as a means of connecting
Marxism more closely with the discourses of continental philosophy
as well as injecting philosophy with a new appreciation for the concrete, the immediate, the routine, the seemingly trivial events of
everyday living.

Lefebvre's critique of everyday life responded not just to the presence of alienation and mystification in capitalist societies but also to
their presumed absence in the "actually practiced" socialism of the
Soviet Union under Stalinism. The everyday world was *everywhere*
being colonized, infiltrated, by a technological rationality that was
extending its effects well beyond the market and the workplace into
the family, the home, the school, the street, the local community; into
the private spaces of consumption, reproduction, leisure, and entertainment. This endemic colonization was filtered through the state
and its bureaucracy in an expanding process that Lefebvre would
later call *spatial planning*, a broader term than even the rich French
phrase for urban and regional planning, *l'aménagement du territoire*. In
a long Foreword to the second edition of the *Critique*, published in
1958 (and included in the 1991 English translation), Lefebvre makes
some provocative connections:

---

[18] Lefebvre rarely used the concept of "place" in his writings, largely because its
richest meaning is effectively captured in his combined use of "everyday life" and
"lived space." Many cultural geographers, in particular, have persistently attempted
to separate the concepts of place and space and to give place greater concreteness,
immediacy, and cultural affect, while space is deemed to be abstract, distanced,
ethereal. As Lefebvre's work demonstrates, this is an unnecessary and misleading
separation/distinction that reduces the meaningfulness of both space and place.

The remarkable way in which modern techniques have pene-
trated everyday life has thus introduced into this backward sector *the uneven development* which characterizes every aspect of
our era. . . . These advances, along with their consequences, are
provoking new structural conflicts within the concrete life of
society. The same period which has witnessed a breathtaking
development in the applications of techniques [to] everyday life
has also witnessed the no-less-breathtaking degradation of
everyday life for large masses of human beings . . . .

Everyday life with all the superior mod cons takes on the distance and remoteness and familiar strangeness of a dream. The
display of luxury to be seen in so many films, most of them
mediocre, takes on an almost fascinating character, and the
spectator is uprooted from his everyday world by an everyday
world *other* than his own. Escape into this illusory but present
everyday world . . . explains the momentary success these films
enjoy.   (1991: 8–10)[19]

For Marxists, the strategic implications of Lefebvre's critique were
clear and powerful, if not so easy to understand and accept.
Everyday life was presented and represented as the place where
alienation and mystification were played out, enacted, concretely
inscribed. It was also, therefore, the place where the struggles to
demystify human consciousness, erase alienation, and achieve true
liberation must be located. What Lefebvre was doing was *substituting everyday life for the workplace* as the primary locus of exploitation,
domination, and struggle; and redefining social transformation and
revolution as intrinsically more socio-cultural (and less economistic)
processes and goals. What Antonio Gramsci and the Frankfurt
School had begun, Lefebvre expanded and developed more fully at a
time when his influence within Western Marxism was still at its
peak. But this was just the beginning of Lefebvre's radical reconceptualization and critique.

In Volume II (1962, subtitled "A Foundation for a Sociology of
Everydayness") and in the new Foreword written for the first volume, Lefebvre moved on from his initial formulations. Depressed by
what he saw as the increasing penetration and control of everyday
life by the state and its surveillant bureaucracy, he began focusing
more specifically on the state and on the sociological aspects of

---

[19] Lefebvre immediately follows this statement with a more positive one: "Happily,
contemporary cinema and theatre have other works to offer which reveal a truth
about everyday life," a "truth" he sees as a source of resistance and a "critique of
everyday life from within." He then goes on to a stimulating, nearly 20-page romp
through the films of Charlie Chaplin and the plays of Bertold Brecht.

Marxism.[20] By 1962, Lefebvre had become a professor of sociology in Strasbourg, after having served two years earlier as director of research for the Centre National de Recherches Scientifiques (CNRS), a state institution that would play a key role in the development of Marxist urban sociology in France and one that he had been involved with since 1948, after his years in the French resistance movement.

Lefebvre's sociological approximation was short-lived (if, unfortunately, long-lasting). Already in the second volume of the critique of everyday life, and then even more forcefully in *Everyday Life in the Modern World*, published in French in 1968, Lefebvre became concerned over what he saw as an emerging "sociologism," filled with the pretensions of scientism, and began regrounding his critique of everyday life and alienation in what he called *the urban condition* and in an accompanying discourse on *modernity*. Flanked by colleagues and students such as Alain Touraine, Henri Raymond, Jean Baudrillard, Manuel Castells, and Daniel Cohn-Bendit after his move to the University of Nanterre in Paris in 1965, Lefebvre initiated a new journey that would take him, especially after the events of May 1968, from sociology to the encompassing specificity of the urban.

Lefebvre's preliminary expositions on such themes as the shift "from the rural to the urban," his forceful promotion of "the right to the city" and the necessity for an "urban revolution" (each phrase the title of a book-length treatment) took many by surprise. To Marxists, and even the Marxist urban sociologists deeply inspired by Lefebvre's writings and teachings, this apparent redefinition of proletarian revolution and social struggle as intrinsically urban in both spirit and substance was not only unacceptably heretical, it seemed

[20] Lefebvre summarized this sociological phase of his approximations most purely in *Sociologie de Marx* (Paris: Presses Universitaires de France, 1966), with translations in Italian (1968), Spanish (1969), Portuguese (Brazil 1969), Japanese (1970), Swedish (1970), German (1972), Danish (1972), Dutch (1973), Greek (1980), and, most significantly, in English (both in London: Penguin Books and New York: Pantheon, 1968). So much is significant – and insignificant – about this book. No other book written by Lefebvre (except *La Survie du capitalisme*, 1973; tr. 1976) was translated into English more quickly, and none except *Le Matérialisme dialectique* (1939) and *Le Marxisme* (1948) were translated into more languages. Of all the pathbreaking books listed previously (and it should be noted that *Sociologie de Marx* is not among them), only *La Survie du capitalisme* and *La Vie quotidienne dans le monde moderne* were translated into English before Lefebvre's death. *The Sociology of Marx*, therefore, has had an extraordinary influence on how Lefebvre was seen and understood in the English-speaking world. Taking nothing away from its contributions to radicalizing North American and British sociology and from Lefebvre's identity as a sociologist, the book is notably outside the story-line being followed in this chapter. It contains little of Lefebvre's critique of everyday life and contains only one passing reference to space (28), and that only to herald the philosophical privileging of time and history. It is not surprising then how incomprehensible and, indeed, invisible Lefebvre's pronounced spatial turn was to anglophones unable to read French or to more fluent francophones who saw Lefebvre only as a sociologist.

to hoist Lefebvre with his own petard, redolent of the ideological mystification, reification, and false consciousness he had so roundly attacked.[21] Despite the important role Lefebvre's ideas played in Paris's short-lived attempt at urban revolution in 1968 – and perhaps also because of it – Lefebvre's heretical geographical journeys came to be seen as going too far from the primacy of the social and the materialist critique of history, and too close to what his critics saw as the fetishism of the spatial.

Thus began a long period in which Lefebvre seemed to recede into the shadows of Parisian intellectual life, especially within French Marxism. His presumed excesses and unbridled spontaneity were blamed by many on the Left for nearly all the failures of the May 1968 uprising, as a new generation of *chapelles* was formed to revive (or to rationalize the total abandonment of) the revolutionary project. Little was heard of Lefebvre in the English-speaking world except through the critical eyes of those who had, often quite respectfully, pushed him into the shadows. Of the nearly 30 books he published between 1970 and 1990, including his autobiographical *Le Temps des méprises* and several critical attacks on the new hegemony of Althusserian structuralism on the Left and the rising cybernetic State Technocracy on the Right, only two were translated into English: the exegetical *Marxist Thought and the City* (1974) and the exuberant *The Survival of Capitalism* (1976), an insightful but confusing preview of *The Production of Space* that was difficult to comprehend without reference to the latter book's more effectively elaborated critique.

With exquisite irony, when viewed from the present, this may have been his most creatively productive period with respect to the conceptualization of spatiality in general and Thirdspace in particular. Moving on from his "urban" approximation, wherein he tried against all odds to project the urban condition as an appropriately grounded metaphor for the specific and by now global geography/spatiality of capitalism, he reached an overlapping moment when, as he described it, "*tout converge dans le problème de l'espace*" (1975: 223) – when "it all came together in the problem of space." In "The Discovery," the opening chapter to *The Survival of Capitalism*, Lefebvre contextualized this convergence in a profound transformation – and spatialization – of dialectical reasoning.

[21] See, for example, the treatment of Lefebvre in Manuel Castells, *La Question urbaine*, Paris: Maspero, 1972, translated in English as *The Urban Question*, London: Edward Arnold, 1977; and in David Harvey, *Social Justice and the City*, Baltimore: Johns Hopkins University Press, 1973. Both Harvey and Castells, while drawing sympathetically from Lefebvre's writings, were unable to accept his assertion of the centrality of urban-cum-spatial revolution in their varying structuralist-Marxist interpretations of urbanism at that time. Both have subsequently softened their structuralisms and their attitudes toward Lefebvre.

The dialectic is back on the agenda. But it is no longer Marx's dialectic, just as Marx's was no longer Hegel's. . . . The dialectic today no longer clings to historicity and historical time, or to a temporal mechanism such as "thesis–antithesis–synthesis" or "affirmation–negation–negation of the negation." . . . *This, then, is what is new and paradoxical: the dialectic is no longer attached to temporality.* Therefore, refutations of historical materialism or of hegelian historicity cannot function as critiques of the dialectic. To recognise space, to recognise what "takes place" there and what it is used for, is to resume the dialectic; analysis will reveal the contradictions of space . . . in the route from mental space to social space . . . [in the] specific contradictions . . . between *centres* and *peripheries* . . . in political economy, in political science, in the theory of urban reality, and in the analysis of all social and mental processes. . . . *We are not speaking of a science of space, but of a knowledge (a theory) of the production of space* . . . this most general of products.   (1976: 14, 17–18)

After 1968, Lefebvre began what in retrospect may turn out to be one of the most significant critical philosophical projects in the 20th century, a radical recasting of critical thought itself, from ontology and epistemology to emancipatory social practice, around a "knowledge," a practical theorization, of the production of space. Over the previous hundred years at least, critical theory and philosophy had been creatively grounded in the interplay of historicality and sociality, in the dynamic relations between the socially conscious and willfull making of history and the historically conscious constitution and reproduction of social relations and social practices. The spatiality of history and social life was, for the most part, frozen into the background as an "external" container, stage, or environment for social action. Lefebvre's project, which remained almost invisible until the 1990s, was to inject a third dimension to the dually privileged dynamics of historicality and sociality: an encompassing and problematic spatiality that demanded at least equivalent attention in critical theory and praxis.

Lefebvre proceeded to fuse a spatial problematic into all his writings: on everyday life, alienation, the urban condition; in his "differentialist manifesto" and theorization of the right to difference; in his development of a critical theory of representations and semiology; in his interpretation of the survival of capitalism and its venue, the reproduction of the social relations of production in space; in his analysis of the State and its growing control over space, knowledge, and power; in his forays into the debates on modernity and modernism, and later, postmodernity and postmodernism. Synthesizing this whirlwind of spatialization was *La Production de l'espace* (1974), the most cryptic yet most revealing of all his works on the spatiality of social life.

In the last chapter of *La Production de l'espace*, translated as "Openings and Conclusions," Lefebvre explicitly grounds and activates his metaphilosophical project around what he insisted on calling a "knowledge" (*connaissance*) of space.[22]

> Space is becoming the principal stake of goal-directed actions and struggles. It has of course always been the reservoir of resources, and the medium in which strategies are applied, but it has now become something more than the theatre, the disinterested stage or setting, of action. Space does not eliminate the other materials or resources that play a part in the socio-political arena, be they raw materials or the most finished of products, be they businesses or "culture". Rather, *it brings them all together* and then in a sense substitutes itself for each factor separately by enveloping it. The result is a vast movement in terms of which space can no longer be looked upon as an "essence", as an object distinct from the point of view of (or as compared with) "subjects", as answering to a logic of its own. Nor can it be treated as result or resultant, as an empirically verifiable effect of a past, a history, or a society. Is space indeed a medium? A milieu? An intermediary? It is doubtless all of these, but its role is less and less neutral, more and more active, both as instrument and as goal, as means and as end. Confining it to so narrow a category as that of "medium" is consequently woefully inadequate.   (1991a: 410–11, emphasis added)

Space, "this most general of products," thus accrues to itself all that has been formerly and familiarly attached to the social production of time as history or social historicality. Space is simultaneously objective and subjective, material and metaphorical, a medium and outcome of social life; actively both an immediate milieu and an originating presupposition, empirical and theorizable, instrumental, strategic, essential. In his rebalanced trialectic of spatiality–historicality–sociality, Lefebvre makes another provocative connection which creatively inverted (and profoundly spatialized) one of the foundational assumptions of historical materialism.

> There is one question which has remained open in the past because it has never been asked: what exactly is the mode of

---

[22] Lefebvre always preferred the term *connaissance*, with its more practical connotation of "understanding," to the synonymous *savoir*, which often suggests a more scholarly and formal notion of "learning." For Lefebvre, this semantic distinction had significant political meaning. It would shape his response to Michel Foucault's emphasis on *savoir/pouvoir* or knowledge/power, as I note briefly in chapter 5.

existence of social relationships? Are they substantial? natural? or formally abstract? The study of space offers an answer according to which the social relations of production have a social existence *to the extent that they have a spatial existence*; they project themselves into a space, becoming inscribed there, and in the process producing the space itself. Failing this, these relations would remain in the realm of "pure" abstraction – that is to say, in the realm of representations and hence of ideology: the realm of verbalism, verbiage and empty words.

(1991a: 129; emphasis added)

The message is clear, but few on the Left have been willing to accept its powerful connotations: that all social relations become real and concrete, a part of our lived social existence, only when they are spatially "inscribed" – that is, *concretely represented* – in the social production of social space. Social reality is not just coincidentally spatial, existing "in" space, it is presuppositionally and ontologically spatial. *There is no unspatialized social reality.* There are no aspatial social processes. Even in the realm of pure abstraction, ideology, and representation, there is a pervasive and pertinent, if often hidden, spatial dimension.

These are complex passages that present the production of space as an all-encompassing worldview and praxis consciously chosen (among other alternatives) for its particular acuity in enabling social and political action to change the world. Lefebvre's argument may seem crudely overstated, especially to those who cling to the emancipatory privileging of historicality and/or sociality. But it is no more (or less) than an assertion that critical thinking about the spatiality of human life is at least comparable to the more familiar, all-encompassing worldview that has centered around the "making of history" and the critical historical-cum-sociological imagination. That "everything" occurs in time and is inherently historical, that our actions always play a part (amidst significant constraints) in constructing sequential temporality and making histories, in the construction of individual and societal "biographies," seems unremarkably true, even if frequently outside of our conscious awareness or submerged in enfolding ideologies. What Lefebvre is arguing for is a similar action-oriented and politicized ontology and epistemology for space: "everything" also occurs in space, not merely incidentally but as a vital part of lived experience, as part of the (social) production of (social) space, the construction of individual and societal spatialities.

It is important to note, for it is so easily misunderstood, that this reconnection of spatiality to historicality and sociality was not meant to deny the political and theoretical significance of critical historiography or sociology. It was instead an effort to open up and enrich the historical and sociological imaginations with a long-neglected or persistently subordinated critical approach to spatiality, an explicitly spatial problematic. No one before or since has so forcefully asserted the significance of space and spatial knowledge in all realms of critical social theory and philosophy.

Arrivals

How did Lefebvre himself explain his provocative spatial turn, his metaphilosophical "convergence" on the spatial problematic? Again, I translate freely from *Le Temps des méprises*.

> Among those who were sympathetic to me, there was great surprise when I started to occupy myself with these questions, speaking of architecture, urbanism, the organization of space.... But this research on space started for me in childhood. I could not comprehend the philosophical separation of subject and object, the body and the world. The boundary between them did not seem so clear and clean. The inner world, was it not also cosmic? The mental and the spatial were splitting European philosophy apart at the seams.... In my sinuous line of development ... through marxist thought ... I arrived at the questions concerning space. (1975: 217)

Some saw in this "discovery" a new geography, others a new sociology, still others a new history or philosophy. For Lefebvre, it was all these things and much more. He cautiously called his spatial perspective *transdisciplinary* as a strategy to prevent spatial knowledge and praxis from being fragmented and compartmentalized (again) as a disciplinary specialty. Space was too important to be left only to the specialized spatial disciplines (Geography, Architecture, Urban Studies) or merely added on as a gap-filler or factual background for historians, social scientists, or Marxist sociologists. The spatiality of human life, like its historicality and sociality, infused every discipline and discourse.

> Some chose other ways to thread through the complexities of the modern world, for example, through literature or the unconscious or language. I chose space. Consistent with the method I have followed, I ploughed into (*creuser*) the concept and tried to see all its implications. (1975: 218)

He would contend with these "other ways to thread through the complexities of the modern world" more directly in *The Production of*

Space, explaining thereby his abiding choice of space. On what today would be called discourse theory, for example, he writes:

> The strategy of centring knowledge on discourse avoids the particularly scabrous topic of the relationship between knowledge and power. It is also incapable of supplying reflective thought with a satisfactory answer to a theoretical question that it itself raises: do sets of non-verbal signs and symbols, whether coded or not, systematized or not, fall into the same category as verbal sets, or are they rather irreducible to them? Among non-verbal signifying sets must be included music, painting, sculpture, architecture, and certainly theatre, which in addition to a text or pretext embraces gesture, masks, costume, a stage, a *mise-en-scene* – in short, a space. Non-verbal sets are characterized by a spatiality which is in fact irreducible to the mental realm. There is even a sense in which landscapes, both rural and urban, fall under this head. To underestimate, ignore and diminish space amounts to the overestimation of texts, written matter, and writing systems, along with the readable and the visible, to the point of assigning to these a monopoly on intelligibility.   (1991a: 62)

In this critique of discourse theory and other comments on alternative critical theoretical perspectives, Lefebvre appears to be harshly dismissive. His aim, however, is not to reject but to *spatialize*, to make sure that a spatial problematic is recognized whatever the theoretical emphasis employed, and to show how such spatialization works against theoretical closure and reductionism whatever interpretive pathway is chosen. For Lefebvre, spatial knowledge is a source and stimulus for radical openness and creativity.

In keeping with his resistance to singular causality, Lefebvre ambles on to other routes and roots to his discovery of an encompassing ontological spatiality, never prioritizing one route/root over the other. Thinking back again to Marx, he reasserts the triad of Land, Labor, and Capital against the tendencies to reduce Marx's critique of capitalism to the categorical binaries of capital–labor, bourgeoisie–proletariat, profit–wages, binaries which too often lead to an estrangement of the critique from nature, the soil, the earth, the production of space. Restoring the capitalist triad by reasserting the significance of Land, "enriches the analysis considerably, for it is not a question of a binary opposition but an incomparably more complex relation across three terms" (1975: 220).

He connects this "trialectical" restoration first to his own largely unpublished work on ground rent and the agrarian question in France, about which he was always exceedingly proud. He extends it in a brief discussion of the old philosophical debates on absolute versus relational space (a favorite source of philosophical legitimacy for more contemporary theorists of space). And then he returns home again to describe his most immediate and personal discovery of the importance of space in the context of the French state's programs for regional development and spatial planning (*l'aménagement de territoire*) in the period between 1960 and 1970.

> Near a village I had passed through many times in my youth one could see being built a new town, Lacq, petroleum, natural gas, sulfur ... the bulldozers moving through the forest, the first building stones set down. It became for me a small laboratory. I started to study urban questions *in vivo*, in *statu nascendi*.... I suspected that this irruption of the urban in a traditional peasant reality was not simply a local hazard but was linked to more global phenomena of urbanization and industrialization. Elsewhere in the region, at the same time, there were great changes taking place in agriculture ... a transformation of the countryside. I followed these events extremely closely, with the idea of telling a story about the birth of a city.
>   At about that time, I began travelling all over the world studying the urban question in New York, Montreal, Teheran, Tokyo, Osaka ... I perceived the global problem to be deeply involved in a complete restructuring of social space.
>   (1975: 222–3)

It is at this point that he writes *tout converge dans le problème de l'espace ... [dans] l'urbanisation de l'espace entier*.

Also in *Le Temps des méprises*, Lefebvre emotionally recalls his postwar encounters with the situationist movement as still another source of his spatial turn. He discusses in detail his controversial relations with Guy Debord, whose best-known work, *The Society of the Spectacle*,[23] he dismissed as "impregnated with sociologism," devoid of attention to the spatial politics and policies of the State, and unaware of the increasing power of globalized capital. Detaching himself from Debord's version of the Situationist International, Lefebvre remembers more warmly his earlier

---

[23] Guy Debord, *La Société du spectacle*, Paris: Buchet-Chastel, 1967, and Éditions Champ Libre, 1971. Translated as *The Society of the Spectacle*, Detroit: Black and Red, 1977. An unpublished translation (1990) was also made by Donald Nicholson-Smith, the translator of Lefebvre's *La Production de l'espace* (1974, tr. 1991). For a recent look at Debord and the situationists, see Sadie Plant, *The Most Radical Gesture: The Situationist International in a Postmodern Age*, London and New York: Routledge, 1992.

encounters with the Amsterdam utopian architect and painter Constant Nieuwenhuis and the COBRA movement, named for the home cities of its major protagonists (Copenhagen, Brussels, and Amsterdam).

COBRA was a neo-surrealist movement of poets, painters, musicians, and urban theorists that focused its revolutionary tactics of "situationism" on architecture and the social production of the built environment. The Amsterdam group in particular was influenced and inspired by Lefebvre's first volume of *The Critique of Everyday Life*, published in 1947. Although COBRA was disbanded in the early 1950s, its remnants continued to be active in Amsterdam, stimulating the Provos (disbanded in 1967) and then the Kabouter (Orange Free State) movements that were to reshape urban life in Amsterdam in ways we are only beginning to recognize today (I pick up these themes again in chapter 9). Lefebvre visited Amsterdam many times between 1958 and 1968 and his look back helps to explain the very special relationships he has had with architects and architecture ever since the 1950s.

> Around 1953, Nieuwenhuis began his projects for a "New Babylon." For him, feelings of emotion (*ambience affective*) are not external to space, nor is space indifferent to emotional feelings. He was able to recapture and elevate a concept from the grand architectural tradition in which space creates something such as a gathering together, a joy, a sadness, a submission; in short, space is active. . . . This is what Constant called an architecture of ambiance . . . the creation of situations . . . the construction of spaces that are the creators of ambiance, emotion, situation, what I called a theory of moments. I find myself in accord with situationists when situationism puts to the forefront such ideas of creation, of the production of new situations. (1975: 157)

Lefebvre's 20th-century adventures did not end with the spatial approximation he consolidated in *La Production de l'espace* (1974). Over the last decade of his life, he continued to explore new dimensions of spatiality, pushing its frontiers beyond the known. He signaled some of these new explorations in the last chapter of *La Production . . .*, translatable as "Openings and Conclusions." "Today," he wrote, "our concern must be with space on a world scale . . . as well as with all the spaces subsidiary to it, at every possible level" (1991a: 412). This globalization process creates a "new situation" in which all places and spatiality itself have "undergone metamorphoses," such radical changes and restructurings that our "languages and linguistic systems need to be dismantled and reconstructed." "This task," he adds, "will be carried out in (spatial) social practice" (1991a: 414).[24]

The primary focus of his last original writings was on "rhythm analysis," which, as noted in the beginning of this chapter, he described as "an idea that may be expected to put the finishing touches to the exposition of the production of space" (1991a: 405). Again anticipating what would become a key issue in contemporary critical cultural studies, Lefebvre turned to the spatio-temporal rhythms of the body, elevating the body to a central position in his metaphilosophy. In many ways his concluding comments on the body begin a deconstruction and reconstitution of all the preceding chapters of *The Production of Space*.

> The whole of (social) space proceeds from the body, even though it so metamorphizes the body that it may forget it alto-gether – even though it may separate itself so radically from the body as to kill it. The genesis of a far-away order [the state? the global?] can be accounted for only on the basis of the order that is nearest to us – namely, the order of the body. Within the body itself, spatially considered, the successive levels consti-tuted by the senses (from the sense of smell to sight, treated as different within a differentiated field) prefigure the layers of social space and their interconnections. The passive body (the senses) and the active body (labour) converge in space. The analysis of rhythms must serve the necessary and inevitable restoration of the total body. This is what makes "rhythm analysis" so important.   (1991a: 405)

He goes further in his review and preview:

> Western philosophy has *betrayed* the body; it has actively partic-ipated in the great process of metaphorization that has *aban-doned* the body; and it has *denied* the body. The living body, being at once "subject" and "object", cannot tolerate such con-ceptual division, and consequently philosophical concepts fall into the category of the "signs of non-body". Under the reign of King Logos, the reign of true space, the mental and the social were sundered, as were the directly lived and the conceived, and the subject and the object. New attempts were forever being made to reduce the external to the internal, or the social to the mental, by means of one ingenious typology or another.

---

[24] These restructuring processes and their impact on urbanization and urban studies will be examined in much greater detail in *Postmetropolis*, the forthcoming compan-ion volume to *Thirdspace*. Lefebvre's brief comment here is suitably precocious and epigraphic.

Net result? Complete failure! Abstract spatiality and practical spatiality contemplated one another from afar, in thrall to the visual realm. (1991a: 407)

Hopeful and confident as ever, Lefebvre asserts that "today the body is establishing itself firmly" beyond philosophy, discourse, and the theory of discourse. A new critical theory and metaphilosophy is "carrying reflexion on the subject and the object beyond the old concepts," and "has re-embraced the body along with space, in space, and as the generator (or producer) of space" (1991a: 407).

Here we have both a conclusion and an opening. Lefebvre's assertive spatialization of philosophy and praxis became an integral part of a significant redirection of the mainstreams of modernist thought and action in the last decades of the 20th century. It was a redirection that would lead circuitously to the irruption of a new cultural politics of identity and difference, a new transdisciplinary field of critical cultural studies, and a radical postmodernism of resistance, all of which revolve today, more than any modern movement of the past century, around the multiplicitous spatialities of social life.

2

The Trialectics of Spatiality

The "imaginary." This word becomes (or better: becomes again) magical. It fills the empty spaces of thought, much like the "unconscious" and "culture." . . .

. . . After all, since two terms are not sufficient, it becomes necessary to introduce a third term. . . . The third term is the *other*, with all that this term implies (*alterity*, the relation between the present/absent other, *alteration-alienation*).

Reflexive thought and hence philosophy has for a long time accentuated dyads. Those of the dry and the humid, the large and the small, the finite and the infinite, as in Greek antiquity. Then those that constituted the Western philosophical paradigm: subject–object, continuity–discontinuity, open–closed, etc. Finally, in the modern era there are the binary oppositions between signifier and signified, knowledge and non-knowledge, center and periphery. . . . [But] is there ever a relation only between two terms. . .? One always has Three. There is always the Other.
(Henri Lefebvre, *La Présence et l'absence*, 1980: 225 and 143)

The recent English translation of *The Production of Space*, and the revived interest in Lefebvre's writings and ideas it has stimulated, presents new opportunities to reconsider Lefebvre's contributions to present-day debates on the theorization of space and social spatiality. In this chapter, I purposefully reappropriate *The Production of Space* to pull from its expansive depths a clearer understanding of the meaning and critical scope of what I have chosen to define as Thirdspace.

Envisioning Thirdspace Through "The Aleph"

After re-reading *The Production of Space*, I find myself drawn once more to "The Aleph," a short story by Jorge Luis Borges, whose distinctive version of the rich Latin-American tradition of "magical realism" resounds so well with Lefebvre's fascination with concrete abstractions, his paradoxically materialist idealism, and his adventurous explorations into the simultaneous worlds of the real-and-imagined.[1] In *Postmodern Geographies*,[2] I used Borges' brilliant evocation of the Aleph as the place "where all places are" to provoke new ways of looking at and understanding contemporary Los Angeles. For the present work, its eye-opening perspective continues to serve a similar purpose, as an introduction and stimulus to the urban essays contained in chapters 7 through 9, each of which defines and redefines Los Angeles in different ways. Here I will use the Aleph again as a point of departure, or better, as a first "approximation" from which to reinterpret *The Production of Space* and recompose its imbricated conceptualizations of Thirdspace.

"The Aleph" was published in Spanish in 1945 and appears in English as the first entry to *The Aleph and Other Stories: 1933–1969*, which also contains an autobiographical essay and commentaries by Borges on each story. On "The Aleph," Borges writes:

> What eternity is to time, the Aleph is to space. In eternity, all time – past, present, and future – coexists simultaneously. In the Aleph, the sum total of the spatial universe is to be found in a tiny shining sphere barely over an inch across.   (Borges, 1971: 189)

Borges introduces "The Aleph" with two classical quotations to amplify his intentions:

> O God! I could be bounded in a nutshell, and count myself a King of infinite space.   (*Hamlet*, II, 2)

> But they will teach us that Eternity is the Standing still of the Present Time, a *Nunc-stans* (as the Schools call it); which neither they, nor any else understand, no more than they would a *Hic-stans* for an Infinite greatness of Place.   (*Leviathan*, IV, 46)

The story begins with the narrator (clearly Borges himself) strolling along the streets of Buenos Aires, reminiscing about the death of a

---

[1] Jorge Luis Borges, "The Aleph," in *The Aleph and Other Stories: 1933–1969*, New York: Bantam Books, 1971: 3–17.
[2] Edward W. Soja, *Postmodern Geographies: The Reassertion of Space in Critical Social Theory*, London and New York: Verso, 1989.

woman he loved. He soon meets a friend, Carlos Argentino Daneri, who excitedly tells him that in Daneri's cellar under the dining room there was an Aleph, "one of the points in space that contains all other points." His friend had discovered it as a child and was now anxious to share his discovery, for the house which contained the Aleph was scheduled for demolition, part of the burgeoning growth of the "pernicious metropolis" of Argentina, the critical urban setting of "The Aleph."

> "One day when no one was home I started down in secret, but I stumbled and fell. When I opened my eyes, I saw the Aleph."
>   "The Aleph?" I repeated.
>   "Yes, the only place on earth where all places are – seen from every angle, each standing clear, without any confusion or blending. I kept the discovery to myself and went back every chance I got . . ."
>   I tried to reason with him. "But isn't the cellar very dark?" I said.
>   "Truth cannot penetrate a closed mind. If all places in the universe are in the Aleph, then all stars, all lamps, all sources of light are in it too."
>   "You wait there. I'll be right over to see it."
>
> Then I saw the Aleph.
>
> I arrive now at the ineffable core of my story. And here begins my despair as a writer. All language is a set of symbols whose use among its speakers assumes a shared past. How, then, can I translate into words the limitless Aleph, which my floundering mind can scarcely encompass? Mystics, faced with the same problem, fall back on symbols: to signify the godhead, one Persian speaks of a bird that is somehow all birds; Alanus de Insulis, of a sphere whose center is everywhere and circumference is nowhere; Ezekiel, of a four-faced angel who at one and the same time moves east and west, north and south. (Not in vain do I recall these inconceivable analogies; they bear some relation to the Aleph.) Perhaps the gods might grant me a similar metaphor, but then this account would become contaminated by literature, by fiction. Really what I want to do is impossible, for any listing of an endless series is doomed to be infinitesimal. In that single gigantic instant I saw millions of acts both delightful and awful; not one of them amazed me more that the fact that all of them occupied the same point in space, without overlapping or transparency. What my eyes beheld was simultaneous, but what I shall now write down will

be successive, because language is successive. Nonetheless, I'll try to recollect what I can.

... I saw a small iridescent sphere of almost unbearable brilliance. At first I thought it was revolving; then I realized that this movement was an illusion created by the dizzying world it bounded. The Aleph's diameter was probably little more than an inch, but all space was there, actual and undiminished. Each thing (a mirror's face, let us say) was infinite things, since I distinctly saw it from every angle of the universe. I saw the teeming sea; I saw daybreak and nightfall; I saw the multitudes of America; I saw a silvery cobweb in the center of a black pyramid; I saw a splintered labyrinth (it was London); I saw, close up, unending eyes watching themselves in me as in a mirror; I saw all the mirrors on earth and none of them reflected me; ... I saw bunches of grapes, snow, tobacco, loads of metal, steam; I saw convex equatorial deserts and each one of their grains of sand; I saw a woman in Inverness that I shall never forget; ... I saw a ring of baked mud on a sidewalk, where before there had been a tree; ... I saw my empty bedroom; I saw in a closet in Alkmaar a terrestrial globe between two mirrors that multiplied it endlessly; ... I saw the coupling of love and the modification of death; I saw the Aleph from every point and angle, and in the Aleph I saw the earth and in the earth the Aleph and in the Aleph the earth; I saw my own face and my own bowels; I saw your face; and I felt dizzy and wept, for my eyes had seen that secret and conjectured object whose name is common to all men but which no man has looked upon – the unimaginable universe.   (1971: 10–14)

"The Aleph" is an invitation to exuberant adventure as well as a humbling and cautionary tale, an allegory on the infinite complexities of space and time. Attaching its meanings to Lefebvre's conceptualization of the production of space detonates the scope of spatial knowledge and reinforces the radical openness of what I am trying to convey as Thirdspace: the space where all places are, capable of being seen from every angle, each standing clear; but also a secret and conjectured object, filled with illusions and allusions, a space that is common to all of us yet never able to be completely seen and understood, an "unimaginable universe," or as Lefebvre would put it, "the most general of products."

Everything comes together in Thirdspace: subjectivity and objectivity, the abstract and the concrete, the real and the imagined, the knowable and the unimaginable, the repetitive and the differential, structure and agency, mind and body, consciousness and the unconscious, the disciplined and the transdisciplinary, everyday life and unending history. Anything which fragments Thirdspace into separate specialized knowledges or exclusive domains – even on the pretext of handling its infinite complexity – destroys its meaning and openness. There is a close connection between this conceptualization of Thirdspace and Lefebvre's nomadic meta-Marxism. Each envisions a complex totality of potential knowledges but rejects any totalization that finitely encloses knowledge production in "permanent structures" or specialized compartments/disciplines. For Lefebvre (and for Borges), spatial knowledge, as a means "to thread through the complexities of the modern world," is achievable only through approximations, a constant search to move beyond (*meta-*) what is known. As Lefebvre notes, others have chosen alternative transdisciplinary perspectives – language and discourse theory, psychoanalysis and the window of the unconscious, literature and critical historiography – to thread through these complexities. Lefebvre, however, was the first to explicitly do so through space, or more specifically the (social) production of (social) spatiality.

This all-inclusive simultaneity opens up endless worlds to explore and, at the same time, presents daunting challenges. Any attempt to capture this all-encompassing space in words and texts, for example, invokes an immediate sense of impossibility, a despair that the sequentiality of language and writing, of the narrative form and history-telling, can never do more than scratch the surface of Thirdspace's extraordinary simultaneities. *The Production of Space* is filled with Aleph-like references to the incapacity of language, texts, discourses, geographies and historiographies to capture fully the meanings of human spatiality, or what Borges, quoting from *Leviathan*, describes as "an Infinite greatness of Place." On the struggle to develop a rigorously analytical "science of space" to meet this daunting task, Lefebvre forcefully expressed his dissatisfaction and despair:

> To date, work in this area has produced either mere descriptions which never achieve analytical, much less theoretical, status, or else fragments and cross-sections of space. There are plenty of reasons for thinking that descriptions and cross-sections of this kind, though they may well supply inventories of what *exists in* space, or even generate a *discourse on* space, cannot ever give rise to a *knowledge of* space. And, without such a knowledge, we are bound to transfer onto the level of discourse, of language *per se* – i.e. the level of mental space – a large portion of the attributes and "properties" of what is actually social space.
>
> ... When codes worked up from literary texts are applied to

spaces – to urban spaces, say – we remain, as may easily be shown, on the purely descriptive level. Any attempt to use such codes as a means of deciphering social space must surely reduce that space itself to the status of a *message*, and the inhabiting of it to the status of a *reading*. This is to evade both history and practice. . . .

As for the above-mentioned sections and fragments, they range from the ill-defined to the undefined – and thence, for that matter, to the undefinable. Indeed, talk of cross-sectioning, suggesting as it does a scientific technique (or "theoretical practice") designed to help clarify and distinguish "elements" within the chaotic flux of phenomena, merely adds to the muddle.   (1991a: 7–8)[3]

The Production of Space is, of course, itself a text, a sequential narrative subject to all the constraints of language and writing and, still further, to the unexpected aporia arising from its translation from French to English. It too is a "reading" rather than an "inhabiting," a "discourse" rather than a practical "knowledge" of space. Like Borges, Lefebvre had to contend with the successive constraints of writing about simultaneities, about the repetitive and the differential, the known and the unknowable, at the same time. How then did Lefebvre attempt to defy some of these discursive inhibitions and to express the multifaceted inclusiveness and simultaneities of lived social space, the term that comes closest to conveying the meanings of Thirdspace? After contemplating this question and remembering his love of music, the rhapsodic, I have become convinced that Lefebvre wrote *The Production of Space* in the form of a fugue, a polyphonic composition based on distinct themes which are harmonized through counterpoint and introduced over and over again in different ways through the use of various contrapuntal devices. Read in this way, every one of the seven chapters is both a repetition and a distinctively different elaboration of the others. As if to emphasize the (counter)point, the fugue ends with "Conclusions" that are also "Openings" or *Ouvertures*.[4]

This makes *The Production of Space* very difficult to comprehend as a conventional text. The ideas are not developed in a straightforward sequential or linear fashion. If one misses the emphatic themes of the first chapter in particular, everything that follows seems to float independently or else seems to contradict erratically

---

[3] Unless otherwise noted, all quoted material is taken from Lefebvre, *The Production of Space*, 1991.
[4] The chapters are, in sequence, Plan of the Present Work, Social Space, Spatial Architectonics, From Absolute Space to Abstract Space, Contradictory Space, From the Contradictions of Space to Differential Space, Openings and Conclusions.

what has gone before. Certain concepts clearly defined at one point either seem to disappear or become confusingly redefined in another way elsewhere.[5] The narrative is distinctly unruly, punctuated with seemingly spontaneous riffs on a range of subjects that is bound to offend the narrow specialist with its quick generalities while challenging even the most perspicacious polymath with its endless scope and specificity.

Like Borges's Aleph, Lefebvre's composition on the production of space, that most general of products, "recollects" a dizzying array of different kinds of spaces. In a recent appreciation of Lefebvre as a forerunner of critical postmodernism, Michael Dear lists these spaces alphabetically.[6] With many of my own additions, they include absolute, abstract, appropriated, architectonic, architectural, behavioral, body, capitalist, conceived, concrete, contradictory, cultural, differentiated, dominated, dramatized, epistemological, familial, fragmented, fresh, geometrical, global, hierarchical, historical, homogeneous, ideological, imagined, impossible, institutional, instrumental, leisure, lived, masculine, mental, natural, neutral, new, opaque, organic, original, perceived, physical, plural, political, possible, pure, real, "real," representational, repressive, sensory, social, socialist, socialized, state, traditional, transparent, true, urban, utopian, and women's space.

What then harmonizes this cacophony of spaces? What gives his composition its thematic structure? What moves *The Production of Space* beyond the string of eclectic recollections in Borges's Aleph? First of all, I suggest, there is Lefebvre's nomadic meta-Marxism and the explicit political and theoretical project – and choice – it engenders.

> The path I shall be outlining here is thus bound up with a strategic hypothesis – that is to say, with a long range theoretical and practical project. Are we talking about a political project? Yes and no. It certainly embodies a politics of space, but at the same time goes beyond politics inasmuch as it presupposes a critical analysis of all spatial politics as of all politics in general. By seeking to point the way towards a different space,

---

[5] In his recent review essay on *The Production of Space*, Andy Merrifield makes a similar argument and suggests that Lefebvre, reflecting the strong influence of Nietzsche on his work, actually tries to subvert the coherence of the introductory chapter. "To my mind, he goes out of his way *not* to follow the "Plan" since he is wary of pinning himself down." See Merrifield, "Lefebvre, Anti-Logos and Nietzsche: An Alternative Reading of *The Production of Space*," *Antipode* 27 (1995), 294–303.

[6] Michael Dear, "Postmodern Bloodlines," in G. Benko and U. Strohmayer, eds, *Space and Social Theory: Geographic Interpretations of Postmodernity*, Cambridge, MA, and Oxford, UK: Blackwell, 1996.

> towards the space of a different (social) life and of a different
> mode of production, this project straddles the breach between
> science and utopia, reality and ideality, conceived and lived. It
> aspires to surmount these oppositions by exploring the dialecti-
> cal relationship between "possible" and "impossible," and this
> both objectively and subjectively.   (1991a: 60)

This project fixes Lefebvre in the present, no matter how far into the
past he roams; and grounds his spatial politics even more inextricably in the social processes of production and reproduction, in the
incessant theme that is summarily captured and intoned in that
repeated phrase: the social production of social space.

But recognizing Lefebvre's theoretical and political project is not
enough to comprehend the conceptualization of Thirdspace that
flows through the pages of his contrapuntal text. To be sure, every
chapter offers bits of sweet music to the ears of those (the present
author included) who share a similar political project; and the force
of his commitment helps to keep the text on track, giving its unruly
composition and florescent detours a semblance of order and discipline amidst its radical openness and polyphony. To proceed further
into a practical and theoretical understanding of Thirdspace, however, requires the exposition of an additional theme, one that is
embedded deeply in *The Production of Space* but never systematically
extrapolated for the reader. I call this, for want of a better term, a
critical strategy of *thirding-as-Othering*, and suggest that it provides
the keynote to Lefebvre's politicized fugue on the meanings and
knowledges of Thirdspace.

Thirding-as-Othering

For Lefebvre, reductionism in all its forms, including Marxist versions, begins with the lure of binarism, the compacting of meaning
into a closed either/or opposition between two terms, concepts, or
elements. Whenever faced with such binarized categories
(subject–object, mental–material, natural–social, bourgeoisie–proletariat, local–global, center–periphery, agency–structure), Lefebvre
persistently sought to crack them open by introducing an-Other
term, a third possibility or "moment" that partakes of the original
pairing but is not just a simple combination or an "in between" position along some all-inclusive continuum. This critical thirding-
as-Othering is the first and most important step in transforming
the categorical and closed logic of either/or to the dialectically open
logic of both/and also. . . .

Thirding-as-Othering is much more than a dialectical synthesis *à la*

Hegel or Marx, which is too predicated on the completeness and temporal sequencing of thesis/antithesis/synthesis. Thirding introduces a critical "other-than" choice that speaks and critiques through its otherness. That is to say, it does not derive simply from an additive combination of its binary antecedents but rather from a disordering, deconstruction, and tentative reconstitution of their presumed totalization producing an open alternative that is both similar and strikingly different. Thirding recomposes the dialectic through an intrusive disruption that explicitly spatializes dialectical reasoning along the lines of Lefebvre's observation quoted in chapter 1. The spatialized dialectic "no longer clings to historicity and historical time, or to a temporal mechanism such as thesis–antithesis–synthesis or affirmation–negation–negation of the negation." Thirding produces what might best be called a cumulative *trialectics* that is radically open to additional othernesses, to a continuing expansion of spatial knowledge.

Stated differently, asserting the third-as-Other begins an expanding chain of heuristic disruptions, strengthening defenses against totalizing closure and all "permanent constructions." Each thirding and each trialectic is thus an "approximation" that builds cumulatively on earlier approximations, producing a certain practical continuity of knowledge production that is an antidote to the hyperrelativism and "anything goes" philosophy often associated with such radical epistemological openness. The "third" term – and Thirdspace as a concept – is not sanctified in and of itself. The critique is not meant to stop at three, to construct a holy trinity, but to build further, to move on, to continuously expand the production of knowledge beyond what is presently known. Lefebvre organizes *The Production of Space* around just such a thirding of his own longstanding interest in the dialectic of the lived and the conceived, the "real" and the "imagined," the material world and our thoughts about it. He produces from this a trialectics of spatiality that at the same time is his most creative contribution to an understanding of social space and the source of an often bewildering confusion on the part of even his most sympathetic interpreters. I refer specifically here to the "conceptual triad" that "emerges" in the first chapter of *The Production of Space*: SPATIAL PRACTICE, REPRESENTATIONS OF SPACE, and SPACES OF REPRESENTATION (a better term than the "representational spaces" that is used in the English translation).

Lefebvre begins this provocative thirding-as-Othering with a different triad of "fields" that he describes as usually "apprehended separately, just as molecular, electromagnetic, and gravitational forces are in physics." This triad of fields is presented twice in one paragraph, in a (deliberately?) non-corresponding order, to suggest the possibility of constructing a "unitary theory" from the "mere bits and pieces of knowledge," the fragments into which spatial knowledge has historically been broken.

> The fields we are concerned with are, first, the *physical* – nature, the Cosmos; secondly, the *mental*, including logical and formal abstractions; and thirdly, the *social*. In other words, we are concerned with logico-epistemological space, the space of social practice, the space occupied by sensory phenomena, including products of the imagination such as projects and projections, symbols, and utopias.   (1991a: 11–12)[7]

After this perplexing introduction, Lefebvre proceeds to fuse (objective) physical and (subjective) mental space into social space through a critique of what he called a "double illusion." This powerful attack on reductionism in spatial thinking is a vital part of the thirding process, working to break down the rigid object–subject binarism that has defined and confined the spatial imagination for centuries, while simultaneously maintaining the useful knowledges of space derived from both of these binary "fields." In this first round of thirding, *social space* takes on two different qualities. It serves both as a separable field, distinguishable from physical and mental space, *and/also* as an approximation for an all-encompassing mode of spatial thinking. Lefebvre continued to use social space in both ways throughout the text. Thirdspace, as I have been defining it, retains the multiple meanings Lefebvre persistently ascribed to social space. It is both a space that is distinguishable from other spaces (physical and mental, or First and Second) and a transcending composite of all spaces (Thirdspace as Aleph).

In *Postmodern Geographies*, I simplified Lefebvre's critique of the double illusion into one of myopia (nearsightedness, seeing only what is right before your eyes and no further) and hypermetropia (farsightedness, seeing so far into the distance that what is immediately before you disappears); and then used these illusions to criticize the epistemological dualism of objectivist-materialist and subjectivist-idealist approaches that has dominated the modern dis-

---

[7] One of the confusions arising from this passage is the possibility that the last phrase, beginning with "including products of the imagination . . . " is attached directly to "the space occupied by sensory phenomena" – or what Lefebvre called spatial practice and *l'espace perçu*. The original French is somewhat clearer. Rather than implying that "*l'imaginaire, les projets et projections, les symboles, les utopies*" (listed quite differently from the English translation) are "included" within the space occupied by sensory phenomena ("*celui qu'occupent les phénomènes sensibles*"), the original seems to be saying that all three of these spaces together should not exclude ("*sans exclure*") these elements. In other words, all spaces should be seen as filled with the products of the imagination, with political projects and utopian dreams, with both sensory and symbolic realities.

cipline of Geography since its origins. Without describing it as such,
I too was involved in a thirding-as-Othering that was similarly
designed to break down and disorder a rigid dichotomy and create a
Thirdspace, an alternative "postmodern geography" of political
choice and radical openness attuned to making practical sense of the
contemporary world. Lefebvre's critique, however, is much broader
and far-ranging. It is useful to examine his treatment of the double
illusion in greater detail.

  The illusion of transparency, for Lefebvre, makes space appear
"luminous," completely intelligible, open to the free play of human
agency, willfulness, and imagination. It also appears innocent, free
of traps or secret places, with nothing hidden or dissimulated,
always capable of being "taken in by a single glance from that
mental eye which illuminates whatever it contemplates"

(1991a: 28).

> What happens in space lends a miraculous quality to thought,
> which becomes incarnate by means of a *design* (in both senses of
> the word). The design serves as a mediator – itself of great
> fidelity – between mental activity (invention) and social activity
> (realization); and it is deployed in space.   (1991a: 27–8)

Approached this way, social space comes to be seen entirely as mental space, an "encrypted reality" that is decipherable in thoughts and
utterances, speech and writing, in literature and language, in discourses and texts, in logical and epistemological ideation. Reality is
confined to "thought things" (*res cogito*) and comprehended entirely
through its representations. What is kept at a distance, unseen and
untouched by this form of reductionism except through the medium
of subjective "design," are actual social and spatial practices, the
immediate material world of experience and realization.

  This overly abstracted "transcendental" illusion is traceable
throughout the entire history of philosophical idealism and post-
Enlightenment rationalism. In many ways, it resembles what
Marxists describe as fetishism, an obsessive fixation on ideas and
ideation emanating from the presumably infinite powers of the
Cartesian *cogito* or the Hegelian Spirit/Mind. It is also present in the
visionary and creative arts among all those who see an immanent
telos or "design" waiting to be discovered. Everything, including
spatial knowledge, is condensed in communicable representations
and re-presentations of the real world to the point that the representations substitute for the real world itself, the "*incommunicable* having no existence beyond that of an ever-pursued residual." Such
subjectivism reduces spatial knowledge to a discourse on discourse
that is rich in potential insights but at the same time filled with illu-

sive presumptions that what is imagined/represented defines the reality of social space.

In contrast, *the realistic illusion* oversubstantiates the world in a naturalistic or mechanistic materialism or empiricism, in which objective "things" have more reality than "thoughts." This illusion of "opacity," the disinclination to see much beyond the surface of things, fills the philosophy of history and the history of philosophy and science. Social space tends to be seen as either natural and naively given (the space of the sculptor or architect "working with nature," the space of the environmental or design determinist); or it is, equally naively, objectively and concretely there to be fully measured and accurately described (the space of the "geometer," the spatial systems analyst,[8] the empirical scientist, the determinedly scientific socialist or social scientist, the idiographic historian or geographer).

The "real" in this realist illusion is reduced only to material or natural objects and their directly sensed relations; the "imagined" is unseen, unmeasurable, and therefore unknowable. For Marxists, who are themselves particularly prone to this illusion of opacity as historical materialists, this resonates with what Marx described as reification, the reduction of the real solely to material objects, to things in themselves. Here, too, one can trace these illusionary tendencies of empiricism, naturalism, economism, environmental and other more social and historical forms of material determinism throughout the fabric of Western philosophy and social theory.

In their purest expression, then, the illusions of transparency and opacity coincide respectively with deterministic forms of subjectivism-idealism and objectivism-materialism. Lefebvre, however, goes on to argue that the double illusion is not always composed in such rigidly antagonistic opposition, "after the fashion of philosophical systems, which armour themselves like battleships and seek to destroy one another." On the contrary, he argues that each illusion often embodies and nourishes the other.

> The shifting back and forth between the two, and the flickering or oscillatory effect that it produces, are thus just as important as either of the illusions considered in isolation. . . . The rational is thus naturalized, while nature cloaks itself in nostalgias which supplant rationality. (1991a: 30)

Through his critical attack on the double illusion, Lefebvre opens the way to a *trialectics of spatiality*, always insisting that each mode of

---

[8] The trenchantly unsystematic Lefebvre was particularly adamant about avoiding the use of systems analysis in thinking about spatiality and the production of social space. The last sentence of the book is quite clear: "And we are concerned with nothing that even remotely resembles a system" (1991a: 423).

thinking about space, each "field" of human spatiality – the physical, the mental, the social – be seen as simultaneously real and imagined, concrete and abstract, material and metaphorical. No one mode of spatial thinking is inherently privileged or intrinsically "better" than the others as long as each remains open to the re-combinations and simultaneities of the "real-and-imagined." This rebalanced and non-illusive trialectics of spatiality, however, is for Lefebvre more an anticipated and desired state than an achieved one. For the present moment, a temporary strategic privileging is necessary to break the hammerlock of binarist logic and to prevent any form of reduction-ism from constraining the free play of the creative spatial imagina-tion. Lefebvre thus begins his critical thirding-as-Othering by focusing attention on *social space*, first as a distinctively different way of thinking about space that has long been obscured by exclusive fix-ations on illusive materialist and/or idealist interpretations; and sec-ond, as an all-inclusive and radically open mode of defining the limitlessly expandable scope of the spatial imagination: the envision-ing of social space as Aleph.

As I think of ways to make this crucial thirding clearer to readers familiar with the epistemology of Marxism, I find it useful to recall that old shibboleth that asks: "Is it consciousness that produces the material world or the material world that produces consciousness?" The answer implied in Lefebvre's trialectics is "yes" to both alterna-tives, and/also something more: a combinatorial and unconfinable third choice that is radically open to the accumulation of new insights, an alternative that goes beyond (*meta*) the mere acceptance of the dualized interrogative. This choice of an-Other alternative is strategically, not presuppositionally, privileged as a means of resist-ing binary closures. It is a thirding that invites further expansion and extension, beyond not just the binary but beyond the third term as well.

The critique of the illusions of transparency and opacity lays the groundwork for the thematic trialectic that is so central to a re-reading of *The Production of Space*, that which inter-relates in a dialectically linked triad:

- Spatial Practice (*espace perçu*, perceived space);
- Representations of Space (*espace conçu*, conceived space);
- Spaces of Representation (*espace vécu*, lived space).

These "three moments of social space" are described twice in the introductory chapter, "Plan of the Present Work," both times as a numbered list with underlined emphasis. As always, Lefebvre modi-fies his descriptions as he moves along, and in subsequent chapters seems either to ignore his earlier formulations or to push them to their limits, ever ready to move on to something else.[9] I will try to capture the meanings of at least his first chapter approximations of this thematic trialectic, referring back to the original French text whenever I think this will help.

(1) *Spatial practice* is defined as producing a spatiality that "embraces production and reproduction, and the particular locations (*lieux specifiés*) and spatial sets (*ensembles*) characteristic of each social formation." It "ensures continuity and some degree of cohesion" and "implies a guaranteed level of *competence* and a specific level of *performance*" (terms he borrows from linguistics but warns should not be seen as subordinating the knowledge of space to its disciplinary hegemony). The spatial practice of a society "secretes that society's space; it propounds (*le pose*) and presupposes it (*le suppose*), in a dialectical interaction; it produces it slowly and surely as it masters and appropriates it." Spatial practice, as the process of producing the material form of social spatiality, is thus presented as both medium and outcome of human activity, behavior, and experience.

"From an analytical standpoint (*À l'analyse*), the spatial practice of a society is revealed (*se découvre*) through the deciphering of its space." To illustrate how this deciphering changes over time, Lefebvre adds a whole paragraph on "Modern" spatial practice under capitalism, which he links to the repetitive routines of everyday life (*la réalité quotidienne*); and to the routes, networks, workplaces, private life, and leisure enjoyments of the urban (*la réalité urbaine*). This materialized, socially produced, empirical space is described as *perceived* space, directly sensible and open, within limits, to accurate measurement and description. It is the traditional focus of attention in all the spatial disciplines and the material grounding for what I redescribe as Firstspace.

(2) *Representations of space* define a "conceptualized space, the space of scientists, planners, urbanists, technocratic subdividers

---

[9] This habitual nomadism has led many of the most sympathetic interpreters of Lefebvre to fix on other examples of the multitude of spaces he addresses in *The Production of Space*, often to the neglect of the keynote trialectic that encompasses them. Especially attractive, it appears, has been Lefebvre's discussion of absolute, relative (or relational), and abstract (or analogical) spaces and their developmental periodization in the history of capitalism. In Derek Gregory's *Geographical Imaginations* (1994), these spaces are listed specifically and in combination are given 34 page references in the Subject Index. Spatial practices, representations of space, and spaces of representation (or representational spaces) do not appear at all in the index, although they are mentioned on p. 403 and frame the much more extensive discussion of Abstract Space and Concrete Space that Gregory diagrams on p. 401. See also Neil Smith and Cindi Katz, "Grounding Metaphor: Toward a Spatialized Politics," in M. Keith and S. Pile, eds, *Place and the Politics of Identity*, London and New York: Routledge, 1993: 67–83, for a similar emphasis on Lefebvre's historiography of absolute, relative, and abstract space at the expense of the keynote triad.

('*découpeurs*' *et* '*agenceurs*'), as of a certain type of artist with a scientific bent – all of whom identify what is lived and what is perceived with what is conceived." This *conceived* space is also tied to the relations of production and, especially, to the order or design that they impose. Such order is constituted via control over knowledge, signs, and codes: over the means of deciphering spatial practice and hence over the production of spatial knowledge.

For Lefebvre, "this is the dominant space in any society (or mode of production)," a storehouse of epistemological power. This conceived space tends, with certain exceptions "towards a system of verbal (and therefore intellectually worked out) signs," again referring to language, discourse, texts, *logos*: the written and spoken word. In these "dominating" spaces of regulatory and "ruly" discourse, these mental spaces, are thus the representations of power and ideology, of control and surveillance. This Secondspace, as I term it, is also the primary space of utopian thought and vision, of the semiotician or decoder, and of the purely creative imagination of some artists and poets.

In a twist that would confuse (or be forgotten by) many readers, Lefebvre did not define the "dominated" space as that of material spatial practices. Instead, he turned to the third space of his triad to exemplify the controlling powers of conceived space.

(3) *Spaces of representation* are seen by Lefebvre both as distinct from the other two spaces and as encompassing them, following his strategic use of social space in his preliminary thirding. Spaces of representation embody "complex symbolisms, sometimes coded, sometimes not." They are linked to the "clandestine or underground side of social life" and also to art, which Lefebvre described as a coding not of space more generally but specifically of the spaces of representation. Clearly an attempt is being made here to retain, if not emphasize, the partial unknowability, the mystery and secretiveness, the non-verbal subliminality, of spaces of representation; and to foreground the potential insightfulness of art versus science (or, for that matter, moral philosophy or semiotics), a key pillar of Lefebvre's metaphilosophy.

Here then is space as directly *lived*, with all its intractability intact, a space that stretches across the images and symbols that accompany it, the space of "inhabitants" and "users." But it is also, Lefebvre takes care to note, inhabited and used by artists, writers, and philosophers – to which he would later add ethnologists, anthropologists, psychoanalysts, and other "students of such representational spaces" – who seek only to *describe* rather than decipher and actively transform the worlds we live in. He follows these references with two key points. First: "this is the dominated – and hence passively experienced (*subi*) or subjected – space which the imagination (verbal but especially non-verbal) seeks to change and appropriate. It overlays (*recouvre*) physical space, making symbolic use of its objects" and tends towards "more or less coherent systems of non-verbal symbols and signs." Second: here we can find not just the spatial representations of power but the imposing and operational power of spatial representations. Combining the real and the imagined, things and thought on equal terms, or at least not privileging one over the other *a priori*, these lived spaces of representation are thus the terrain for the generation of "counterspaces," spaces of resistance to the dominant order arising precisely from their subordinate, peripheral or marginalized positioning. With its foregrounding of relations of dominance, subordination, and resistance; its subliminal mystery and limited knowability; its radical openness and teeming imagery, this third space of Lefebvre closely approximates what I am defining as Thirdspace.

Both Thirdspace and Lefebvre's most encompassing notion of social space are comprised of all three spatialities – perceived, conceived, and lived – with no one inherently privileged *a priori*. And yet, there is an implied preference in all of Lefebvre's (and my) spatial trialectics and thirdings that derives not from ontological privilege or priority but from that political choice that is so central to Lefebvre's spatial imagination. It is political choice, the impetus of an explicit political project, that gives special attention and particular contemporary relevance to the spaces of representation, to *lived space as a strategic location* from which to encompass, understand, and potentially transform all spaces simultaneously. Lived social space, more than any other, is Lefebvre's limitless Aleph, the space of all inclusive simultaneities, perils as well as possibilities: the space of radical openness, the space of social struggle.

For Lefebvre, the spaces of representation teem with symbols, hence the tendency of some to see him primarily as a semiologist and to describe lived space as "symbolic" space. These spaces are also vitally filled with politics and ideology, with the real and the imagined intertwined, and with capitalism, racism, patriarchy, and other material spatial practices that concretize the social relations of production, reproduction, exploitation, domination, and subjection. They are the "dominated spaces," the spaces of the peripheries, the margins and the marginalized, the "Third Worlds" that can be found at all scales, in the corpo-reality of the body and mind, in sexuality and subjectivity, in individual and collective identities from the most local to the most global. They are the chosen spaces for struggle, liberation, emancipation.

> Representational space is alive: it speaks. It has an affective kernel (*noyau*) or centre: Ego, bed, bedroom, dwelling, house; or: square, church, graveyard. It embraces the loci of passion, of action, of lived situations, and this immediately implies time. Consequently it may be qualified in various ways: it may be directional, situational or relational, because it is essentially qualitative, fluid and dynamic. (1991a: 42)

Spaces of representation contain all other real and imagined spaces simultaneously. In different ways, appropriately understood, so too do spatial practices and representations of space, but only in so far as one can escape from the double illusion of objectivism and subjectivism that weakens their insights into the workings of power.[10] Lefebvre never made explicit his strategic "preference" for the spaces of representation, in part, I believe, because he either assumed it would be implicitly understood; or avoided an explicit recognition of it for fear that it would be construed as unchallengeably fixed, a complete answer, a permanent construction that would divert too much attention away from other modes of spatial thinking (the more likely explanation).

Lefebvre was always preparing himself to move on, and move on he did after 1974, never again elaborating his spatial approximation so thoroughly yet keeping it alive in promises of further adventures, such as the *rhythmanalyse* that he proposed would complete his exposition on the production of space. In *La Présence et l'absence* (1980), however, Lefebvre's most extensive discussion and critique of the theory of representations, there are not only further explanations for his emphasis on the power-filled spaces of representation but also a veritable fountain of triplicities similar to that described for social space. In what he called a *Tableau des Triades (associées)*, he divides his list of triads into *Libido* and *Pensée*, and separates them with a mediating row described as *Représentant–Représenté–Représentation*, or roughly Representing–Represented–Representation, a triad (or trialectic) that is close to signifier–signified–signification, with the third term different yet encompassing and partially dependent upon the other two. Lefebvre refused to tabularize his triads into neat First, Second, and Third columns or to present them in triangular form, but instead lists them vertically. The following is a translated selection from the Tableau. Each one can be read as a model of trialectical thirding – and as a summation of most of the arguments presented in this chapter.

---

[10] How these spaces of representation are echoed in the heterotopias and heterotopology of Michel Foucault is discussed in chapter 5.

TOTALITY	CENTRALITY	DISPLACEMENT	HOMOGENEITY	
CONTRADICTION	PERIPHERY	SUBSTITUTION	FRAGMENTATION	
POSSIBILITY	MEDIATION	REPRESENTATION	HIERARCHY	
THING-*chose*	PRESENCE	SUBJECT	IDENTITY	
PRODUCT-*produit*	ABSENCE	OBJECT	CONTRADICTION	
WORK-*oeuvre*	REPRESENTATION	UNITY	DIFFERENCE	
FIRST NATURE	HISTORY	MELODY	PHILOSOPHY	
SECOND NATURE	SPACE	RHYTHM	SCIENTISM	etc.
PRODUCTION	GLOBALITY	HARMONY	METAPHILOSOPHY	

Discomforted by this table (a form he rarely used), Lefebvre immediately follows with one of the many poems to life, death, love, and his nomadic travels that are sprinkled throughout the book (which begins with a long unsent letter to Octavio Paz).

His discomfort aside, Lefebvre's trialectics are infused with increasing power in this galaxy of triads, each with its strategic preference for the third term but always as a transcending inclusion of the other two. The third term never stands alone, totally separate from its precedents or given absolute precedence on its own. This is the key point of Lefebvre's "dialectics of triplicity" and of what I have chosen to describe as trialectical thinking and, more specifically, the trialectics of spatiality.

Summarizing Again/Before Moving On

Thinking trialectically is a necessary part of understanding Thirdspace as a limitless composition of lifeworlds that are radically open and openly radicalizable; that are all-inclusive and transdisciplinary in scope yet politically focused and susceptible to strategic choice; that are never completely knowable but whose knowledge none the less guides our search for emancipatory change and freedom from domination. Trialectical thinking is difficult, for it challenges all conventional modes of thought and taken-for-granted epistemologies. It is disorderly, unruly, constantly evolving, unfixed, never presentable in permanent constructions.

With these warnings in mind, the first two chapters of *Thirdspace* can be graphically summarized around two simplified diagrams that serve as convenient road maps to all the journeys that will follow. The first presents an *ontological trialectic*, a statement of what the world must be like in order for us to have knowledge of it. It is a crude picture of the nature of social being, of human existence, and also of the search for practical knowledge and understanding.

Although primarily an ontological assertion, the trialectics of Spatiality, Historicality, and Sociality (summary terms for the social production of Space, Time, and Being-in-the-world) apply at all levels of knowledge formation, from ontology to epistemology, theory building, empirical analysis, and social practice. At all these levels, however, there has been a persistent tendency during at least the past century to over-privilege, in another "double illusion," the dynamic relations between the "making" of Historicality and the "constitution" of social practices or Sociality. Built into the arguments of *The Production of Space* as I have excavated them is a critical thirding-as-Othering that involves the reassertion of Spatiality against this pronounced tendency in Western philosophy, science, historiography, and social theory (including its most critical variants) to bifocalize on the interactive Historicality and Sociality of being.[11]

When the Trialectics of Being is reduced to the relations between Historicality and Sociality, Spatiality tends to be peripheralized into the background as reflection, container, stage, environment, or external constraint upon human behavior and social action. Muted efforts to reactivate space more centrally can be found in the writings of many critical philosophers, from Kant and Hegel to Heidegger and Sartre; and in the work of critical social theorists (Simmel, Kracauer, Benjamin, Giddens, Harvey, to name only a few). But no one has so

---

[11] Chapter 6 contains a more detailed discussion of this over-privileging of Historicality and Sociality.

forcefully and successfully activated Spatiality and rebalanced the trialectic as Henri Lefebvre.

Applying Lefebvre's critique, the Historicality–Sociality, or more simply, history–society link has too often been conceptualized in the form of an all-inclusive ontological and epistemological dyad, with the "shifting back and forth between the two" and the "flickering effect" that this produces creating often illusory knowledges that "embody" and "nourish" each other. In the wake of this circumscribed oscillation, Spatiality in nearly all its forms is unproblematically silenced, pushed to the periphery, to the margins of critical intellectual inquiry. With a profound voice from the periphery, Lefebvre deconstructs the dualism with an-Other term, and in so doing reconstitutes a social ontology that is radically opened to Spatiality in at least two ways: through what I once called the "socio-spatial dialectic" (Spatiality–Sociality) and through the problematic interplay between space and time, the making of historical geographies or geohistories (Spatiality–Historicality).

In the first opening, the social and the spatial are seen as mutually constructed, with neither deterministically privileged *a priori*. Sociality, both routinely and problematically, produces spatiality, and *vice versa*, putting to the forefront of critical inquiry a dynamic socio-spatial dialectic that by definition is also intrinsically historical. A similar trialectical logic infuses the spatio-temporal structuration of Sociality. Historicality and Spatiality, or more familiarly history and geography, intertwine in a simultaneously routine and problem-filled relation that evokes another crucial field of inquiry and interpretation in the spatio-temporal or geohistorical dialectic. The Trialectics of Being thus generates three ontological fields of knowledge formation from what for so long has only been one (Historicality–Sociality).

The three moments of the ontological trialectic thus contain each other; they cannot successfully be understood in isolation or epistemologically privileged separately, although they are all too frequently studied and conceptualized this way, in compartmentalized disciplines and discourses. Here again, however, the third term, Spatiality, obtains a strategic positioning to defend against any form of binary reductionism or totalization. The assertion of Spatiality opens the Historicality and Sociality of human lifeworlds to interpretations and knowledges that many of its most disciplined observers never imagined, while simultaneously maintaining the rich insights they provide for understanding the production of lived space. It is in this sense that Lefebvre's trialectics contains a deep, if indirect, critique of historicism (the excessive confinement in knowledge production that arises from interpreting the world only through its Historicality); and the similar overemphasis on the evolutionary and revolutionary power of social will and political consciousness (freed from spatial or environmental constraints) that enveloped both the liberal social sciences and radical scientific socialism beginning in the last half of the 19th century.[12]

All excursions into Thirdspace begin with this ontological restructuring, with the presupposition that being-in-the-world, Heidegger's *Dasein*, Sartre's *être-là*, is existentially definable as being simultaneously historical, social, and spatial. We are first and always historical-social-spatial beings, actively participating individually and collectively in the construction/production – the "becoming" – of histories, geographies, societies. Our essential historicity and sociality have long been recognized in all the human sciences. The project begun by Lefebvre in the 1960s, and only now beginning to be understood and realized, was nothing less than to reassert the equally existential spatiality of life in a balanced trialectic that ranges from ontology through to a consciousness and *praxis* that are also simultaneously and presuppositionally social, historical, *and spatial*.[13]

A second diagram (triagram?), building off the first, moves us further into the multiple meanings of Thirdspace. Here the emphasis shifts from an existential ontology (statements about what the world must be like in order for us to exist as social beings) to a more specific discussion of the epistemology of space (how we can obtain accurate and practicable knowledge of our existential spatiality). Again, for at least the past century, there has been a "double illusion" bracketing the accumulation of spatial knowledge within the

---

[12] Still the best treatment of the origins of the liberal (or "bourgeois") social sciences is H. S. Hughes, *Consciousness and Society: The Reconstruction of European Thought, 1890–1930*, New York: Knopf, 1958.

[13] Much of the contemporary literature interpreting Lefebvre's work seems incapable of engaging in this balanced form of trialectical thinking. Instead, one form or another of rigidly categorical logic is applied to argue that Lefebvre was fundamentally a sociologist or a geographer or a historian. In their recent reappropriation of Lefebvre as an urban sociologist, for example, Kofman and Lebas distinguish his *Writings on Cities* from his more "purely spatial" theorizing, missing in this misconstrual not only Lefebvre's presuppositional definition of the urban as spatial but also his insistence that space should never be theorized purely, apart from its historicality and sociality (and so too for history and sociology). Similarly, they take his sensitivity to time and history as a sign of his supposedly deeper prioritization of historicality versus spatiality and chide me and other "Anglo-American geographers" for falsely appropriating Lefebvre to "dethrone" history and "crown" geography as the master discipline, "at the expense of an impoverished historical understanding and simplification of the richness of temporalities and their significance for lived experience" (1996: 47–8). Derek Gregory, in his *Geographical Imaginations* (1994: 359–60), goes even further to label Lefebvre a "historicist," a secretive Hegelian using spatial tropes to construct a determinative "history of the present." This sort of categorical labelling and construction of neat either/or oppositions between spatiality, historicality, and sociality seriously distorts Lefebvre's nomadic metaphilosophy of approximations and his assiduously historical and social interpretation of the production of space.

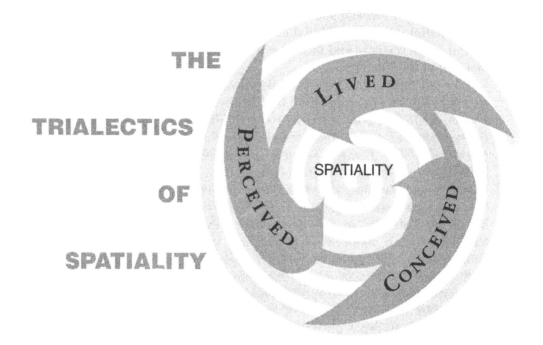

oscillation between two contrasting epistemes, two distinctive modes of producing spatial knowledge. Lefebvre's trialectics help us to break open this dualism to a third alternative, to other ways of making practical sense of the spatiality of social life.

Figure 2b is, of course, deceptively simplified. As with the ontological trialectic in figure 2a, each term appropriately contains the other two, although each is distinguishable and can be studied in splendidly specialized isolation. No one of the three forms of spatial knowledge is given a priori or ontological privilege, but again there is a strategic privileging of the third term, in this case Thirdspace, as a means of combating the longstanding tendency to confine spatial knowledge to Firstspace and Secondspace epistemologies and their associated theorizations, empirical analyses, and social practices. At the risk of oversimplification, each of these modes of accumulating spatial knowledge and what might be called their characteristic "mentalities" can now be briefly and tentatively described.

Firstspace epistemologies and ways of thinking have dominated the accumulation of spatial knowledge for centuries. They can be defined as focusing their primary attention on the "analytical deciphering" of what Lefebvre called Spatial Practice or perceived space, a material and materialized "physical" spatiality that is directly comprehended in empirically measurable *configurations*: in the absolute and relative locations of things and activities, sites and situations; in patterns of distribution, designs, and the differentiation of a multitude of materialized phenomena across spaces and places; in the concrete and mappable geographies of our lifeworlds, ranging from the emotional and behavioral space "bubbles" which invisibly surround our bodies to the complex spatial organization of social practices that shape our "action spaces" in households, buildings, neighborhoods, villages, cities, regions, nations, states, the world economy, and global geopolitics.

Firstspace epistemologies tend to privilege objectivity and materiality, and to aim toward a formal science of space. The human occupance of the surface of the earth, the relations between society and nature, the architectonics and resultant geographies of the human "built environment," provide the almost naively given sources for the accumulation of (First)spatial knowledge. Spatiality thus takes on the qualities of a substantial text to be carefully read, digested, and understood in all its details. As an empirical text, Firstspace is conventionally read at two different levels, one which concentrates on the accurate description of surface appearances (an indigenous mode of spatial analysis), and the other which searches for spatial explanation in primarily exogenous social, psychological, and biophysical processes.

Over the years, various schemas have been indigenously devised in the spatial disciplines to describe the virtually limitless factual scope and "scape" of Firstspace knowledge. One that I remember from my own education as a geographer consisted of two matrices, with the rows of both representing locations or places. In the first matrix, the columns itemized a potentially endless list of "attributes;" while the columns of the second matrix contained an also potentially endless list of "interactions" or flows connecting different places. Spatial analysis would then consist of quantitative measurement, pattern recognition, and correlative modeling of the row and column variables and their "regional" configurations. Another memorable schema for Firstspatial analysis defined five nested tiers of pattern finding, from the flowing "movements" of people, goods, and information, to the formation of regular "networks" of such movements, to the growth of "nodes" in these networks, to the hierarchical patterning of these nodes of different sizes and functions, and finally to composite "surfaces" that summarize the uneven development over space of all these patternings together.

In the field of Geography, these and other schemas coagulated to form the conceptual foundation for a fundamentally positivist "spatial science" based primarily on the quantitative and mathematical description of these spatial data patternings of Firstspace. Mathematical functions were "fit" into the point patterns of settlement nodes to explain their distributional configuration; a form of "social physics" based on the gravity model and other measures of the "friction of distance" was used to describe human interactions over space and the concentric zoning of land uses in cities and around markets; graph theory and mathematical topology was applied to measure accessibility levels in transportation and communications networks; multivariate statistical ecologies were employed to describe complex urban and regional geographies; two- and three-dimensional languages and shape grammars were developed to describe all sorts of spatial forms, structures, and designs.[14]

Today, spatial science is being increasingly focused on Geographical Information Systems (GIS) and remote sensing by satellite photography to collect and organize massive data banks describing the empirical content of Firstspace. Many of these techniques have proved useful in such fields as architecture and urban planning, archeology and anthropology, petroleum geology and military science. What these techniques provide are more sophisticated and objectively accurate ways to do what most geographers, spatial analysts, and, for that matter, the colonial adventurers and cartographers in the Age of Exploration, had been doing all along: accumulating and mapping what was presumed to be increasingly accurate "factual" knowledge about places and the relations between places over the surface of the earth. The key difference brought about by this so-called quantitative "revolution" in geography was the presumption that these increasingly accurate empirical descriptions of geographical "reality" also contained the intrinsic sources of spatial *theory*.

When captivated by such realist illusions and impelled by the presumptions of scientism, Firstspace epistemologies become fixated on the material form of things in space: with human spatiality seen primarily as outcome or *product*. Explanation and theory-building in turn derived essentially from the material form and covariation of spatial patterns, with one or more "independent" geographical configurations "explaining" the "dependent" configuration or outcome in increasingly complex equations and causal chains. In Geography as well as in the more scientific approaches to other spatial disciplines, this positivist epistemology, in one form or another, continues to dominate mainstream spatial thinking and analysis.

Responding to the epistemological confinement of spatial science, more advanced (and less illusory) forms of Firstspace analysis have developed well beyond such descriptive matrices to explore the historicality and sociality of spatial forms. In this more exogenous mode of Firstspatial analysis, human spatiality continues to be defined primarily by and in its material configurations, but explana-

---

[14] My own doctoral dissertation was inspired by this spatial science and its enhancements of more traditional approaches to geographical research. See Edward W. Soja, *The Geography of Modernization in Kenya*, Syracuse, NY: Syracuse University Press, 1968.

tion shifts away from these surface plottings themselves to an inquiry into how they are socially produced. A rich tradition of historical geography, for example, relocates the sources of geographical understanding and explanation from positivist science to the powers of the historical imagination and narrative historiography. The social production of Firstspace is treated as a historical unfolding, an evolving sequence of changing geographies that result from the dynamic relations between human beings and their constructed as well as natural environments.

Relatively new fields of behavioral and/or critical human geography – often significantly influenced by the varied forms of structuralism that have developed in the 20th century – seek their sources of understanding the social production of Firstspace either in individual and collective psychologies or, more directly, in the social processes and practices presumed to be underlying and structuring the production of material spatialities. Marxist geographers of Firstspace, for example, explain the material worlds of human geography and geographically uneven development through appeals to class analysis, the labor theory of value, and the evolving historical effects of the interplay between social relations of production and the development of the productive forces. The normal operations of the market, backed by the presumption of rational profit-seeking or cost-minimizing behavior, are used by other economic geographers and spatial economists to explain the same spatial outcomes; while more humanistic cultural geographers, drawing upon phenomenology and hermeneutics, seek to root the material patternings of space in the imprint of cultural beliefs and commitments and the free play of human nature.

In these and other attempts to study and understand the production of Firstspace exogenously, explanation and theory-building tend to derive from presumably non-spatial variables, behaviors, and social activity such as historical development, class consciousness, cultural preferences, and rational economic choice. The flow of causality in this epistemology thus tends to run primarily in one direction, from historicality and sociality to the spatial practices and configurations. This has resulted in an increasingly rigorous and insightful understanding of how Firstspace is socially produced, as well as a welcome exploration by geographers and other spatial analysts of a wide range of "non-spatial" disciplines and ideas.

For many different reasons, however, relatively little attention is given to the causal flow in the other direction, that is, to how material geographies and spatial practices shape and affect subjectivity, consciousness, rationality, historicality, and sociality. The impact of Firstspace on history and society in the broadest sense has either been ignored by those who seek "external" and presumably non-spatial explanations of material spatial configurations; or else it is consciously avoided for fear that Firstspatial analysis will fall prey to the presumed dangers of environmental or spatial determinism. Such Firstspace determinisms were rampant within the spatial disciplines in the 19th and early 20th centuries, and played an important role in the formation of the liberal social sciences and Marxian scientific socialism (as well as related critical theories of history and society) that were explicitly constructed to free social will and social consciousness from such "external" Firstspatial determinations. Although various forms of spatial causality with regard to the sociality and historicality of life are suggestively implied in more contemporary Firstspatial analyses, there continues to be a prevalent inhibition that constrains attempts to explore the trialectics of spatiality–historicality–sociality in its fullest complexity and interdependency. This ontological confinement is only one of the reasons why Firstspace epistemologies, despite their broad purview and impressive accumulation of accurate spatial knowledge, are fundamentally incomplete and partial.

In the long history of spatial thinking, *Secondspace epistemologies* have tended to arise in reaction to the excessive closure and enforced objectivity of mainstream Firstspace analysis, pitting the artist versus the scientist or engineer, the idealist versus the materialist, the subjective versus the objective interpretation. In part through these ideational and epistemological critiques of Firstspace approaches, the boundaries between First- and Secondspace knowledge have become increasingly blurred, especially with the intermixing of positivist, structuralist, poststructuralist, existential, phenomenological, and hermeneutic ideas and methods. Firstspace analysts now frequently adopt Secondspace epistemologies for their purposes, and Secondspace interpretations often extend themselves specifically to address actual material spatial forms.

Using Lefebvre's words again, there are times when these two epistemological worlds "armour themselves like battleships and seek to destroy one another;" but there are also times when each "embodies and nourishes the other," with shiftings back and forth marking the historical relations between Firstspace and Secondspace. Buried beneath these oscillations, however, has been a presumption of epistemological completeness that channels the accumulation of spatial knowledge into two main streams or some selective combination of both. Little room is left for a lateral glance beyond the long-established parameters and perimeters that map the overlapping terrains of Firstspace and Secondspace.

Despite these overlappings, Secondspace epistemologies are immediately distinguishable by their explanatory concentration on conceived rather than perceived space and their implicit assumption that spatial knowledge is primarily produced though discursively devised representations of space, through the spatial workings of the mind. In its purest form, Secondspace is entirely ideational, made up of projections into the empirical world from conceived or imagined geographies. This does not mean that there is no material reality, no Firstspace, but rather that the knowledge of this material reality is comprehended essentially through thought, as *res cogito*, literally "thought things." In so empowering the mind, explanation becomes more reflexive, subjective, introspective, philosophical, and individualized.

Secondspace is the interpretive locale of the creative artist and artful architect, visually or literally re-presenting the world in the image of their subjective imaginaries; the utopian urbanist seeking social and spatial justice through the application of better ideas, good intentions, and improved social learning; the philosophical geographer contemplating the world through the visionary power of scientific epistemologies or the Kantian envisioning of geography as way of thinking or the more imaginative "poetics" of space; the spatial semiologist reconstituting Secondspace as "Symbolic" space, a world of rationally interpretable signification; the design theorist seeking to capture the meanings of spatial form in abstract mental concepts. Also located here are the grand debates about the "essence" of space, whether it is "absolute" or "relative" and "relational," abstract or concrete, a way of thinking or a material reality.

However its essence is defined, in Secondspace the imagined geography tends to become the "real" geography, with the image or representation coming to define and order the reality. Actual material forms recede into the distance as fixed, dead signifiers emitting signals that are processed, and thus understood and explained when deemed necessary, through the rational (and at times irrational) workings of the human mind. Here too, as with Firstspace, there are at least two levels of conceptualization, one introverted and indigenous and the other more extroverted and exogenous in its perspective.

The more internal or indigenous approaches to Secondspace knowledge – ranging from the egocentric, self-explanatory genius of the "starchitect" as masterful creator of spaces to the mental mappers of human spatial cognition and the epistemological referees enforcing their control over spatial knowledge production – are particularly subject to illusory insights. To illustrate, I remember an intense flurry of interest in Geography over the elucidation of "cognitive maps," mental images of space that we all carry with us in our daily lives. Many studies were done to elicit such mental maps across gender, race, and class lines by asking individuals to draw maps of the city in which they lived. Various techniques were used to summarize these imagined urban depictions and the resultant composite mappings were then compared.

Often, some very interesting insights about human spatiality were produced. But equally often the interpretation abruptly ended with naive categorical idealizations, such as "men's mental maps are extensive, detailed, and relatively accurate" while women's were "domicentric" (centered on the home), more compact, and less accurate in terms of urban details; or, the poor have highly localized mental maps in contrast to the wealthy, whose mental maps come close to reproducing a good road map from the gas station. Readers were left with the impression that the conceived space defined an urban reality on its own terms, the mental defined and indeed produced and explained the material and social worlds better than precise empirical descriptions. In such illusions of transparency, as Lefebvre called them, Firstspace collapses entirely into Secondspace. The difference between them disappears. Even more significantly, also lost in the transparency of space are its fundamental historicality and sociality, any real sense of how these cognitive imageries are themselves socially produced and implicated in the relations between space, power, and knowledge.

The more advanced and exogenous approaches to Secondspace derive either directly from idealist philosophies (e.g. Hegelianism) or from what might be called the idealization of epistemology, its confident representation as a masterful and complete ordering of reality. Lefebvre was particularly concerned with the hegemonic power often ascribed to (and by) this idealized and elevated spatial epistemology. More than anything else, it made the Representations of (Second)Space what he called the dominant space, surveying and controlling both spatial practices and the lived spaces of representation.

Looking back over the past three decades, this seemingly overdrawn concentration on the power of epistemology and on the "dominance" of Secondspace explanatory perspectives (which in so many other ways were much less important in the spatial disciplines than was Firstspace analysis) makes much more sense. Here was the most powerful blockage to the creative rethinking of spatiality, to the trialectical reassertion of Spatiality in ontological conjunction with Sociality and Historicality, to the struggle against all forms of spatial reductionism and disciplinary fragmentation. The broader philosophical hegemony of what I will call modernist epistemologies and their tacit silencing of other knowledges would become the primary focus for a series of post-prefixed (postmodern, poststructuralist, post-Marxist, postcolonial) and related feminist critiques of modernism more generally. Lefebvre's targeting of Secondspace epistemologies as dominant and dominating was thus an important precursor to these more recent developments, which will be explored at greater length in later chapters.

Thirdspace epistemologies can now be briefly re-described as arising from the sympathetic deconstruction and heuristic reconstitution of the Firstspace–Secondspace duality, another example of what I have called thirding-as-Othering. Such thirding is designed not just to critique Firstspace and Secondspace modes of thought, but also to reinvigorate their approaches to spatial knowledge with new possibilities heretofore unthought of inside the traditional spatial disciplines. Thirdspace becomes not only the limitless Aleph but also what Lefebvre once called the city, a "possibilities machine;" or, recasting Proust, a madeleine for a *recherche des espaces perdus*, a remembrance-rethinking-recovery of spaces lost ... or never sighted at all.[15]

The starting-point for this strategic re-opening and rethinking of new possibilities is the provocative shift back from epistemology to ontology and specifically to the ontological trialectic of Spatiality–Historicality–Sociality. This ontological rebalancing act induces a radical skepticism toward all established epistemologies, all traditional ways of confidently obtaining knowledge of the world. Many, especially in the explicitly postmodern discourse, have reacted to this epistemological crisis by unleashing a free-wheeling anything-goes eclecticism or hyperrelativism, almost always without addressing the new ontological issues being raised. Addressing them directly, seeking to rebalance the ontological foundations of knowledge formation, however, makes a significant difference. Such ontological restructuring, at least for the present moment, re-centers

---

[15] The reference to Proust brings to mind an interesting passage in Joseph Frank's "Spatial Form in Modern Literature," originally published in 1945 (*Sewanee Review* 53) and reissued in *The Idea of Spatial Form* (New Brunswick and London: Rutgers University Press, 1991). Against the grain of most literary treatments of Proust as infatuated with time *per se*, Frank writes:

> To experience the passage of time, Proust had learned, it was necessary to rise above it and to grasp both past and present simultaneously in a moment of what he called "pure time." But "pure time," obviously, is not time at all – it is perception in a moment of time, that is to say, space. And by the discontinuous presentation of character Proust forces the reader to juxtapose disparate images spatially, in a moment of time, so that the experience of time's passage is communicated directly to his sensibility. Ramon Fernandez has acutely stressed this point in some remarks on Proust and Bergson. "Much attention has been given to the importance of time in Proust's work ... but perhaps it has not been sufficiently noted that he gives time the value and characteristics of space.... [T]he reactions of his intelligence on his sensibility, which determine the trajectory of his work, would orient him rather toward a *spatialisation* of time and memory." (1991: 26–7)

Remember these words, for their meaning will re-appear in a different guise in chapter 6.

knowledge formation first around the long-submerged and subordinated spatiality of existential being and becoming, and then in the spatialization of historicality and sociality in theory-formation, empirical analysis, critical inquiry, and social practice.

This far-reaching spatialization, I believe, was Lefebvre's primary intent in *The Production of Space*. As he persistently demonstrated, such knowledge is not obtained in permanent constructions confidently built around formalized and closed epistemologies, but through an endless series of theoretical and practical approximations, a critical and inquisitive nomadism in which the journeying to new ground never ceases. Accordingly, I leave the discussion of Thirdspace epistemologies radically open. We must always be moving on to new possibilities and places.

3

Exploring the Spaces that Difference Makes: Notes on the Margin

Postmodern culture with its decentered subject can be the space where ties are severed or it can provide the occasion for new and varied forms of bonding. To some extent, ruptures, surfaces, contextuality, and a host of other happenings create gaps that make space for oppositional practices which no longer require intellectuals to be confined to narrow separate spheres with no meaningful connection to the world of the everyday ... a space is there for critical exchange ... [and] this may very well be "the" central future location of resistance struggle, a meeting place where new and radical happenings can occur.

(bell hooks, *Yearning*, 1990: 31)

Distinctive features of the new cultural politics of difference are to trash the monolithic and homogeneous in the name of diversity, multiplicity, and heterogeneity; to reject the abstract, general and universal in light of the concrete, specific and particular; and to historicize, contextualize and pluralize by highlighting the contingent, provisional, variable, tentative, shifting and changing. ... [T]hese gestures are not new in the history of criticism ... yet what makes them novel – along with the cultural politics they produce – is how and what constitutes difference, the weight and gravity it is given in representation and the way in which highlighting issues like exterminism, empire, class, race, gender, sexual orientation, age, nation, nature, and region at this historical moment acknowledges some discontinuity and disruption from previous forms of cultural critique. To put it bluntly, the new cultural politics of difference consists of creative responses to the

precise circumstances of our present moment ... in order to empower
and enable social action.
(Cornel West, "The New Cultural Politics of Difference," 1990: 19–20)

These introductory remarks from bell hooks and Cornel West re-
open the exploration of Thirdspace with voices from the community
that has generated some of its most intrepid and insightful contemporary adventurers: African-American cultural critics, writers, and
philosophers; and more broadly, people (especially women) of color.
The pathways into Thirdspace taken by these itinerant explorers lead
us to the specific terrain of postmodern culture and into discussions
of the new cultural politics of difference and identity that is re-awakening the contemporary world to the powerfully symbolic spaces of
representation, to struggles over the right to be different, to a new
politics of location and a radical spatial subjectivity and praxis that is
postmodern from the start. The pathways chosen move forward
many of the journeys begun by Henri Lefebvre, re-routing and re-
rooting them in the experienced immediacy of the present.

Both bell hooks and Cornel West are primarily concerned with
reconceptualizing radical African-American subjectivity in a way
that retains and enhances the emancipatory power of blackness, but
is at the same time innovatively open to the formation of multiple
communities of resistance, polyvocal political movements capable of
linking together many radical subjectivities and creating new "meeting places" and real-and-imagined "spaces" for diverse oppositional
practices. As Black intellectuals, hooks and West have been multiply
marginalized, made peripheral to the mainstreams of American
political, intellectual, and everyday life; but they have also consciously chosen to envelop and develop this *marginality*, as hooks
puts it, as a space of radical openness, a context from which to build
communities of resistance and renewal that cross the boundaries and
double-cross the binaries of race, gender, class, and all oppressively
Othering categories. Like Lefebvre, they obtain a particular centrality
– and an abiding globality – in their purposeful peripheralness, a
strategic positioning that disorders, disrupts, and transgresses the
center–periphery relationship itself.

Hear bell hooks in *Yearning*:[1]

As a radical standpoint, perspective, position, "the politics of
location" necessarily calls those of us who would participate in

---

[1] bell hooks, *Yearning: Race, Gender, and Cultural Politics*, Boston, MA: South End
Press, 1990. See also hooks, *Ain't I a Woman: Black Women and Feminism*, Boston, MA:
South End Press, 1981; *Feminist Theory: From Margin to Center*, Boston: South End
Press, 1984; *Talking Back: Thinking Feminist, Thinking Black*, Boston, MA: South End
Press, 1989; and *Outlaw Culture: Resisting Representations*, New York and London:
Routledge, 1994.

the formation of counter-hegemonic cultural practice to identify the spaces where we begin the process of re-vision. . . . For many of us, that movement requires pushing against oppressive boundaries set by race, sex, and class domination. Initially, then, it is a defiant political gesture. (1990, 145). . . . For me this space of radical openness is a margin – a profound edge. Locating oneself there is difficult yet necessary. It is not a "safe" place. One is always at risk. One needs a community of resistance.   (1990: 149)

Hear also Cornel West describing bell hooks in their dialogue, *Breaking Bread*.[2]

bell hooks's bold project locates her on the margins of the academy and the Black community – in search of a beloved community whose members will come from these very same margins. Her kind of Black feminism – or womanism – puts a premium on reconstructing new communities of Black people (of whatever gender, sexual orientation, and class) and progressives (of whatever race) regulated by thoroughly decolonized visions, analyses and practices.   (Hooks and West, 1991: 61)

Their politics of location concentrates on the de-colonizing role of the African-American intellectual (as cultural critic and philosopher) in the search for wider spheres of participation in the "world of the everyday" and in the enablement of radical social action everywhere in the world, from the personal to the planetary. It is a politics and spatial positioning that is explicitly but cautiously and critically postmodern. In her essay on "Postmodern Blackness," hooks presents a powerful critique of conventional postmodern discourse for its white male exclusivity and especially its persistent separation of a generalized "politics of difference" from the more specified and lived "politics of racism." But hooks follows her critique by reasserting, with no regrets, her own commitment to radical postmodernism.

The overall impact of postmodernism is that many other groups now share with black folks a sense of deep alienation, despair, uncertainty, loss of a sense of grounding even if it is not

---

[2] bell hooks and Cornel West, *Breaking Bread: Insurgent Black Intellectual Life*, Boston: South End Press, 1991. See also West, "The New Cultural Politics of Difference," in R. Ferguson, M. Gever, T. T. Minh-ha and C. West, eds, *Out There: Marginalization and Contemporary Cultures*, Cambridge MA: MIT Press and New York: The New Museum of Contemporary Art, 1990; and *Race Matters*, New York: Vintage Books, 1994.

informed by shared circumstance. Radical postmodernism calls attention to those shared sensibilities which cross the boundaries of class, race, gender, etc., that could be fertile ground for the construction of empathy – ties that would promote recognition of common commitments, and serve as a base for solidarity and coalition. . . . To change the exclusionary practice of postmodern critical discourse is to enact a postmodernism of resistance. (1990: 27 and 30)

She concludes with the passage selected to begin this chapter, on "postmodern culture" as a (Third)space of political choice.

I will continue to draw upon bell hooks for the remainder of this chapter (and beyond), for her voice more than any other I have heard brings to Thirdspace a creatively postmodern conceptualization of the cultural politics of difference, a radically open and openly radical critique that infuses Thirdspace with explicit attention to the multiplicity of spaces that difference makes around the "issues" listed earlier by Cornel West: exterminism, empire, class, race, gender, sexual orientation, age, nation, nature, and region.

On the Differences that Postmodernity Makes

The paths to the new cultural politics, to a radically open postmodernism of resistance, and to the multiplicitous Thirdspaces of bell hooks start off by covering some old ground. We must pass again, for example, through the trialectical critique of modernist binarisms and explore, in a different way, the representations of power and the power of representations that were charted out by Lefebvre. What we will get to eventually in the new cultural politics is that familiar intersection of space/knowledge/power that defines the context of postmodern radical spatial subjectivity and praxis.[3]

On the Workings of Power

The cultural politics of difference, whether labeled old or new, arise primarily from the workings of power in society and on space in their simultaneously perceived, conceived, and lived worlds. In other words, power – and the specifically cultural politics that arise

---

[3] For much of what follows in this chapter and the next, I owe a great deal to Barbara Hooper, who co-authored the essay from which these two chapters are primarily drawn and who deserves, perhaps more than any other, my deep appreciation for the students who have taught me. See Edward Soja and Barbara Hooper, "The Spaces That Difference Makes: Some Notes on the Geographical Margins of the New Cultural Politics," in Michael Keith and Steve Pile, eds, *Place and the Politics of Identity*, London and New York: Routledge, 1993, 183–205.

from its workings – is contextualized and made concrete, like all social relations, in the (social) production of (social) space. As Michel Foucault has argued, these links between space, knowledge, power, and cultural politics must be seen as both oppressive and enabling, filled not only with authoritarian perils but also with possibilities for community, resistance, and emancipatory change.

The multisidedness of power and its relation to a cultural politics of difference and identity is often simplified into hegemonic and counter-hegemonic categories. Hegemonic power, wielded by those in positions of authority, does not merely manipulate naively given differences between individuals and social groups, it actively *produces and reproduces difference* as a key strategy to create and maintain modes of social and spatial division that are advantageous to its continued empowerment and authority. "We" and "they" are dichotomously spatialized and enclosed in an imposed territoriality of apartheids, ghettos, barrios, reservations, colonies, fortresses, metropoles, citadels, and other trappings that emanate from the center–periphery relation. In this sense, hegemonic power universalizes and *contains* difference in real and imagined spaces and places.

Those who are territorially subjugated by the workings of hegemonic power have two inherent choices: either accept their imposed differentiation and division, making the best of it; or mobilize to resist, drawing upon their putative positioning, their assigned "otherness," to struggle against this power-filled imposition. These choices are inherently spatial responses, individual and collective reactions to the ordered workings of power in perceived, conceived, and lived spaces.

The outcomes of this socio-spatial differentiation, division, containment, and struggle are cumulatively concretized and conceptualized in spatial. practices, in representations of space, and in the spaces of representation, for all three are always being profoundly shaped by the workings of power. It is useful to see these presuppositions, processes of social production, and outcomes of the workings of power as conflated, historically and geographically, in *uneven development*: the composite and dynamic spatio-temporal patterning of socially constructed differences at many different spatial scales, from the body and the home to the nation and the world economy. As Lefebvre has creatively argued, geohistorically uneven development has become increasingly fragmented, homogenized, and hierarchically structured in capitalist societies, making the simple division between hegemonic and counter-hegemonic power, centers and peripheries, more and more complex and difficult to decipher. Pushed one step further, this leads us to the conditions of postmodernity as they are being expressed in the contemporary world. But we jump too quickly ahead.

Returning to Foucault for the moment, the struggles against (and for) hegemonic power can be seen to operate in three overlapping are(n)as, around what he termed subjection, domination, and exploitation. Subjection, the least well studied and understood he claimed, has two expressions, reflecting the double meaning of the word *subject*: subject to control by and dependency on someone else, usually implying some form of submission; and subject as asserted identity and self-knowledge, i.e. individual or collective subjectivity. Struggles against subjection are thus struggles over subjectivity and submission, over who defines the individual or collective subject. Foucault argued that "nowadays, the struggle against the forms of subjection – against the submission of subjectivity – is becoming more and more important," although he hastened to add that struggles against domination (ethnic, social, religious) and exploitation (against alienation, the socio-spatial relations that separate individuals from what they produce) continue to prevail as primary targets of resistance.[4] Cultural politics broadly defined include all three forms of resistance struggle, but in the "new" cultural politics of difference and identity, subjection has assumed a much more central role, especially with regard to gender, race, and the far-reaching relation between the "colonizer" and the "colonized."

Growing out of this view of power, differences that are ascribed to gender, sexual practice, race, class, region, nation, etc., and their expression in social space and geohistorically uneven development, are appropriately seen as "brute fashionings."[5] Like social space itself, they are neither transhistorical nor "natural" (in the sense of being naively or existentially given, as in "human nature"). This brute fashioning, as the social and spatial production (and strategic reproduction) of difference, becomes the catalyst and the contested space for both hegemonic (conservative, order-maintaining) and counter-hegemonic (resistant, order-transforming) cultural and identity politics: the most general form of the center–periphery relation defined (and deconstructed) by Lefebvre.

On Radical Subjectivities

Counter-hegemonic cultural politics, as it has conventionally been defined, has typically followed one of two paths, at least within Western capitalist societies. The first and most "acceptable" is rooted in the post-Enlightenment development of liberal humanism and

---

[4] Michel Foucault, "Afterword: The Subject and Power," in H. Dreyfus and P. Rabinow, eds, *Michel Foucault: Beyond Structuralism and Hermeneutics*, Chicago: University of Chicago Press, 1982, 208–26.

[5] Joan Cocks, *The Oppositional Imagination: Feminism, Critique and Political Theory*, London: Routledge (1989), 20.

modernism. It has traditionally based its opposition on the assertion of universal principles of equality, human rights, and democracy, seeking to reduce to a minimum the negative effects of difference, whatever their origins. A second form of counter-hegemonic politics, not always completely separable from the first, arises from more radical contestation over the many axes along which socially constructed power differentials have historically and geographically polarized. Rather than aiming above all to erase differences or to "even things out," it uses difference as a basis for community, identity, and struggle against the existing power relations at their source.

Without excluding the liberal alternative entirely, what is defined here as the *modernist cultural politics* of difference and identity refers primarily to this tradition of more radical subjectivity and resistance as it has developed since the mid-19th century around such categories of cultural consciousness as class, race, ethnicity, nationality, colonial status, sexuality, and gender. More particularized and fragmented than the universal humanism of the liberal alternative, modernist cultural politics usually preserve inter-categorical distinctions while seeking to eliminate differences in power along each "internal" axis.

Even when rejecting Marxian categories and explicitly revolutionary ideology, modernist cultural politics and the various social and cultural movements associated with it have often tended to develop along the lines initiated more than a century ago with respect to the formation of anti-capitalist class consciousness and the revolutionary struggle against economic exploitation. While varying greatly in their specificities, these modern movements have generally followed analogous trajectories based on a similar praxis of refusal and resistance that parallels the bipolar logic of class struggle: capital versus labor, bourgeoisie versus proletariat: a struggle defined around a deep structural dichotomy that "orders" differential power into two primary social categories, one dominant and the other subordinate.[6]

Each separate sphere of modernist cultural politics has typically mobilized its version of radical subjectivity around a fundamentally

[6] These observations must not be taken as privileging class struggle as paradigmatic or as a model which other forms of modernist resistance simply mimicked. Such privileging is largely the result of Eurocentric and masculinist interpretations of social reality, opposition, and resistance as they developed (and became hegemonic as "Marxist") in the 19th and 20th centuries. It is precisely this essentialist privileging of the working class, the radical (male) intellectual, and later the Party as *the* agents "making history" – and the attendant deployment of Marxism as an all-inclusive, self-sufficient grand narrative of social revolution – that radical postmodernism and the new identity politics (as well, it might be added, as some 19th-century feminisms and anti-racisms) have worked to deconstruct. Even drawing parallels and analogies must be done with care, for the emphasis on exploitation that is so central to class struggle is significantly different from the emphasis on subjection and domination that motivates most other forms of cultural politics.

epistemological critique of the binary ordering of difference that is particular to it: capital/labor, self/other, subject/object, colonizer/colonized, white/black, man/woman, majority/minority, heterosexual/homosexual. This critique of knowledge formation is aimed at "denaturalizing" the origins of the binary ordering to reveal its social and spatial construction of difference as a means of producing and reproducing systematic patterns of domination, exploitation, and subjection. As socially constructed, context-specific "technologies" of power (to use another Foucauldian term), these binary structures become subject to social and cultural transformation via a politics of identity and a radical subjectivity that builds upon the empowerment of the "subaltern" against the "hegemon" (to use the most general terms covering the various oppositional forms of the cultural politics of difference).[7]

Each major social movement within modernist cultural politics characteristically projects its particular radical subjectivity, defined within its own oppressive binary structure, as overarchingly (and often universally) significant. Whether or not this (bounded) totalization and essentialism is actually believed, its powerful mobilizing effect is used strategically in attempts to consolidate and intensify counter-hegemonic consciousness "for itself" and on its identified "home ground." In both theory and practice, therefore, a significant degree of closure and exclusiveness is embedded within the strategies and tactics of modernist cultural politics. Even when one counter-hegemonic movement avows its openness to alliance with others, it is usually open only on the former's terms and under its primary strategic guidance. The result has been the production of parallel, analogous, and segregated channels of radical political consciousness and subjectivity, each designed and primed to change its own discrete binarized world of difference.[8]

Under these ordered conditions, *fragmentation* (in the very real

[7] The generalized bipolarity of the relation between hegemon and subaltern is directly comparable to the relation between core (or center) and periphery (or margin). In both, the binary ordering of difference is socially constructed (not natural or transhistorical) and arises from the workings of power in a complex process of geohistorically uneven development. The homology (rather than mere analogy) of these two expressions of binary ordering is another important stimulus to the conceptual and practiced spatialization of the cultural politics of difference that has been occurring over the past several decades.

[8] Here is where the universalist politics of liberal humanism (and its attendant individualism) frequently complicates the theory and praxis of radical identity politics by providing an alluring all-embracing alternative viewpoint. This categorically simplified choice, occupy only one "radical" channel or "liberally" choose them all, especially when backed by the similar categorical binarism of "if you are not our friend you are our enemy," imposes a rigid logic which intensifies the need for an "epistemological police" to guard boundaries and punish deviation, especially for those who seek a different set of choices.

form of complex multiple subjectivities, with and without overlapping), becomes an endemic problem in modernist cultural politics, especially for those social movements which theorize either a universalist encompassing of other radical subjectivities (e.g. substituting "woman" for all women, or asserting the transcendental unity of the "workers of the world" to encompass multiracial, multiethnic, and otherwise diverse men and women workers); or alternatively, recognize differences but none the less theorize and strategize from the assumed primacy and privileging of one or another set of agents in the process of radical social change. In the extreme case, as with most orthodox forms of modern Marxism and some forms of radical feminism and black nationalism, these essentialist tendencies present a barrier to cross-cutting alliances of political significance by attributing "false consciousness" or subordinate identity to all radical subjectivities other than that emanating from the privileged bipolarity.[9]

When the primacy of one binary opposition is viewed as competing with the privileging of another, the prospects for flexible and cooperative alliance and "empathy" (a key term for bell hooks) are likely to be dim. While there have been many fruitful dialogues between various radical movements (between Marxism and feminism, for example, and between both and those struggling against racism and colonial oppression), the deeply engrained essentialisms of modernist identity politics have tended to create a competitive exclusivity that resists, even rejects, seeing a "real" world populated by multiple subjects with many (often changeable) identities located in varying (and also changeable) subject positions. Hence, modernist cultural politics, in its fear and rejection of a fragmented reality, has often tended to create and intensify political divisiveness rather than working toward a multiple, pluralized, and yet still radical conceptualization of agency and identity.

Modernist identity politics has always had to face both a forceful reactionary resistance from those in power and the temptations of reformist liberal diversion and co-optation. Over the past two decades, however, new theoretical and philosophical critiques (especially of taken-for-granted epistemologies and, in particular, the politically-divisive tendencies toward master-narrative essentialism and binary totalization) and actual political events (ranging from the global restructurings of capital, labor, and ideology to the apparent

---

[9] Further intensifying this tendency to fragmentation (and expanding the role of the epistemological police) is an internal competition for primacy, driven by the same "urge to unity" that pervades modernist identity politics. Here we find the effort to exclude, to separate the pure from the impure, the deviation from the essence: a demarcation of categorical distinctions between *inside and outside* that actively represses difference and otherness. For a similar view on the inside–outside relation, see the work of Diana Fuss (1991) cited in the next chapter (p. 118, see footnote 12).

defeat and retreat of governments inspired by modernist Marxism-Leninism and its dominant form of identity politics) have ushered in an extraordinary period of deep questioning that strikes more disruptively than ever before at the very foundations of modernist political practices – and simultaneously opens up new possibilities for radical resistance to all forms of hegemonic subordination. I am speaking, in particular, of the critiques that have been associated with the development of a radically open and openly radical postmodernism, from Lefebvre and Foucault to bell hooks and Cornel West, with all their similarities and differences.

On the Disordering of Difference

Postmodernism lives. Legions of detractors and years of intellectual debate have done nothing to arrest its expansion or reduce its impact, and scores of usurpers have failed miserably in stultifying its scope. Despite or because of being profanely ambivalent and ambiguous, rejoicing in consumption and celebrating obsessions, ignoring consistency and avoiding stability, favoring illusions and pleasure, postmodernism is the only possible contemporary answer to a century worn out by the rise and fall of modern ideologies, the pervasion of capitalism, and an unprecedented sense of personal responsibility and individual impotence.

Whether one likes it or not, postmodernism is a state of things. It is primarily determined by an extremely rapid and freewheeling exchange to which most responses are faltering, impulsive, and contradictory. What is at stake is the very constitution of being – the ways we perceive ourselves and others, the modes of experience that are available to us, the women and men whose sensibilities are shaped by urban exposure.   (Olalquiaga, 1992: xi)

There is no doubt that some who hasten to proclaim the death of modernism and announce the emergence of a new postmodern era from its ashes are motivated by the same old impulses of hegemonic reempowerment and/or liberal diversion. More will be said of this reactionary postmodernism in later chapters. But there is also emerging – in the postmodern blackness and post-colonialist critiques of bell hooks, Cornel West, Celeste Olalquiaga, Gayatri Spivak, Arjun Appadurai, Trinh T. Minh-ha, Edward Said, Chandra Mohanty, Homi Bhabha; in the increasingly spatialized postmodern feminisms of Iris Marion Young, Jane Flax, Judith Butler, Diana Fuss, Meaghan Morris, Rosalyn Deutsche, Donna Haraway; and in the anti-essentialist critiques of various postmodern Marxist scholars such as

Ernesto Laclau and Chantal Mouffe – a polyvocal postmodernism that maintains a political commitment to radical social change while continuing to draw (selectively, but sympathetically) from the most powerful critical foundations of modernist cultural politics. The intent behind these radical postmodernisms of resistance and redirection is to deconstruct (not to destroy) the ebbing tide of modernist radical politics, to renew its strengths and shed its weaknesses, and to reconstitute an explicitly postmodern radical politics, a new cultural politics of difference and identity that moves toward empowering a multiplicity of resistances rather than searching for that one "great refusal," the singular transformation that somehow must precede and guide all others.

The *disordering* of difference from its persistent binary structuring and the reconstitution of difference as the basis for a new cultural politics of multiplicity and strategic alliance among all who are peripheralized, marginalized, and subordinated by the social construction of difference (especially in its strictly dichotomized forms) are key processes in the development of radical postmodern subjectivity. Whether this revisioning of radical subjectivity requires a major transformation or merely a significant reform of modernist identity politics is still being contested. But it is clear that politics as usual can no longer be practiced as it was in the past, at least among those who take seriously the conditioning effects of postmodernity.[10]

In the wake of this continuing debate, a new breed of radical anti-postmodernists has emerged in force to "spin-doctor" the critical discourse toward either reformist or "total" revolutionary solutions, and away from any deep but sympathetic deconstruction and reconstitution of modernist traditions. Given the intellectual elitism, masculinism, racism, and neoconservative politics that have dominated so much of the postmodernist discourse, there is ample ammunition for the radical anti-postmodernist project. After all, the deconstructive challenge raised by postmodernism often sounds suspiciously like the old hegemonic strategies of opposition, co-optation, and diversion; and the affirmedly prefixed "post-" seems too literally to signal the irrevocable "end of" all progressive modernist projects rather than their potentially advantageous reconstitution and renewal. Moreover, to most radical modernist critics, the multiplicity of resistances continues to be seen as inevitably leading to a

---

[10] It is important to note the emergence of an increasingly powerful reactionary form of postmodern politics that also engages in a "disordering of difference" and has learned effectively how to create political advantage from fragmentation, ruptures, discontinuities, and the apparent disorganization of capitalist (and formerly socialist) political economies and cultures. In many ways, the hegemon has learned to adapt to the conditions of postmodernity more rapidly and creatively than the subaltern, especially when the latter remains frozen between modernist and postmodernist radical subjectivities.

politically debilitating fragmentation and the abandonment of long-established forms of counter-hegemonic struggle. Under these presumed circumstances, the promise of eventual emancipatory reconstitution rings hollow, if not cruelly deceptive, especially at a time when nearly all radical modern movements are either in crisis or massive retreat.

The tendency to homogenize postmodernism and to generalize (i.e. ascribe to all postmodernisms) certain negative and oppressive political practices associated with postmodernity, makes the construction of a radical postmodern alternative even more difficult.[11] Theoretically suggestive aphorisms arising from the postmodern and related poststructuralist critiques ("there is no reality outside the text," "the death of the subject," "anything goes") are now routinely set up as straw-objects to demonstrate that all postmodern politics are abstract and immaterial, unrelated to everyday life, inherently reactionary – and hence, immanently nihilistic with respect to "real" radical politics. A forbidding wall of categorical totems and taboos has thus been raised to hide the very possibility of radical postmodernism, making it all but invisible to outsiders, especially on the Left.

This forbidding wall has, however, tended to materialize around the same modernist conditioning and rigid either/or logic that has become so central a target for contemporary cultural criticism and proponents of the new politics of difference: the infatuation with clean and orderly binary oppositions; the intolerance of ambiguity, disordering, multiplicity, fragmentation; the urge to unity enforced by epistemological closure and essentialism. The arguments that have been outlined as a critique of modernist identity politics are thus also a critique of the flourishing new anti-postmodernism. It is a critique that calls for a new way of looking at and making practical political sense of what Cornel West described as "the precise circumstances of our present moment," drawing insight from the realization that postmodernity has made more significant differences to our real and imagined political worlds than it has reinforced repressive continuities with the past.

Perhaps the most significantly revealing difference arising from radical postmodernism (as a critical metaphilosophy) and poststructuralism (as its primary methodological support) has been a profoundly introspective ontological, epistemological, and discursive

---

[11] One might argue as well that a similar tendency can be found in my earlier critique of modernist identity politics. I hope that I have not been so categorically dismissive of modernisms and have not homogenized modernist identity politics to the degree that is present in the contemporary radical backlash against postmodernism. I recognize, however, that there is a widespread tendency to be simplistically and categorically anti-modernist (more specifically, anti-Marxist and anti-feminist) among many who proclaim themselves to be postmodern.

deconstruction of modernist political theory. And the most powerful exposition of this difference in ways of thinking about politics has come from feminism. In their introduction to *Feminists Theorize the Political*,[12] Judith Butler and Joan W. Scott "open up a political imaginary for feminism" that is situated in what they call a poststructuralist critique of "positionings," an unsettling critique that deconstructs all presumptions of cognitive "mastery." Their comments effectively summarize what has been described here as the postmodern disordering of difference.

> Poststructuralism is not, strictly speaking, *a position*, but rather a critical interrogation of the exclusionary operations by which "positions" are established. In this sense, a feminist poststructuralism does not designate a position from which one operates, a point of view or standpoint which might be usefully compared with other "positions" within the theoretical field. This is, of course, the conventional grammar of political debate: "I am an _____ , and I believe _____ ." And surely there are strategic occasions when precisely that kind of grammar is important to evoke. But the questions raised in this volume concern the *orchestration* of that kind of position taking, not simply within academic discourse, but within feminist political discourse. What are the political operations that constrain and constitute the *field* within which positions emerge? What *exclusions* effectively constitute and naturalize that field? Through what means are women positioned within law, history, political debates on abortion and rape, contexts of racism, colonization, postcoloniality? Who qualifies as a "subject" of history, a "claimant" before the law, a "citizen"? Indeed, what qualifies as "reality," "experience" and "agency." ... Through what differential and exclusionary means are such "foundational" notions constituted? And how does a radical contestation of these "foundations" expose the silent violence of these concepts as they have operated not merely to marginalize certain groups, but to erase and exclude them from the notion of "community" altogether, indeed, to establish exclusion as the very condition and possibility of "community"?   (Butler and Scott, 1992: xiv)

Butler and Scott are quick to point out that this interrogation of the primary terms of conventional political discourse, or what I have called modernist cultural politics, is in no sense intended "to censor their usage, negate them, or to announce their anachronicity. On the

---

[12] Judith Butler and Joan W. Scott, eds, *Feminists Theorize the Political*, New York and London: Routledge, 1992.

contrary, this kind of analysis *requires* that these terms be reused and rethought, exposed as strategic instruments and effects, and subjected to a critical reinscription and redeployment."

These interrogations turn what has been a primarily deconstructive critique of modernism into a more constructive program for a radical postmodernist cultural politics that revolves around disordering difference and empowering multiplicity. To situate the debate within Thirdspace, however, requires a return to other sources.

In Thirdspace with bell hooks

In *Yearning: Race, Gender, and Cultural Politics*, and particularly in the chapters on "Postmodern Blackness" and "Choosing the Margin as a Space of Radical Openness," bell hooks attempts to move beyond modernist binary oppositions of race, gender, and class into the multiplicity of *other* spaces that difference makes; into a re-visioned spatiality that creates from difference new sites for struggle and for the construction of interconnected and non-exclusionary communities of resistance. In so doing, she opens up in these real-and-imagined other spaces a Thirdspace of possibilities for a new cultural politics of difference and identity that is both radically postmodern and consciously spatialized from the beginning.

This creative spatialization involves more than wrapping texts in appealing spatial metaphors. It is a vital discursive turn that both contextualizes the new cultural politics and facilitates its conceptual re-visioning around the empowerment of multiplicity, the construction of combinatorial rather than competitively fragmented and separated communities of resistance. It also leads to a new spatial conception of social justice based on the politics of location and the right to difference within the revised situational contexts of postmodernity.

Early work on the spatialization of cultural politics can be found in the writings of Siegfried Kracauer and Walter Benjamin, Frantz Fanon and Simone de Beauvoir, Michel Foucault and Henri Lefebvre. In addition, there has developed over the past decade a substantial body of literature that has been specifically concerned with disordering modernist binaries and promoting a new and radical cultural politics. In bell hooks's writings, there has been a particularly lucid, expressive, and accessible convergence of these discourses, both old and new, making her *Yearnings* an especially useful place from which to re-explore Thirdspace.

Choosing Marginality

Hooks finds her place, positions herself, first of all as an African-American woman and then by the simultaneously political and geo-graphical act of *choosing marginality*. This positioning of identity and subjectivity is purposefully detached from the "narrow cultural nationalism masking continued fascination with the power of the white [and/or male] hegemonic order." The latter identification, hooks suggests, would not be choosing marginality but accepting its imposition by the more powerful, binary Other, a submission to the dominant, order-producing, and unremittingly modernist ideology and epistemology of difference. Instead of totally ignoring cultural nationalisms, however, such an assertion of recentered identity "is evoked as a stage in a process wherein one constructs radical black subjectivity" (or what she has more recently called "wildness").[13] By extension and adjustment, choosing marginality becomes a critical turning-point in the construction of other forms of counter-hegemonic or subaltern identity and more embracing communities of resistance.

As an initial stage, categorical identity as subaltern is crucial and necessary. But hooks's construction of radical black subjectivity pushes the process of identity formation beyond exclusionary struggles against white racism and on to a new terrain, a "space of radical openness" where the key question of *who we can be and still be black* can be politically re-imagined and practiced. "Assimilation, imitation, or assuming the role of rebellious exotic are not the only options and never have been," hooks notes in rejecting the conventional choices that liberal modernist discourse has frequently imposed upon the activist black subject. Instead, she chooses a space that is simultaneously central and marginal (and purely neither at the same time), a difficult and risky place on the edge, filled with contradictions and ambiguities, with perils but also with new possibilities: a Thirdspace of political choice.

"Fundamental to this process of *decentering the oppressive other* and claiming our right to subjectivity," she writes, "is the insistence that we must determine how we will be and not rely on colonizing responses to determine our legitimacy." In a similar avowal, Pratibha Parmar argues that creating identities as black women is not done " 'in relation to,' 'in opposition to,' or 'as a corrective to,' ... *but in and for ourselves*. Such a narrative thwarts that binary hierarchy of centre and margin: the margin refuses its authoritative

---

[13] bell hooks, *Black Looks: Race and Representation*, Boston: South End Press, 1992. All the other quoted references to bell hooks are taken from "Choosing the Margin as a Space of Radical Openness," in *Yearning*, 1990: 145–53.

emplacement as 'Other'." Significantly, hooks cites Parmar in "Choosing the Margin" to confirm their shared presupposition of the political significance of spatiality. After stating that spaces "can be real and imagined … can tell stories and unfold histories … can be interrupted, appropriated, and transformed," hooks adds: "As Pratibha Parma[r] notes, 'The appropriation and use of space are political acts.' "[14]

This evocative process of choosing marginality reconceptualizes the problematic of subjection by deconstructing and disordering both margin and center. In those restructured and recentered margins, new spaces of opportunity and action are created, the new spaces that difference makes. For hooks, and by extension and invitation, all others involved in this spatial disordering of difference, there is a "definite distinction between the marginality which is imposed by oppressive structure and that marginality one chooses as site of resistance, as location of radical openness and possibility." She adds, more extensively:

> It was this marginality that I was naming as a central location for the production of a counter-hegemonic discourse that is not just found in words but in habits of being and the way one lives. As such, I was not speaking of a marginality one wishes to lose, to give up, but rather as a site one stays in, clings to even, because it nourishes one's capacity to resist. It offers the possibility of radical perspectives from which to see and create, to imagine alternatives, new worlds. …
>
> Understanding marginality as position and place of resistance is crucial for oppressed, exploited, colonized people. If we only view the margin as a sign, marking the condition of our pain and deprivation, then certain hopelessness and despair, a deep nihilism penetrates in a destructive way. … I want to say that these margins have been both sites of repression and sites of resistance. … (1990: 149–51)

She concludes with a resounding invitation:

> This is an intervention. A message from that space in the margin that is a site of creativity and power, that inclusive space where we recover ourselves, where we move in solidarity to erase the category colonizer/colonized. Marginality is the space of resistance. Enter that space. Let us meet there. Enter that space. We greet you as liberators. (1990: 152)

---

[14] Pratibha Parmar, "Black Feminism: The Politics of Articulation," in J. Rutherford, ed., *Identity: Community, Culture, Difference*, London: Lawrence and Wishart, 1991: 101.

In choosing marginality as a space of radical openness, hooks contributes significantly to a powerful revisioning not only of the cultural politics of difference but also of our conceptualization of human geographies, of what we mean by the politics of location and geohistorically uneven development, of how we creatively combine spatial metaphor and spatial materiality in an assertively spatial *praxis*. By recontextualizing spatiality, she engages in a cognitive remapping of our many real-and-imagined worlds – from the most local confines of the body, the geography closest in, to the nested geographical worlds that are repeated again and again in an expanding sequence of scales reaching from the "little tactics of the habitat" to the "great strategies" of global geopolitics.[15]

For hooks, the political project is to occupy the (real-and-imagined) spaces on the margins, to reclaim these lived spaces as locations of radical openness and possibility, and to make within them the sites where one's radical subjectivity can be activated and practiced in conjunction with the radical subjectivities of others. It is thus a spatiality of inclusion rather than exclusion, a spatiality where radical subjectivities can multiply, connect, and combine in polycentric communities of identity and resistance; where "fragmentation" is no longer a political weakness but a potential strength: the spatiality searched for but never effectively discovered in modernist identity politics.

Hooks first makes her Thirdspace consciousness explicit in the Preface to *Feminist Theory: From Margin to Center*. Like Lefebvre's, hooks's consciousness is rooted in the everyday life experiences of youth, home, and family.

> To be in the margin is to be part of the whole but outside the main body. As black Americans living in a small Kentucky town, the railroad tracks were a daily reminder of our marginality. Across those tracks were paved streets, stores we could not enter, restaurants we could not eat in, and people we could not look directly in the face. Across these tracks was a world we could work in as maids, as janitors, as prostitutes, as long as it was in a service capacity. We could enter that world but we could not live there. We had always to return to the margin, to cross the tracks, to shacks and abandoned houses on the edge of town.

---

[15] The full quote from Foucault comes from "The Eye of Power," reprinted in *Power/Knowledge: Selected Interviews and Other Writings 1972–1977*, C. Gordon, ed., New York: Pantheon, 1980: 149. Foucault writes: "A whole history remains to be written of *spaces* – which would at the same time be the history of *powers* (both terms in the plural) – from the great strategies of geopolitics to the little tactics of the habitat." The same quote reappears as the lead-in to Daphne Spain, *Gendered Spaces*, Chapel Hill: University of North Carolina Press, 1992.

There were laws to ensure our return. To not return was to risk being punished. Living as we did – on the edge – we developed a particular way of seeing reality. *We looked both from the outside in and from the inside out. We focused our attention on the center as well as on the margin. We understood both.* This mode of seeing reminded us of the existence of a whole universe, a main body made up of both margin and center. Our survival depended on an ongoing public awareness of the separation between margin and center and an ongoing private acknowledgement that we were a necessary, vital part of that whole.

This sense of wholeness, impressed upon our consciousness by the structure of our daily lives, provided us an oppositional world view – a mode of seeing unknown to most of our oppressors, that sustained us, aided us in our struggle to transcend poverty and despair, strengthened our sense of self and our solidarity.   (1984: ix; emphasis added)

The familiar "other side of the tracks" story is here transformed in a trialectical twist, a thirding that constructs an Other world from the simple binary opposition of center and periphery, inside and outside. Hidden from the view of the oppressor, this third-world of political choice ensures the survival of the oppressed, nurtures resistance, and provides openings "on the edge," in the chosen context of marginality, to build larger communities of survival and resistance. This Thirdspace on the margin is hooks's point of departure for a sympathetic critique of feminist theory and the exclusionary practices and liberal individualism that often emerge from its white mainstreams. It is also an invitation to all progressive women and men to be themselves – and at the same time be black feminists, participating inhabitants of that radically open and openly radical world on the margin.

The Margin as a Space of Radical Difference

In *Feminist Theory*, there is little else in the text that is explicitly spatial, but everything follows from the evocations of Thirdspace in the subtitle and the Preface. In *Yearning*, however, hooks is resplendently spatial, not only in her use of metaphors but in her imaginative material grounding of the powerful force of black cultural criticism. The introductory chapter on "Liberation Scenes" begins with a comment on Lorraine Hansberry's play, *A Raisin in the Sun*, as a "counter-hegemonic cultural production" which interrogates and challenges the fear of black people that "being out of our place" (not conforming to the rules of white supremacy) would lead to destruction, even death. "On a basic level," hooks adds, "the play was about housing – the way racial segregation in a capitalist society meant that black folks were discriminated against when seeking places to live.... Mama reminds her family that black people survived the holocaust of slavery because they had oppositional ways of thinking, ones that were different from the structures of domination determining so much of their lives" (1990: 1–2).

Hooks goes on in this chapter to criticize more conventional black cultural criticism for focusing too narrowly on black "images" in the media and failing to draw upon the counter-hegemonic cultural criticism "honed and developed in black living rooms, kitchens, barber shops, and beauty parlors" (1990: 4). "Replaced by an ethic in complete contradiction to those values stressed by Mama in *A Raisin in the Sun*, the emphasis was on finding work for black folks in the culture industry" (1990: 5). Hooks concludes: "Cultural critics who are committed to a radical cultural politics (especially those of us who teach students from exploited and oppressed groups) must offer theoretical paradigms in a manner that connects them to *contextualized political strategies*" (1990: 8; emphasis added).

Hooks turns next to "The Politics of Radical Black Subjectivity," declaring the inspiration she draws from the work of Paulo Freire and positioning herself more clearly on the margin, in a "homeplace" that is much more than ghetto or shantytown.

> Perhaps the most fascinating constructions of black subjectivity (and critical thinking about the same) emerge from writers, cultural critics, and artists who are poised on the margins of various endeavors. I locate myself within this group, imagining I reside there with a wild crowd of known and unknown folks. We share commitment to left politics (yes, we critique capitalism and explore the revolutionary possibilities of socialism); we are concerned with ending domination in all its forms; we are into reading and deeply concerned with aesthetics (I mean, I have my own bumper sticker firmly stuck on my heart: "I die for style"); we are into all kinds of culture and do not fear losing our blackness; we see ourselves as one of the people, while simultaneously acknowledging our privileges, whatever they may be.... The ground we stand on is shifting, fragile, and unstable. We are avant-garde only to the extent that we eschew essentialist notions of identity, and fashion selves that emerge from the meeting of diverse epistemologies, habits of being, concrete class locations, and radical political commitments. We believe in solidarity and are working to make spaces where black women and men can dialogue about everything, spaces where we can engage in critical dissent without violating one

another. We are concerned with black culture and black identity. (1990: 19)

She criticizes more hard-line leftist, black nationalist, white feminist, and even some postcolonial deconstructionist perspectives in her espousal of an inclusive – and flexible – politics of "relocation."

> Cultural identity has become an uncool issue in some circles. One of the very crowd I mentioned earlier suggests "the emphasis on culture is a sign of political defeat," whereas it seems to me a practical gesture to shift the scene of action if in fact the location of one's political practice does not enable change. We return to "identity" and "culture" for relocation, linked to political practice – identity that is not informed by a narrow cultural nationalism masking continued fascination with the power of the white hegemonic other.   (1990: 20)

> Often when black subjects give expression to multiple aspects of our identity, which emerge from a different location, we may be seen by white others as "spectacle." For example, when I give an academic talk without reading a paper, using a popular, performative, black story-telling mode, I risk being seen by the dominating white other as unprepared, as just entertainment. Yet their mode of seeing cannot be the factor which determines style of representation or the content of one's work. Fundamental to the process of decentering the oppressive other and claiming our right to subjectivity is the insistence that we must determine how we will be and not rely on colonizing responses to determine our legitimacy. We are not looking to that Other for recognition. . . . [Much more than with a critical confrontation with "texts"], I am moved by that confrontation with difference which takes place on new ground, in that counter-hegemonic marginal space where radical black subjectivity is *seen*, not overseen by any authoritative Other claiming to know us better than we know ourselves. (1990: 22)

Hooks follows her discussion of radical black subjectivity with a chapter on "Postmodern Blackness," the last paragraph of which is the source for the richly Thirdspatial quotation that opens the present essay and helps us in our efforts "to enact a postmodernism of resistance." The next two chapters, "The Chitlin Circuit: On Black Community," and "Homeplace: A Site of Resistance," further explore the particularities of African-American Thirdspace, the world of "Southern, rural, black growing up . . . a world where we had a history." "For me," hooks says, "this experience, of growing up in a segregated small town, living in a marginal space where black people (though contained) exercised power, where we were truly caring and supportive of one another, was very different from the nationalism I would learn about in black studies classes or from the Black Muslims selling papers at Stanford University my first year there" (1990: 35). Special attention is reserved for the homeplace and its social construction and nurturing, especially by women, as a "community of resistance."

> Throughout our history, African-Americans have recognized the subversive value of homeplace, of having access to private space where we do not directly encounter white racist aggression. Whatever the shape and direction of black liberation struggle (civil rights reform or black power movement), domestic space has been a crucial site for organizing, for forming political solidarity. Homeplace has been a site of resistance. (1990: 47)

Hooks carries these arguments forward in a wonderful essay on "An Aesthetic of Blackness: Strange and Oppositional." The essay begins and ends with a story about the house in which she grew up.

> *This is the story of a house. It has been lived in by many people. Our grandmother, Baba, made this house a living space. She was certain that the way we lived was shaped by objects, the way we looked at them, the way they were placed around us. She was certain that we were shaped by space. From her I learn about aesthetics, the yearning for beauty that she tells me is the predicament of heart that makes our passion real. A quiltmaker, she teaches me about color. Her house is a place where I am learning to look at things, where I am learning how to belong in space. In rooms full of objects, crowded with things, I am learning to recognize myself. She hands me a mirror, showing me how to look. The color of wine she has made in my cup, the beauty of the everyday. Surrounded by fields of tobacco, the leaves braided like hair, circles and circles of smoke fill the air. We string red peppers fiery hot, with thread that will not be seen. They will hang in front of a lace curtain to catch the sun. Look, she tells me, what the light does to color! Do you believe that space can give life, or take it away, that space has power? These are the questions she asks which frighten me. Baba dies an old woman, out of place. Her funeral is also a place to see things, to recognize myself. How can I be sad in the face of death, surrounded by so much beauty? Death, hidden in a field of tulips, wearing my face and calling my name. Baba can make them grow. Red, yellow, they surround her body like lovers in a swoon, tulips everywhere. Here a soul on fire with beauty burns and passes, a soul touched by flame. We*

see her leave. She has taught me how to look at the world and see beauty. She has taught me "we must learn to see."

(1990: 103; italics in the original)

One of my five sisters wants to know how it is I come to think about these things, about houses, and space. She does not remember long conversations with Baba. She remembers her house as an ugly place, crowded with objects. My memories fascinate her. She listens with astonishment as I describe the shadows in Baba's house and what they meant to me, the way the mood entered an upstairs window and created new ways for me to see dark and light. After reading Tanizaki's essay on aesthetics "In Praise of Shadows," I tell this sister in a late night conversation that I am learning to think about blackness in a new way. Tanizaki speaks of seeing beauty in darkness and shares this moment of insight: "The quality we call beauty, however, must always grow from the realities of life, and our ancestors, forced to live in dark rooms, presently came to discover beauty in shadows, ultimately to guide shadows toward beauty's end." My sister has skin darker than mine. We think about our skin as a dark room, a place of shadows. We talk often about color politics and the ways racism has created an aesthetic that wounds us, a way of thinking about beauty that hurts. In the shadows of late night, we talk about the need to see darkness differently, to talk about it in a new way. In that space of shadows we long for an aesthetic of blackness – strange and oppositional.   (1990: 113)

There are few writings anywhere in which life's intimate spatiality is made so explicitly and brilliantly revealing. Throughout *Yearning*, more so than in any of her writings before or since, hooks creatively and consistently grounds her work in what I have described as the trialectics of spatiality, an alternative mode of understanding space as a transdisciplinary standpoint or location from which to see and to be seen, to give voice and assert radical subjectivity, and to struggle over making both theoretical and practical sense of the world.

Hooks makes no direct reference to Henri Lefebvre anywhere in *Yearning*, but his inspiration is tacitly there in so many places. In a hotel by the beach in Santa Monica, I had breakfast once with bell hooks and we spoke about Lefebvre. She easily admitted his influence on her thinking about space and warmly remembered his stay

at the University of California at Santa Cruz in 1984 and the gathering of spatial thinkers at a conference organized in his honor by Fred
Jameson and the History of Consciousness Program – which I at the
time thought might be renamed, at least temporarily in honor of
Lefebvre's presence, the Spatiality of Consciousness Program. She
smiled at the thought and spoke freely of the spatialization of her
own consciousness in the homeplaces of her youth, her interests in
design and architecture, and in those spaces of representation
screened through films and television. And she joked about a new
demand on her time: invitations from Geography and Architecture
departments to speak about space.

The last paragraph in "Choosing the Margin as a Space of Radical
Openness," perhaps the most Lefebvrean of the essays in *Yearning*,
provides an apt conclusion to this brief journey into Thirdspace with
bell hooks – and another point of departure for further and future
explorations.

> I am located in the margin. I make a definite distinction
> between that marginality which is imposed by oppressive struc-
> tures and that marginality one chooses as site of resistance – as
> location of radical openness and possibility. This site of resis-
> tance is continually formed in that segregated culture of opposi-
> tion that is our critical response to domination. We come to this
> space through suffering and pain, through struggle. We know
> struggle to be that which pleasures, delights, and fulfills desire.
> We are transformed, individually, collectively, as we make radi-
> cal creative space which affirms and sustains our subjectivity,
> which gives us a new location from which to articulate our
> sense of the world.   (1990: 153)

4

Increasing the Openness of Thirdspace

Lefebvre's analysis of the spatial exercise of power as a construction and conquest of difference, although it is thoroughly grounded in Marxist thought, rejects economism and predictability, opening up possibilities for advancing analysis of spatial politics into realms of feminist and anti-colonialist discourse and into the theorization of radical democracy. More successfully than anyone of whom I am aware, Lefebvre has specified the operations of space as ideology and built the foundation for cultural critiques of spatial design as a tool of social control.

(Rosalyn Deutsche, "Uneven Development," 1988: 122)

The West is painfully made to realize the existence of a Third World in the First World, and vice versa.

(Trinh T. Minh-ha, *Woman, Native, Other*, 1989: 98)

I want to end by asking for a geography that acknowledges that the grounds of its knowledge are unstable, shifting, uncertain and, above all, contested.

(Gillian Rose, *Feminism and Geography*, 1993: 160)

Bell hooks has not been alone in choosing Thirdspace, or something very much like it, as a strategic location for exploring postmodern culture and seeking political community among all those oppressively peripheralized by their race, class, gender, erotic preference, age, nation, region, and colonial status. And both hooks and Lefebvre would be the last to give permanence and essentiality to their particular formulations and journeys. Thirdspace must always be kept radically open (and yes, openly radical) for its interpretive insights and strategic power to be grasped and practiced. Although hooks and Lefebvre provide significant points of departure and destination, other openings, pathways, and practices must be mapped in advance of our peripatetic journeys to Los Angeles and other geographically real-and-imagined places.

Two rather expansive and heterogeneous literatures are explored here for their particular-and-general contributions to a constantly evolving conceptualization of Thirdspace. The first is feminist, and so too, in a significant way, is a major segment of the second, what is described today as the postcolonial discourse. My particular selections from these two literatures are guided by the accomplished journeys of Lefebvre and hooks. What is foregrounded, therefore, is a critical scholarship that is (1) explicitly, if not assertively, spatial in its modes of thinking, writing, and interpretation; (2) in some recognizable way attuned to the rebalanced trialectics of Spatiality–Historicality–Sociality (and not, therefore, historicist or deterministically – versus strategically – spatialist); and (3) is at least moving toward a Thirdspace perspective and away from the narrowed channels of Firstspace and Secondspace modes of knowledge formation.

This foregrounding carries with it an unsettling epistemological and theoretical critique that revolves around disruptions and disorderings: of difference, of confidently centered identities, and of all forms of binary categorization. It seeks instead a multiplicitous "alterity," a transgressive "third way" that is more than just the sum or combination of an originary dualism. At its best, such critical spatial thinking seeks to undermine its own authority by a form of textual and political practice that privileges uncertainties, rejects authoritative and paradigmatic structures that suggest permanence or inviolability, invites contestation, and thereby keeps open the spatial debate to new and different possibilities.

The Spatial Feminist Critique

There is a long heritage of critical feminist analysis of the social production of space, and especially of the social space of the city, that has only recently been critically un-silenced and given voice in the canons and corridors of Western social and cultural theory and philosophy. As far back as *The Book of the City of Ladies*, for example, written by Christine de Pizan, a Frenchwoman born in Venice in 1364, an explicitly and strategically spatial perspective was applied to an allegoric idealization of the medieval city both to reveal its

oppressive gendering and to imagine new sites and spaces of resistance from which to redesign cityspace for women.[1]

With the development and diffusion of capitalism, the most radical feminist critiques increasingly shifted away from the city and
spatiality to that encompassing web defined by the relatively space-
less dialectic between history and society (writ large as Historicality
and Sociality). A rich tradition of socialist feminism emerged, not
only in the mainstreams of Marxist historical materialism but also,
even more creatively, within the evolving traditions of utopian and
libertarian (anarchist) socialism, where issues of space, gender, and
nature were more openly addressed. Just how spatial this socialist
feminism was is difficult to ascertain, in part because this question
has rarely been asked in the historiography of socialist thought.

In the more contemporary reinterpretations of the history of
explicitly spatial feminism, class analysis and the socialist critique of
capitalism continue to be relevant. But the center of attention has
shifted back to urbanism, to everyday life within the built (socially
produced) environment of the city, and to the philosophy, epistemology, and cultural politics of spatiality. Indeed, the language and
critical scholarship of feminism – especially among women who
experience multiple sources of oppression and marginalization and
explicitly accept the possibility, if not the necessity, of a radical postmodernist politics – have become more thoroughly, creatively, and
strategically spatialized than those of any other comparable group of
critical scholars. And from this prolific spatialization new terrains
and itineraries are being opened for the exploration of Thirdspace.

Gendering Cityspace: The Feminist Critique of Urbanism

An appropriate starting-point for discussing the contemporary
development of a feminist critique of urban life and citybuilding is
the work of the architectural and urban historian, Dolores Hayden.
Hayden's writings, significantly influenced by Lefebvre's critique of
everyday life, critically explore the historical geography of the built
environment in the United States. Their subject matter ranges from
the imprint of 19th- and early 20th-century material feminists on the
design of American homes, neighborhoods, and cities; to the gendering of space and daily life in experimental American utopian socialist communities; to redesigning the "American Dream" with regard

---

[1] Christine de Pizan, *The Book of the City of Ladies*, tr. E. J. Richards, New York:
Persea Books, 1982. For a tracing of the feminist critique of urbanism and the "gen-
dering" of the capitalist city (and of urban and regional planning theory), see
Barbara Hooper, " 'Split at the Roots': A Critique of the Philosophical and Political
Sources of Modern Planning Doctrine," *Frontiers: A Journal of Women Studies* 13
(1992), 45–80.

to a more feminist future in housing, work, and family life; to a more recent concern for the "power of place" in preserving urban memories of social resistance and creative achievement among those subordinated by class, race, and gender. Raised again in the retrospective and prospective specificity of Hayden's work is the old question posed by Christine de Pizan: "What would a non-sexist city look like?"[2]

Hayden was not alone. Beginning in the late 1970s, a broadly based feminist urban critique arose from within the spatial disciplines (Architecture, Urban Planning, Geography), producing wide-ranging contemporary analyses of gendered spaces and the reproduction of this masculinist and phallocentric gendering through the contextualizing effects of patriarchy on the built environment of the city.[3] For the most part, this early round of critical spatial feminism remained characteristically modernist, in the sense of channeling its critical power and emancipatory objectives around the gendered binary Man/Woman. Little was said about the theorization of space itself, except to promote a spatial perspective as a useful way of historically unraveling the impact of oppressive social processes and practices.

Urban spatiality was seen as oppressively gendered in much the same way that the cityscape was shown to be structured by the exploitative class relations of capitalism and the discriminatory geographical effects of racism, the two other major channels of radical modernist urban critique developing over the same period in the spatial disciplines. Another important project, however, was begun in these writings and made more explicit by the spatial feminists

---

[2] The last question is the title of a chapter in *Women and the American City*, Catherine Stimpson, Else Dixler, Martha Nelson, and Kathryn Yatrakis, eds, Chicago: University of Chicago Press, 1980: 167–84. See also Hayden's *Seven American Utopias: The Architecture of Communitarian Socialism (1790–1975)*, Cambridge MA: MIT Press, 1976; *The Grand Domestic Revolution: A History of Feminist Designs for American Homes, Neighborhoods, and Cities*, Cambridge MA: MIT Press, 1982; *Redesigning the American Dream: The Future of Housing, Work, and Family Life*, Cambridge MA: MIT Press, 1984; and, most recently, *The Power of Place: Urban Landscapes as Public History*, Cambridge MA: MIT Press, 1995. I will return to Hayden's work on The Power of Place in chapter 7.

[3] A sampling of this literature would include Dolores Hayden and Gwendolyn Wright, "Architecture and Urban Planning," *Signs: A Journal of Women in Culture and Society*, 1 (1976), 923–33; the special issue of *Signs* on "Women and the American City," 5 (1980); Gerde Werkele, Rebecca Peterson, and David Morley, *New Space for Women*, Boulder: Westview Press, 1980; Dolores Hayden: *The Grand Domestic Revolution*, Cambridge MA: MIT Press, 1982; Shirley Ardener, ed., *Women and Space: Ground Rules and Social Maps*, New York: St Martin's Press, 1981; Mary Ellen Mazey and David R. Lee, eds, *Her Space, Her Place: A Geography of Women*, Washington, D.C.: Association of American Geographers, 1983; Matrix Collective, *Making Space: Women and the Man-Made Environment*, London: Pluto Press, 1984; Women and Geography Study Group of the Institute of British Geographers, ed., *Geography and Gender: An Introduction to Feminist Geography*, London: Hutchinson, 1984.

than by the radical urban critics focusing on race and class. It aimed at an active and intentional "remapping" of the city as a space of radical openness, a space where, as in hooks's margin, ties are severed and subjection abounds but also, at the same time, a strategic location for recovery and resistance, a meeting place where new and radical happenings can occur beyond the centered domain of the patriarchal urban order.

The spatialization of patriarchal power, or what can be termed the specific geography of patriarchy, was seen not only in the design of buildings (the home, the office, the factory, the school, the public monument, the skyscraper) but in the very fabric of urbanism and everyday life in the city. Suburbanization and urban sprawl became materially symbolic of the marginalization or, as it came to be called, the "entrapment" of women, their purposefully designed isolation from the workplace and public life in gadgeted homes and modern lifestyles that facilitated subservience to the male breadwinner and his cohorts. More broadly, what urban theorists described as urban morphology, the presumably innocent if not "natural" geographical patterning of urban land uses, was critically reinterpreted by spatial feminists as a veiled cartography of power and exploitation, not just in conjunction with class and race but also with respect to gender. The social production of cityspace and the institutionalized citybuilding processes that drive it thus became contested terrain, filled with new spaces and places of community, resistance, and critique.

More recently, primarily through the development of an openly postmodern feminism, new directions are being taken that expand the earlier feminist urban critique and its revealing recognition that space makes a difference into a creative exploration of the prolific multiplicity of spaces that difference makes. What is new about this development, that is, how the postmodern feminist critique of urban spatiality can be distinguished from its more modernist forms, is represented well in Cornel West's observations at the beginning of chapter 3. Along with the cultural politics it produces, what is novel is the emphasis given not to space or gender *per se*, or simply to (Firstspace) spatial practices and (Secondspace) spatial representations, but to "how and what constitutes difference, the weight and gravity it is given in representation," especially in what Lefebvre called the lived spaces of representation, and I describe as Thirdspace. In short, there was a shift from an *equality* to a *difference* model, and along with it a shift from an emphasis on material spatial forms to a more real-and-imagined urban spatiality.

The Postmodern Spatial Feminist Critique

Postmodern spatial feminism brings us into a more encompassing theorization and politics of difference and identity built on the opening of new spaces for critical exchange and creative radical responses "to the precise circumstances of our present moment." It relocates us not in the past or in the tacitly built environment of the city, but in the marginality and overlapping psychological, social, and cultural borderlands of contemporary *lived* spaces, and particularly in those chosen spaces of radical openness where difference is critically disordered in a deconstruction and reconstitution of both modernist binary oppositions and the universalist principles of liberal democracy.

In contrast to the earlier modernist critiques, postmodern spatial feminism is no longer so bound up with the Man/Woman binary. Gillian Rose, drawing upon the work of Teresa de Lauretis, captures the meanings of this unbinding of "the subject of feminism."[4]

> Teresa de Lauretis insists that, in order to break out of the masculinist field of knowledge, feminism must think beyond sexual difference: in order to challenge the patriarchal claim that the field of Man/Woman is exhaustive, the subject of feminism must be positioned in relation to social relations other than gender. The subject of feminism is thus constituted "not by sexual difference alone, but rather across languages and cultural representations; a subject en-gendered in the experiencing of race and class, as well as sexual, relations; a subject, therefore, not unified but rather multiple, and not so much divided as contradicted.

Rose quotes other key passages from de Lauretis.

> It is the elsewhere of discourse here and now, the blind spots or the space-off, of its representations. I think of it as spaces in the margins of hegemonic discourses, social spaces carved in the interstices of institutions and in the chinks and cracks of the power-knowledge apparati.

> It is a movement between the (represented) and what the representation leaves out or, more pointedly, makes unrepresentable.

---

[4] Gillian Rose, *Feminism and Geography: The Limits of Geographical Knowledge*, Cambridge, UK: Polity Press, 1993, 138–40. The passages from de Lauretis are from *Technologies of Gender: Essays on Theory, Film and Fiction*, London: Macmillan, 1987: 2, 25, 26. Neither Rose nor de Lauretis, I suspect, would feel entirely comfortable with being called postmodern feminists. Their observations are acutely relevant here none the less.

It is a movement between the (represented) discursive space of the positions made available by hegemonic discourses and the space-off, the elsewhere, of those discourses. ... These two kinds of spaces are neither in opposition to one another or strung along a chain of signification, but they coexist concurrently and in contradiction. The movement between them, therefore, is not that of a dialectic, of integration, of a combinatory, or of *différance*, but is the tension of contradiction, multiplicity, and heteronomy.

As illustrated in these passages, the movement between modern and postmodern feminism is being pointedly expressed in spatial terms and tropes. This is not merely a metaphorical ploy but rather a consciously political grounding of critique and resistance in the spatiality of social life. Today, more so than ever in the past, the spatial problematic has become increasingly transdisciplinary, extending well beyond the traditionally spatial disciplines. And in the widening orbit of such critical spatial thought and praxis, a more openly postmodern feminism has played a particularly forceful and reconstitutive role.

In this growing spatial feminist literature,[5] many different ways of looking at and experiencing urban spatiality are being explored. At the core of these explorations has been a revived interest in the body as the most intimate of personal-and-political spaces, an affective microcosm for all other spatialities. The spatiality and sensuality of the body is being given a central positioning in the critical interpretation of the real-and-imagined geographies of everyday life in and outside the city. In Elizabeth Grosz's words, "the city is made and made over into the simulacrum of the body, and the body, in its turn, is transformed, 'citified,' urbanized as a distinctively metropolitan body" (in Colomina, 1992: 242).

Through this material and symbolic embodiment, urban spatiality has become recharged with multiple sexualities, eroticisms, and desires – what Sue Golding described as "the whole damn business

[5] In 1992, for example, at least four notable (and very different) books appeared from US publishers exploring the spaces that difference makes. Daphne Spain, in *Gendered Spaces*, (see footnote 15, p. 99), critically synthesizes the modernist feminist literature and develops further the strategy of degendering spaces in dwellings, schools, and workplaces. *Sexuality and Space*, edited by Beatriz Colomina, Princeton: Princeton Papers on Architecture, 1992, foregrounds sexuality and the body in the postmodern politics of space; while Celeste Olalquiaga, in *Megalopolis: Contemporary Cultural Sensibilities*, Minneapolis: University of Minnesota Press, 1992, infuses the "culturescapes" and "cyberspaces" of the city with wildly postmodern (and postcolonial) sensibilities. Finally, Leslie Kanes Weisman explores "bodyscape as landscape" and urban spatiality as a "patriarchal symbolic universe," in *Discrimination by Design: A Feminist Critique of the Man-Made Environment*, Urbana and Chicago: University of Illinois Press, 1992.

of sweat and blood and pleasure and death."[6] Cityspace is no longer just dichotomously gendered or sexed, it is literally and figuratively transgressed with an abundance of sexual possibilities and plea-sures, dangers and opportunities, that are always both personal and political and, ultimately, never completely knowable from any sin-gular discursive standpoint.

Lesbians can be said to have taken the lead in exploring the "queer spaces" of the city, but, as Golding so effectively notes, even the term "lesbian" or "gay" denotes too narrow and exclusive a view of erotic preferences and identity. The key point being raised by Golding and others is that identity, sexual or otherwise, is unstable, shifting, mul-tiplicitous, situational, refractory, hybridizable, always being negoti-ated and contested, never static or fixed. In an ironic twist, Golding makes analogies to quantum physics in her pungent essay on "Quantum Philosophy" and "Impossible Geographies."

> To put this slightly differently, we have now before us not only a fluid (and yet discrete) concept of time, but also a dynamic concept of space, a concept, to be even more precise, that debunks any notion of an "eternally infinite" spatiality, while simultaneously refusing its uniqueness as if only geographi-cally "singular", that is to say, "closed", "totalized", "homoge-neous" and there-with, always-already, "fixed". We have here, in other words, "space-time", an imaginary, but real, and utterly *dynamic* fourth dimension.   (1993: 211)

On a more broadly political palette, Barbara Hooper brilliantly elaborates on Lefebvre's insistent argument that "the whole of (social) space proceeds from the body." In her ongoing project on "Bodies, Cities, Texts," Hooper takes this elaboration directly into a thirdspatial journey to contemporary Los Angeles.[7]

---

[6] Sue Golding, "Quantum Philosophy, Impossible Geographies and a Few Small Points About Life, Liberty and the Pursuit of Sex (All in the Name of Democracy)," in Keith and Pile, eds, *Place and the Politics of Identity*, 1993: 206–19. The preceding quote is from p. 212.

[7] The unpage-referenced quotes that follow are taken from Barbara Hooper, "Bodies, Cities, Texts: The Case of Citizen Rodney King," unpublished manuscript, February 1994, 80 pp. A preliminary version of this manuscript was presented at the annual meetings of the Association of American Geographers in Atlanta, 1992; and much more extensive extracts will appear in the final chapter of Edward W. Soja, *Postmetropolis*, the forthcoming companion volume to *Thirdspace*. Hooper's work on Rodney King is part of a larger dissertation/book project under the general title *Bodies, Cities, Texts*, in which she also explores these interlocked themes in the Athenian polis and nineteenth-century Paris. On the latter, see Hooper, "The Poem of Male Desires: Female Bodies, Modernity, and 'Paris, Capital of the Nineteenth Century'," *Planning Theory* 13, 1995: 105–29.

Henri Lefebvre suggests that power survives by producing space; Michel Foucault suggests that power survives by disciplining space; Gilles Deleuze and Felix Guattari suggest that to reproduce social control the state must reproduce spatial control. What I hope to suggest is that the space of the human body is perhaps the most critical site to watch the production and reproduction of power.

Hooper inscribes the body–city–text in what she calls *the order in place* and the representational practice of *somatography*, or body-writing, "a hierarchical differentiating of flesh that began millennia ago with the division of body and mind and that, like geography, earth-writing, orders ambiguous substances of matter as political meanings and territories." The human body here becomes more than just a product of culture or the creation of biology.

> It is a concrete physical space of flesh and bone, of chemistries and electricities; it is a highly mediated space, a space transformed by cultural interpretations and representations; it is a lived space, a volatile space of conscious and unconscious desires and motivations — a body/self, a subject, an identity: it is, in sum, a social space, a complexity involving the workings of power and knowledge *and* the workings of the body's lived unpredictabilities.

In her forays into the production of citizen-bodies in classical Athens, nineteenth-century Paris, and especially contemporary Los Angeles (through a critical reading of the political emanations from the body of Citizen Rodney King), Hooper acts spatially to *disorder* the (b)orderlands of bodies, cities, texts.

> Body and the body politic, body and social body, body and city, body and citizen-body, are intimately linked productions.... The practice of using the individual body as a metaphor for the social body, of deploying it as a sign of the health or disease of the social body, develops in the Athenian polis with ideas of democracy and reason and continues into the present. Body and city are the persistent subjects of a social/civic discourse, of an imaginary obsessed with the fear of unruly and dangerous elements and the equally obsessive desire to bring them under control: fears of pollution, contagions, disease, things out of place [for the ancient Greeks, the definition of "pollution"]; desires for controlling and mastering that [become] the spatial practice of enclosing unruly elements within carefully guarded spaces. These acts of differentiation, separation, and enclosure involve material, symbolic, and lived spaces ... bodies and cities and texts ... and are practiced as a politics of difference,

as segregation and separation.

She continues her unruly border work to set the stage for an interpretation of "The Case of Citizen Rodney King" and the events of spring 1992 in Los Angeles. After citing Foucault's observation that, "in the first instance, discipline proceeds from the distribution of individuals in space," she writes:

> In times of social crisis — when centers and peripheries will not hold — collective and individual anxiety rise and the politics of difference become especially significant. The instability of the borders heightens and concern with either their transgression or maintenance is magnified. When borders are crossed, disturbed, contested, and so become a threat to order, hegemonic power acts to reinforce them: the boundaries around territory, nation, ethnicity, race, gender, sex, class, erotic practice, are trotted out and vigorously disciplined. At the same time, counter-hegemons are working to harness the disorder ... for political use. In these periods, bodies, cities, and texts become key sites of hegemonic and counter-hegemonic contestations.... In the late twentieth century, it is the global megacity with its restless (teeming, breeding) populations, and (in the US) the sensationalized, demonized bodies of black males and urban gangs who have taken on the role of representing social disorder and pathology.

Moving circuitously outward from the body without ever losing touch with it, the alternative spaces of the visual, kinetic, and aesthetic imagination – in films, photography, advertising, fashion, museum exhibitions, murals, poems, novels, but also in shopping malls and beaches, factories and streets, motels and theme parks – are being creatively evoked by other spatial feminists as ways of seeing, hearing, feeling, interpreting, and changing the city. In the works of such urban cultural critics as Rosalyn Deutsche, Meaghan Morris, and Elizabeth Wilson, the construction of these spaces (as *oeuvres*, works of art, as well as *produits*, manufactured or "fashioned" products of a culture industry) is connected directly to the material dynamics of geohistorically uneven development and the political economy of contemporary capitalism, bringing culture performatively into the realm of class politics.[8]

[8] Rosalyn Deutsche, "Uneven Development: Public Art in New York City," *October* 47 (1988) and "Alternative Space," in *If You Lived Here: The City in Art, Theory, and Social Activism*, B. Wallis, ed., Seattle: Bay Press, 1991. Meaghan Morris, "At Henry Parkes Motel," *Cultural Studies* 2 (1988) and "Things to Do with Shopping Centres," in *Grafts: Feminist Cultural Criticism*, S. Sheridan, ed., London: Verso, 1988. Elizabeth Wilson, *The Sphinx and the City: Urban Life, the Control of Disorder, and Women*, London: Virago Press, 1991.

Across another spatial plane, Iris Marion Young and Nancy Fraser, among others, have engaged effectively in deconstructing and reconstituting the ideology of urban community and the old modernist binary of public versus private space in an explicitly radical postmodern and feminist politics of difference; while Donna Haraway has added nature and cybernetic high technology to this ideological deconstruction and reconstitution.[9] In their wake, cityspace is remapped around "unassimilated otherness," "fractured scenes," and "deprivileged spaces" rather than the overvalorized category of shared community identity and its exaggerated privileging of face-to-face communications.

Donna Haraway describes this critical remapping as a process of "diffraction" in her cartographic essay, "The Promises of Monsters."

Diffraction is a mapping of interference, not of replication, reflection, or reproduction. A diffraction pattern does not map where differences appear, but rather maps where the effects of differences appear. Tropically, for the promises of monsters, the first invites the illusion of essential, fixed position, while the second trains us to a more subtle vision.   (1992: 300)

She ends with a call to confront the "scary issues" of racism, sexism, historical tragedy, and technoscience with a remapping of possibility: "an inescapable possibility for changing maps of the world, for building new collectives"   (1992: 327).

In these postmodern recontextualizations of contemporary life, the great modernist narratives that connected "fixed" community (whether identified by class, race-ethnicity, gender, or mere propinquity) with emancipation (if not revolution) are shattered. Another spatiality is recognized, one which cannot be so neatly categorized and mapped, where the very distinction between mind and body, private and public space, and between who is inside or outside the boundaries of community, is obliterated and diffracted in a new and different cultural politics of real-and-imagined everyday life. In place of the community ideal, for example, Young advocates a radical democratic cultural pluralism that makes spatiality itself a primary field of struggle against all forms of oppression, wherever they may be located.

[9] Iris Marion Young, "The Ideal of Community and the Politics of Difference," in *Feminism/Postmodernism*, L. J. Nicholson, ed., London and New York: Routledge (1990); Nancy Fraser, "Rethinking the Public Sphere: A Contribution to the Critique of Actually Existing Democracy," *Social Text* 25/26 (1990); Donna Haraway, *Simians, Cyborgs, and Women: The Reinvention of Nature*, London and New York: Routledge (1991) and "The Promises of Monsters: A Regenerative Politics for Inappropriate/d Others," in *Cultural Studies*, L. Grossberg, C. Nelson, and P. Treichler, eds, New York and London: Routledge, 1992: 295–337.

In her spatial critique of everyday urban life in the postmodern world, Anne Friedberg goes *Window Shopping* to explore the spectacles and "spectatorship" of Thirdspace in the arcades of Paris, the oldest and newest theme parks, world exhibitions and fairs, the department store, and the shopping mall (through which she wanders as a self-conscious *flâneuse du mall*, a postmodern feminist recomposition of Baudelaire and Walter Benjamin). Friedberg's *flânerie* focuses on those "machines of virtual transport" that break us out of our constraining spatio-temporal containers, starting with the panorama and the diorama and ending with the "virtual tourism" provided by cinema and its extensions, most notably the television and the VCR.[10] With her, we enter an engrossing world of urban *hyperreality*, always there in the past but now teeming with the new perils and possibilities of postmodernity.

Of particular importance, here and now, in these "other" spaces and different geographies being opened up by the postmodern feminist cultural critics of spatiality, is the insight provided on how fragmentation, ruptures, deviation, displacements, and discontinuities can be politically transformed from liability and weakness to a potential source of opportunity and strength, a project which helps define again the boundary between adaptive late modernism and creatively critical postmodernism. In the new postmodern cultural and geographical politics of difference, as hooks has passionately shown us, we position ourselves first by subjectively choosing *for ourselves* our primary "marginal" identities as feminist, black, socialist, anti-colonialist, gay and lesbian activist, and so on. But we do not remain rigidly confined by this "territorial" choice, as was usually the case in modernist identity politics. We seek instead to find more flexible ways of being other than we are while still being ourselves, of becoming open to coalitions and coalescences of radical subjectivities, to a multiplicity of communities of resistance, to what Trinh T. Minh-ha has called "the anarchy of difference."[11]

Trinh develops a strategy of *displacement* (as opposed to a strategy of reversal) that adds significantly to hooks's reconceptualization of marginality and Hooper's disorderly notion of border work. "Without a certain work of displacement," she writes, "the margins can easily recomfort the center in goodwill and liberalism." The margins are "our fighting grounds" but also "their site for pilgrimage . . .

---

[10] Anne Friedberg, *Window Shopping: Cinema and the Postmodern*, Berkeley and Los Angeles: University of California Press, 1993. Friedberg locates many of her contemporary examples in the Los Angeles region, where she currently lives.
[11] Trinh T. Minh-ha, *When the Moon Waxes Red: Representation, Gender, and Cultural Politics*, New York and London: Routledge, 1991: 120. See also her *Woman, Native, Other: Writing Postcoloniality and Feminism*, Bloomington: Indiana University Press, 1989.

while we turn around and claim them as our exclusive territory, they happily approve, for the divisions between margins and center should be preserved, and as clearly demarcated as possible, if the two positions are to remain intact in their power relations." By actively displacing and disordering difference, by insisting that there are "no master territories," one struggles to prevent "this classifying world" from exerting its ordered, binary, categorical power.

Diana Fuss adds further insight to the caution needed and dangers involved in choosing marginality, and helps to defend this strategic choice against those who feel uneasy when persons of substantial power and status subjectively position themselves in the margins.[12] Fuss returns us to the tensions of being both inside and outside at the same time, and to the necessity for deconstruction and reconstitution, for disordering difference, for bell hooks's "wildness" and Lefebvre's anarchically centered peripheralness.

> Does inhabiting the inside always imply cooptation? ... And does inhabiting the outside always and everywhere guarantee radicality? The problem, of course, with the inside/outside rhetoric, if it remains undeconstructed, is that such polemics disguise the fact that most of us are both inside and outside at the same time. Any displaced nostalgia for or romanticization of the outside as a privileged site of radicality immediately gives us away, for in order to realize the outside we must already be, to some degree, comfortably on the inside. We really only have the leisure to idealize the subversive potential of the power of the marginal when our place of enunciation is quite central.   (1991: 5)

What Trinh, Fuss, hooks, and other feminist critics are capturing here are the political and personal tensions inherent in choosing marginality as a strategic standpoint, and even more so with adopting a postmodernist perspective in so doing. As Jane Flax notes in her essay, "The End of Innocence":[13]

> Postmodernism, especially when combined with certain aspects of psychoanalysis, necessarily destabilizes the (literal and

---

[12] Diana Fuss, "Inside/Out," in Diana Fuss, ed., *Inside/Out: Lesbian Theories, Gay Theories*, New York and London: Routledge, 1991: 5. As Fuss makes clear, there will always be significant tensions and dangers in choosing marginality as a space of resistance, not the least of which is the temptation toward "romancing the margins," assuming the margins to be the sole "authentic" location for the voices of truth and justice. That the dangers exist even after deconstruction must also be recognized.
[13] Jane Flax, "The End of Innocence," in *Feminists Theorize the Political*, Butler and Scott, eds, New York and London: Routledge, 1992: 445–63.

figurative) grounds of feminist theorizing, just as it does many other forms of Western philosophy. If one takes some of its central ideas seriously, even while resisting or rejecting others, postmodernism is bound to induce a profound uneasiness, or threatened identity, especially among white Western intellectuals, whose consciousness and positions are among its primary subjects of critical analysis.   (1992: 447)

Emerging from these postmodern spatial (and some not so spatial) feminist critiques is a further elaboration and exploration of a Thirdspace perspective on the complex contextuality of human existence. The brief selections reviewed here have been coming primarily from radical women of color, from gay and lesbian activists, from artists, architects, urbanists, writers, poets, literary critics, film critics, photographers, philosophers, and critical journalists. I turn next to put a different light on the spatial feminist critique that has developed in the field I know best: Geography.

The Paradoxical Space of Feminism and Geography

Despite the development of a vigorous feminist movement and the centrality of space in the discipline, there have been relatively few contributions by geographers to what I have described as the postmodern spatial feminist critique. This may be partially explained by the peculiar institutional and intellectual history of the modern discipline of Geography, its insular isolation on the margins of academia, and especially the related absence, until quite recently, of a strong critical-theoretical tradition. When a critical human geography did arise, after decades of concentration on just-the-facts descriptions, it was deeply shaped by a Marxism that had for a hundred years excluded spatiality (as well as gender) from its most incisive critical perspectives.

There were two primary arenas of critical debate for feminist geographers. The first was centered on the institutional and professional makeup of Geography itself, that is, on the entrenchment of masculinist authority in the definition and production of geographical knowledge and in the gender composition of the profession and its leading figures. This introspective critique of modern Geography mobilized a strong equity politics aimed at affirmative action for women, paralleling similar movements in other disciplines. It was expressed, in particular, in arguments that women have been silenced or "left out" of the geographical "project," an allusion to the Marxist or socialist approaches that had become so influential in Geography in the 1970s.

The second arena of struggle was similarly modernist in its

concentration on equity politics and affirmative action, but here the focus was on empirical research and theory-building. A growing body of feminist research on such issues as male versus female differentials in work, wages, and social status and their patterning across urban and regional spaces as well as within the social and spatial divisions of labor in both First World and Third World contexts, reflected the prevailing political economy approach to radical geography. It also linked up to the growing importance within the discipline of the subfields of economic and industrial geography, where the spatial restructuring of Fordism and the rise of what has come to be called a "postfordist regime of flexible accumulation" have become a primary nexus for geographical theory-building and rigorous empirical analysis. As participants in these debates, feminist economic geographers have contributed significantly to understanding the embeddedness of gender and sexism in the material forms emerging from this local, urban, regional, national, and global restructuring process.[14]

For the most part, feminist geographical research on the political economy of urbanization and regional and international development, as well as in other branches of the new critical human geography, was essentially modernist in the sense that it revolved primarily around the Man/Woman dualism and its partitioning of gendered spaces. More pertinent to the present argument, it also tended to remain tightly cocooned within the discipline of Geography, especially within its traditional oscillation between Firstspace and Secondspace perspectives. Feminist geographers worked effectively to describe and explain the objective material expressions of gender, sexism, patriarchy, and women's discriminatory oppression in (First)space; and to unmask the deep imprint of masculinism in subjective geographical thinking and in the representations of (Second)space that filled the writings of male (and some female) geographers. But, until very recently, there was little attempt to contribute to the development of feminist theory and criticism outside the discipline by introducing directly to it the feminist geographer's own critical geographical imagination.

When the powerful epistemological critique of modernism and the assertion of postmodern critical perspectives entered Geography, most feminist (as well as non-feminist) geographers seemed to be taken aback. As would happen in other spatial disciplines such as architecture and urban planning, many of the most prominent geog-

---

[14] The most prominent figure in these developments has been Doreen Massey, a longtime professor at the Open University in the UK. Her best-known work is *Spatial Divisions of Labor: Social Structures and the Geography of Production*, London: Macmillan, 1985. For a collection of her earlier writings as well as her more recent ventures into feminist critical theory, see Massey, *Space, Place and Gender*, Oxford and Cambridge, UK: Blackwell and Polity Press, 1994.

raphers became frozen into an ambiguous middle-ground, unable or unwilling to take sides between defending the long-established modernist traditions and adopting the seemingly faddist attractions of postmodernism. For feminists, the dilemma was particularly acute. It had become increasingly clear that modern Geography was intrinsically and perhaps indelibly masculinist, as was being demonstrated for other modernist disciplines. But at the same time, the new postmodern Geography as it was developing appeared only to continue the domination of powerful white males drawing on other, mostly French, white males for their theoretical inspiration.

A temptation grew among feminist geographers, especially in Great Britain, to declare a pox on both the old and the new houses of Geography and their leading proponents.[15] In "Flexible Sexism," Doreen Massey (1991: 33) led the attack by packaging her critical review of David Harvey's *The Condition of Postmodernity* (1989) and my *Postmodern Geographies* (1989) in the assertion that "the nature of the response to the crisis [of modern Geography] is such as to find, somehow, a way of hanging on to the intellectual hegemony, or at least of not letting anyone else have it." Similarly, Linda McDowell (1992: 65) adds: "displacement from the center [among other post-structuralist strategies] allows those with power to counter the increasingly insistent challenge from those at the margins;" and Liz Bondi (1990: 5) concludes: "postmodernism may be understood as a crisis in the experience of modernity among white, western men, and as a response centred on that experience."[16] More recently,

---

[15] In the US, the response among feminist geographers to the postmodern debate was much less antagonistic than in Great Britain. The research and writings of Cindi Katz, Sallie Marston, Margaret FitzSimmons, Susan Christopherson, among others, while often critical of the excesses of postmodernism, has for the most part been sympathetic and supportive of the radical postmodernist critique in and outside Geography. Why such a difference arose between the US and Great Britain is difficult to explain. Was it due to the significantly different political situations and cultures? Was more at stake in Britain, where Geography is a more prestigious and prominent field than in the US? Is British Geography more pervasively masculinist? It is perhaps best to leave the question open for more careful analysis, although I might add that it is interesting that the four American feminist geographers I mentioned have not been employed exclusively in university departments of Geography but rather in positions with close contact with other spatial disciplines such as urban planning.

[16] Doreen Massey, "Flexible Sexism," *Environment and Planning D: Society and Space* 9 (1991), 31–57; Linda McDowell, "Multiple Voices: Speaking From Inside and Outside 'the Project,' " *Antipode* 24 (1992), 56–72; Liz Bondi, "On Gender Tourism in the Space Age: A Feminist Response to *Postmodern Geographies*," paper presented at the annual meeting of the Association of American Geographers in Toronto, 1990, quoted in Massey, 1991: 33. In addition to my *Postmodern Geographies*, the major focus for these feminist geographical critiques was David Harvey, *The Condition of Postmodernity: An Inquiry into the Origins of Cultural Change*, Cambridge, MA, and Oxford, UK: Blackwell, 1989. I note in passing that the feminist critique so dominated the reaction to *Postmodern Geographies* in the UK that to my knowledge not one major British geographical journal ever published a specific review of the book.

Patricia Price-Chalita reasserts a similar critical position in her attempt to explore the possible compatibility of feminist and post-modern geographical approaches:[17]

> It appears that some scholars have embarked on a wholly patri-archal attempt to "spatialize," and I argue, to thus know and control, the Other-that-is woman. This presents feminist geog-raphers with yet another interestingly spatial gesture, one that negates rather than sets in motion the liberating possibility inherent in so-called postmodern work. Thus in spite of their claim to work within some sort of postmodern framework, a great many scholars continue to deny the voices of difference. (1994: 245)

All these responses to the postmodernization of Geography raise cogent criticisms of the style and substance contained in the first round of writings about geography and postmodernism.[18] But at the same time, it is important to note that this forceful critique signifi-cantly deflected the impact of the postmodern spatial feminist writ-ings that were developing outside Geography and delayed the creation of an alternative feminist approach to the critical retheoriza-tion of space from within the discipline. The long-established disci-plinary hold of the Firstspace-Secondspace dualism continued to be virtually unseen and unquestioned, severely constraining the explo-ration of Thirdspace perspectives and the participation of geogra-phers in the wider cultural studies debates on spatiality.

In the past few years, however, the once rigid anti-postmodernist stance has softened significantly and a different feminist geography has begun to emerge, not yet avowedly postmodern but clearly more open to alternative critical spatial perspectives. In my view, the best statement of this repositioning is *Feminism and Geography: The Limits of Geographical Knowledge* (1993), written by Gillian Rose. Rose's book maintains the British tradition of almost entirely dismissing the con-tributions of authoritative male geographers (perhaps to make them understand better how it feels to be left out of the project). But it also breaks important new ground in the debates on geography and spa-tiality, and on what I have called the trialectics of space–power–knowledge. At the risk of appearing to appropriate her work

---

[17] Patricia Price-Chalita, "Spatial Metaphor and the Politics of Empowerment: Mapping a Place for Feminism and Postmodernism in Geography?" *Antipode* 26 (1994), 236–54.
[18] I might add that the most insightful and constructive critiques of this first round of writings came from outside the discipline of Geography. See, especially, Rosalyn Deutsche, "Boys Town," *Environment and Planning D: Society and Space* 9 (1991), 5–30; and Meaghan Morris, "The Man in the Mirror: David Harvey's 'Condition' of Postmodernity," *Theory, Culture and Society* 9 (1992), 253–79.

against her will, I will try, with admiration and respect, to link *Feminism and Geography* to the evolving explorations of *Thirdspace*.

As she writes in the Introduction, "this is not a book about the geography of gender but about the gender of geography." (1993: 4–5) In the first six chapters, Rose explores the masculinism of Geography and the limits it imposes on geographical knowledge more incisively and creatively than anyone else before her. In her narrative, one can see more clearly the build-up of intellectual, political, professional, and personal frustrations that angrily exploded in the late 1980s and early 1990s, especially among British feminist geographers facing what they understandably saw, in the new literature on postmodernity, as yet another new, and if not masculinist then certainly not feminist, twist on the internal trajectory of modern Geography.

In the final chapter, however, Rose moves out of the disciplinary cocoon to explore "the politics of paradoxical space" in a wonderful meandering tour through the spatial feminist critique that has been lying beyond the geographer's disciplinary pale. With one minor exception, no geographer, male or female, is referenced in the more than 90 footnotes to this chapter. Instead, Rose ventures into the very best of feminist spatial scholarship: Teresa de Lauretis on *Technologies of Gender* and "eccentric subjects," Adrienne Rich on *Blood, Bread and Poetry* and the body as "the geography closest in," Chandra Mohanty on Third World women and the "cartographies of struggle," Diana Fuss's *Inside/Out; Lesbian Theories, Gay Theories*, Virginia Woolf on having *A Room of One's Own*, Gloria Anzaldúa's spiraling borderlands and consciousness of the *mestiza*, Donna Haraway's remappings and "reinvention of nature," Marilyn Frye's angry "domestic geographies," Iris Marion Young breaking out from conventional spatial confinements, Trinh T. Minh-ha *Writing Postcoloniality and Feminism*, Elspeth Probyn's embodiment of geographies, the Combahee River Collective's "Black feminist statement" and the empowering *Black Feminist Thought* of Patricia Hill Collins and Pratibha Parmar, and ultimately, with particular appreciation, bell hooks's *Yearning* for a "space of radical openness."

En route, Rose charts out a project in her own words.

> I want to explore the possibility of a space which does not replicate the exclusions of the Same and the Other. I examine a spatiality imagined by some feminists which can acknowledge the difference of others. ... I want to end this book by arguing, then, that the spatial images which structure one kind of feminist imagination offer us one new way of thinking geography. They offer a new kind of geography which refuses the exclusions of the old.... Feminism, I think, through its awareness of

the politics of the everyday, has always had a very keen awareness of the intersection of space and power – and knowledge.
As de Lauretis says, there is "the *epistemological* priority which
feminism has located in the personal, the subjective, the body,
the symptomatic, the quotidian, as the very site of material
inscription of the ideological."   (1993: 137, 141, 142)

On the "geometrics of difference and contradiction:"

Social space can no longer be imagined simply in terms of a territory of gender. The geography of the master subject and the
feminism complicit with him has been ruptured by the diverse
spatialities of different women. So, a geographical imagination
is emerging in feminism which, in order to indicate the complexity of the subject of feminism, articulates a "plurilocality."
In this recognition of difference, two-dimensional social maps
are inadequate. Instead, spaces structured over many dimensions are necessary.   (1993: 151)

On "the paradox of being within the Same/Other and also elsewhere," Rose enters bell hooks's space of radical openness:

The manipulation of the field of the Same/Other, being both
separate and connected, the simultaneous occupation of both
the centre and the margin, being at once inside and outside: all
these discursive spaces depend on a sense of an "elsewhere" for
their resistance. The subject of feminism has to feel that there is
something beyond patriarchy in order to adopt these strategies
of subversion. Thus the paradoxes described in the previous
subsection themselves depend on a paradoxical space which
straddles the spaces of representation and unrepresentability.
This space of unrepresentability can acknowledge the possibility of radical difference.   (1993: 154)

The subject of feminism, then, depends on a paradoxical geography in order to acknowledge both the power of hegemonic
discourses and to insist on the possibility of resistance. This
geography describes that subjectivity as that of both prisoner
and exile; it allows the subject of feminism to occupy both the
centre and the margin, the inside and the outside. It is a geography structured by the dynamic tension between such poles, and
it is also a multidimensional geography structured by the
simultaneous contradictory diversity of social relations. It is a
geography which is as multiple and contradictory and different
as the subjectivity imagining it ... [it is] a different kind of

space through which difference is tolerated rather than
erased.   (1993: 155)

Rose concludes the chapter with the exemplary work of two feminists "who invoke this [paradoxical] geography in their own under-
standings of their lives in order to think about a politics of
resistance." The choice is itself paradoxical: two American women
from the South, one (Minnie Bruce Platt) a white Christian lesbian,
the other bell hooks, writing about "Choosing the Margin as a Space
of Radical Openness." Rose dwells primarily on hooks's interpretations of "home" as a paradoxical geography, but also travels with
her very close to the Thirdspace I have evoked from her writings in
the previous chapter.

No geographer, male or female, who has read *Feminism and
Geography* will look at the discipline in the same way they may have
before. What is still missing from this critical rethinking of
Geography, however, is an appreciation for the wider range of othernesses and marginalities found in those polycentric coalitional communities of resistance that hooks speaks about as a vital part of the
new cultural politics. Geography and critical human geographers
have an opportunity to play a significant role in these communities of
resistance, drawing on the discipline's continued peripheralization
within the long-established academic and intellectual division of
labor. Here there is another chance to strategically choose this marginality as a space of radical openness, a place for critical rethinking,
re-envisioning, and more effective resistance to all forms of subordination and oppression both inside and outside the academy. I am not
saying that identity as a "marginalized" geographer is the same thing
as identity defined along gender, race, or class lines, but it must not
be entirely overlooked as a "home" for a community of critical scholarship and spatial praxis that reaches out to join with other such communities in the new cultural politics of postmodernity.

The Postcolonial Critique[19]

As has already been indicated, there is a significant overlapping of
the spatial feminist debate into the postcolonial critique that has

[19] The tense ambiguity of the term "postcolonial," as opposed to the, at first
thought, clearer "anti-colonial" is worthwhile preserving for several reasons. First,
the critique is not simply anti-colonial, for colonization is, like Foucault's notion of
power, both oppressive and enabling at the same time, a complex form of governmentality (or govern mentality). The use of "post-" also resonates with connections
to the postmodern and postmodernism. In this sense, the postcolonial critique is a
move beyond modernist anti-colonialism and an adaptive adjustment to the particular circumstances of the present moment.

been developing equally rapidly over the past 20 years. The post-colonial critique has many connections with the "postmodern black-ness" described by bell hooks and other explorations of open and multi-layered radical subjectivities. Although emerging from a simi-lar rootedness in the intertwining questions of race, gender, and class, the postcolonial critics tend also to give particular attention to several additional issues from Cornel West's list: to exterminism, empire, nation, nature, and region. Foregrounded in all these con-texts is an emphasis on colonization and the relations between the colonizer and the colonized, another variation on the themes of sub-jection and marginalization. And in contrast to the urban-focused feminist critique, the postcolonial critics roam more globally in the spaces created by the workings of geohistorically uneven develop-ment.

I find it useful to understand the postcolonial critique as a product of still another critical thirding-as-Othering, an assertively different and intentionally disruptive way of (re)interpreting the relation between the colonizer and the colonized, the center and the periph-ery, Firstworlds and Thirdworlds. More specifically, the critique addresses two metanarratives that have overarchingly dominated the Firstworld discourse on coloniality: a fundamentally capitalist metanarrative of *development* that wraps world history in the neces-sity for continuous progress and modernization; and a predomi-nantly Marxist or socialist metanarrative of *social justice* that requires radical if not revolutionary transformation for social justice to be achieved.

In the best of the postcolonial critiques, neither of these two encompassing historical metanarratives is dismissed entirely. Instead, both are creatively interrogated to expose first of all their totalizing and persistently Eurocentric co-hegemony and their equally persistent silencing of peripheral voices and alternative points of view. Allowing the "subaltern" to speak, to assert an-Other voice, pushes the discourse on to a different plane and into a recre-ative space of radical openness where both development and social justice can be revisioned together, along with their histories and geo-graphies, not as an either/or choice but in the limitless expansion of the both/and also ... As with Thirdspace, the critique does not come to a full stop at this third way; nor is it a simple additive or "in-between" positioning that can be marked with dogmatic assurity. It is instead an invitation to continuous deconstruction and reconstitu-tion, to a constant effort to move beyond the established limits of our understanding of the world.

What follows are rather hurried glimpses of five variations on the critical theme of postcoloniality chosen for their rich and varied con-tributions to a specifically postmodern and Thirdspatial approach to understanding the new cultural politics of difference, identity, and radical subjectivity. As is also true with the preceding discussions, I draw for the most part on those interpretations that are most relevant for our eventual journeys to the real-and-imagined spaces of Los Angeles.

The Borderlands of Gloria Anzaldúa

A special place within the overlapping spheres of the spatial feminist and postcolonial critiques is occupied by Gloria Anzaldúa, whose major works, *Borderlands/La Frontera: The New Mestiza* and *Making Face/Making Soul: Haciendo Caras*, explore the poetics of the "borderlands" as a space of radical openness, a Thirdspace filled with the perils and possibilities that infiltrate the chosen marginality of bell hooks.[20]

> The actual physical borderland that I'm dealing with ... is the Texas–U.S. Southwest/Mexican border. The psychological borderlands, the sexual borderlands and the spiritual borderlands are not particular to the Southwest. In fact, the Borderlands are physically present wherever two or more cultures edge each other, where people of different races occupy the same territory, where under, lower, middle and upper classes touch, where the space between two individuals shrinks with intimacy.
>
> I am a border woman, I grew up between two cultures, the Mexican (with a heavy Indian influence) and the Anglo (as a member of a colonized people in our own territory). I have been straddling that *tejas*-Mexican border, and others, all my life. It's not a comfortable territory to live in, this place of contradictions. Hatred, anger and exploitation are the prominent features of this landscape.
>
> However, there have been compensations for this *mestiza*, and certain joys. Living on borders and in margins, keeping intact one's shifting and multiple identity and integrity, is like trying to swim in a new element, an "alien" element ... [that] has become familiar – never comfortable, not with society's clamor to uphold the old, to rejoin the flock, to go with the herd. No, not comfortable but home. (1987: unpaged preface)

[20] Gloria Anzaldúa, *Borderlands/La Frontera: The New Mestiza*, San Francisco: Spinsters/Aunt Lute, 1987; and *Making Face/Making Soul: Haciendo Caras*, Gloria Anzaldúa, ed., San Francisco: Spinsters/Aunt Lute, 1990. Anzaldúa was also co-editor, with Cherríe Moraga, of an influential collection of writings that helped define feminism as practiced by radical women of color in the US: *This Bridge Called My Back: Writings by Radical Women of Color*, New York: Kitchen Table: Women of Color Press, 1981.

Anzaldúa positions herself in The Homeland, Aztlan, *El otro Mexico*, the "other" Mexico, *"una herida abierta* where the Third World grates against the First and bleeds. And before a scab forms it hemorrhages again, the lifeblood of two worlds merging to form a third country – a border culture." (1987: 3) She shapes this culture with a poem. (1987: 2–3)

> I press my hand to the steel curtain –
> chainlink fence crowned with rolled barbed wire –
> rippling from the sea where Tijuana touches San Diego
> unrolling over mountains
> and plains
> and deserts,
> this "Tortilla Curtain" turning into *el río Grande*
> flowing down to the flatlands
> of the Magic Valley of South Texas
> its mouth emptying into the Gulf.

> 1,950 mile-long open wound
> dividing a *pueblo*, a culture,
> running down the length of my body,
> staking fence rods in my flesh,
> splits me   splits me
> *me raja*   *me raja*

> This is my home
> this thin edge of
> barbwire.

> But the skin of the earth is seamless.
> The sea cannot be fenced,
> *el mar* does not stop at borders.
> To show the white man what she thought of his
> arrogance,
> *Yemaya* blew that wire fence down.

> This land was Mexican once,
> was Indian always
> and is.
> And   will be again.

In the borderlands is born(e) a new consciousness, the consciousness of the *mestiza*. "As a *mestiza*," Anzaldúa writes, "I have no country, my homeland casts me out; yet all countries are mine because I am every woman's sister or potential lover. (As a lesbian I have no race, my own people disclaim me; but I am all races because there is the queer of me in all races.) ... I am an act of kneading, of uniting and joining that not only has produced both a creature of darkness and a creature of light, but also a creature that questions the definitions of light and dark and gives them new meanings" (1987: 80–1). In the postmodern culture of the borderlands, with its "decentered subjects," space is created for "oppositional practices," for "critical exchange," for "resistance struggle," for "new and radical happenings" (I am now quoting hooks). Anzaldúa offers her own spatialization in the Preface to *Making Face.* ...

> We need theories that will rewrite history using race, class, gender and ethnicity as categories of analysis, theories that cross borders, that blur boundaries. ... Because we are not allowed to enter discourse, because we are often disqualified or excluded from it, because what passes for theory these days is forbidden territory for us, it is *vital* that we occupy theorizing space, that we not allow white men and women solely to occupy it. By bringing in our own approaches and methodologies, we transform that theorizing space. ...
>
> We are articulating new positions in these in-between, Borderland worlds of ethnic communities and academies, feminist and job worlds. In our literature, social issues such as race, class and sexual difference are intertwined with the narrative and poetic elements of a text, elements in which theory is embedded.   (1990: xxv–xxvi)

Chicanisma/Chicanismo: Reworlding the Border

While Anzaldúa may be the leading spatial theoretician of the borderlands and *mestizaje*, there has been developing over the past ten years a remarkable Chicana/o-Latina/o literature that has opened the space of the US–Mexico border to innovative new forms of (Third)spatial interpretation.[21] Chicana feminists such as Norma Alarcón, Sandra Cisneros, Terri de la Peña, and María Lugones, to name but a few, expand our understanding of the intersections of space and language, *sitios y lenguas*, the production of space and the production of texts by exploring in novels, poetry, and cultural criticism the multiple mean-

---

[21] See the forthcoming dissertation by Mary Pat Brady, *Re-Mapping the Terrain of Power: Capitalism, Spatiality and Cultural Response in Chicana and Chicano Cultural Production*, English, UCLA. I wish to thank Mary Pat for making me aware of this literature and its creative extensions of the new theorizations of spatiality, and for her very constructive and stimulating critique of earlier drafts of chapters 3 and 4.

ings of *la linea* – the line, the border, the boundary – as lived space.[22] I cannot hope to capture the richness of this literature here, but a few passages from María Lugones's "Playfulness, 'World'-Traveling, and Loving Perception" are indicative.

> The paper describes the experience of "outsiders" to the mainstream White/Anglo organization of life in the U.S., and stresses a particular feature of the outsider's existence: the acquired flexibility in shifting from the mainstream construction of life to other constructions of life where she is more or less "at home." This flexibility is necessary for the outsider but it can also be willfully exercised by those who are at ease in the mainstream. I recommend this willful exercise which I call "world"-traveling and I also recommend that the willful exercise be animated by a playful attitude. . . .
>
> I can explain some of what I mean by a "world." I do not want the fixity of a definition because I think the term is suggestive and I do not want to lose this. A "world" has to be presently inhabited by flesh and blood people. That is why it cannot be a utopia. It may also be inhabited by some imaginary people. It may be inhabited by people who are dead or people that the inhabitants of this "world" met in some other "world" and now have this "world" in imagination. . . .
>
> A "world" may be incomplete. . . . Given lesbian feminism, the construction of "lesbian" in "lesbian community" (a "world" in my sense) is purposefully and healthily still up in the air, in the process of becoming. To be Hispanic in this country is, in a dominant Anglo construction, purposefully incomplete. Thus one cannot really answer questions like "What is a Hispanic?" "Who counts as a Hispanic?" "Are Latinos, Chicanos, Hispanos, black Dominicans, white Cubans, Korean-Columbians, Italian-Argentinians Hispanic?" What it means to be a "Hispanic" in the varied so-called Hispanic communities in the U.S. is also up in the air. We have not yet decided whether there are any "Hispanics" in our varied "worlds. . . ."
>
> One can "travel" between these "worlds" and one can inhabit more than one of these "worlds" at the very same time. I think that most of us who are outside the mainstream U.S. construc-

---

[22] Norma Alarcón, "The Theoretical Subject(s) of *This Bridge Called My Back* and Anglo-American Feminism," in *Criticism in the Borderlands*, Héctor Calderón and José David Saldívar, eds, Durham NC: Duke University Press, 1991: 28–39; Sandra Cisneros, *Woman Hollering Creek*, New York: Random House, 1991, and *The House on Mango Street*, Houston TX: Arte Publico Press, 1993; Terri de la Peña, *Margins*, Seattle WA: Seal Press, 1992; María Lugones, "Playfulness, 'World'-Traveling, and Loving Perception," in *Making Face, Making Soul*, Gloria Anzaldúa, ed., San Francisco: Spinsters/Aunt Lute, 1990: 390–402.

tion ... are "world-travelers" as a matter of necessity and survival. ... Those of us who are "world"-travelers have the distinct experience of being different in different "worlds" and ourselves in them. ... The shift from being one person to being a different person is what I call "travel. ...."

When considering the value of "world"-traveling and whether playfulness is the loving attitude to have while traveling, I recognized the agonistic attitude as inimical to traveling across "worlds." The agonistic traveler is a conqueror, an imperialist. Given the agonistic attitude one *cannot* travel across "worlds," though (one) can kill other "worlds" with it. ...

So for people who are interested in crossing racial and ethnic boundaries, an arrogant western man's construction of playfulness is deadly. One needs to give up such an attitude if one wants to travel. Huizinga in his classic book on play [*Homo Ludens*, 1968], interprets Western civilization as play. That is an interesting thing for Third World people to think about. Western civilization has been interpreted by a white western man as play in the agonistic sense of play: he reviews western law, art, and many other aspects of western culture and sees agon, contest, in all of them.   (1990: 390, 395, 396, 400)

Lugones's "worlds" are artfully lived spaces, metaphorical and material, real and imagined, contested and loving, dangerous and playful, local and global, knowable and incomplete, necessary for survival yet often still "up in the air," as limitless as Borges's magically real Aleph yet intimate and comforting at the same time. Here we have another version of a space of radical openness, a space filled with the representations of power and the power of representations.

Even more playful in his "world"-travels is Guillermo Gómez-Peña, a writer (as he is known in Mexico, his homeland) and performance artist (as he is known in the US and Europe). In his performances as a "Warrior for Gringostroika," Gómez-Peña, a self-described "interdisciplinary troublemaker," explores the border, *la linea*, within the dynamic of deterritorialization–reterritorialization, as a zone of crossing, alterity, opposition, symbiosis, fissure, suture, permanent exile, a way of life, a mode of expression, a broken line, a political esthetic, a performance laboratory, a community of resistance.[23] Under "Involuntary Postmodernism," he describes Mexico

---

[23] Guillermo Gómez-Peña, *Warrior for Gringostroika: Essays, Performance Texts, and Poetry*, St Paul MN: Graywolf Press, 1993; Introduction by Roger Barra, Douglas Keiller, Associate Editor. While a Master's student in Urban Planning at UCLA and suffering through some of my earlier lectures on Thirdspace, Doug Keiller insisted that I take a look at the creatively spatial work of Gómez-Peña and lent to me his personal copy of this book. Thank you, Doug. I hope I remember to return your book.

City as "the 'postmodern' city par excellence," a "mega-pastiche" filled with "a vernacular postmodern sensibility," "cross-cultural fusion," "dualities ... perceived ... as overlapping realities," *mestizaje* (1993: 18). In a brief performance poem entitled "Punctured Tire (Highway 5, California, 1988)" (1993: 167), he describes life across the border:

> I am stuck in the middle of the journey
> a highway without human activity
> a text without visible structure
> life on this side of the border
> on your side ...
> I no longer know who I am
> but I like it

In "The Border Is ... (A Manifesto)" (1993: 43–4) Gómez-Peña engages in descriptions of "border culture" that resound with brilliant evocations of Borges's Aleph. I recompose his performance text as a selective listing:

> BORDER CULTURE IS a polysemantic term ... [it] means boycott, complot, ilegalidad, clandestinidad, transgresíon ... hybrid art forms for new contents-in-gestation: spray mural, techno-altar, poetry-in-tongues, audio graffiti, punkarachi, video corrido, anti-bolero, anti-todo ... to be fluid in English, Spanish, Spanglish, and Ingleñol ... transcultural friendship and collaboration among races, sexes, and generations ... creative appropriation, expropriation, and subversion of dominant cultural forms ... a new cartography: a brand-new map to host the new project; the democratization of the East; the socialization of the West; the Third-Worldization of the North and the First-Worldization of the South ... a multiplicity of voices away from the center, different geo-cultural relations among more cultural akin regions: Tepito-San Diejuana, San Pancho-Nuyorrico, Miami-Quebec, San Antonio-Berlin, your home town and mine, digamos, a new internationalism ex centris ... to return and depart again [for] to arrive is just an illusion ... a new terminology for new hybrid identities and métiers constantly metamorphosing: sudacá, not sudaca; Chicarrican, not Hispanic; mestizaje, not miscegenation; social thinker, not bohemian; accionista, not performer; intercultural, not postmodern ... to develop new models to interpret the world-in-crisis, the only world we know ... to push the borders of countries and languages or, better said, to find new languages to express the fluctuating borders ... experimenting with the fringes

between art and society, legalidad and illegality, English and
español, male and female, North and South, self and other; and
subverting these relationships ... to speak from the crevasse,
desde acá, desde el medio. The border is the juncture, not the
edge, and monoculturalism has been expelled to the margins ...
it also means glasnost, not government censorship ... to ana-
lyze critically all that lies on the current table of debates: multiculturalism, the Latino "boom," "ethnic art," controversial art,
even border culture ... to question and transgress border culture [for what] today is powerful and necessary, tomorrow is
arcane and ridiculous; what today is border culture, tomorrow
is institutional art, not vice versa ... But it also means to look at
the past and the future at the same time. 1492 was the beginning of a genocidal era. 1992 will mark the beginning of a new
era: America post-Colombina, Arteamérica sin fronteras. Soon a
new internationalism will have to gravitate around the spinal
cord of this continent – not Europe, not just the North, not just
white, not only you, compañero del otro lado de la frontera, el
lenguaje y el océano.

An excerpt from the performance poetry of "Border Brujo" (1993: 71)
amplifies *la linea* in another way:

> standing on the corner of Broadway & Hell
> where English meets Spanish
> & Death performs the last striptease of the day
> El Johnny longs for his inner carnales
> al Sueño lo torcieron
> al Misterioso lo balacearon
> & a Susy la Sad Girl
> la trituraron los medios de información
> "Mexico is sinking
> California is on fire
> & we are all getting burned
> aren't we? we're just
> a bunch of burning myths!"
> – he begins to yell at the gringos –
> "but what if suddenly
> the continent is turned upside down?
> what if the U.S. was Mexico?
> what if 200,000 Anglosaxicans
> were to cross the border each month
> to work as gardeners, waiters
> musicians, movie extras
> bouncers, babysitters, chauffeurs

syndicated cartoons, featherweight
boxers, fruit-pickers & anonymous poets?
what if they were called waspanos,
waspitos, wasperos or waspbacks?
what if we were the top dogs?
what if literature was life?
what if yo were you
& tú fueras I, Mister?"

Gómez-Peña is the bard of the borderlands, a maestro of *mestizaje*, a playfully serious poet-*lineate* whose "tangential" attitude and post-modern stance cries out for the "borderization of the world" and a continual rethinking of Thirdspace.

The Reworldings of Gayatri Spivak

Also deserving a special place among the postcolonial critics and those, like bell hooks, Gloria Anzaldúa and Guillermo Gómez-Peña, who are revisioning Thirdspace from the regional borderlands defined by both the micro-workings of power and the macrogeographies of uneven development, is Gayatri Chakravorty Spivak. A major translator and (re)interpreter of Jacques Derrida, Spivak is the hard-hitting Derridean theoretician of postcoloniality. Her geographical imagination sparkles occasionally, but it rarely becomes the self-conscious center of attention. Nevertheless, she too adds new dimensions to a specifically global understanding of Thirdspace.

Spivak faces directly, in her writings, the challenges of choosing marginality and asserting the radical alterity of subaltern voices. Her discursive practice of a locational politics of difference is filled with acts of deconstruction, subversion, displacement, and repositioning. Spivak and other key figures in postcolonial studies also move beyond theoretical deconstruction and regional repositioning to begin another remapping of explicitly global proportions. This remapping for Spivak, is not just an opening up of the texts of Marx to non-Eurocentric readings, it is a move "beyond a homogeneous internationalism, to the persistent recognition of heterogeneity," to a new "worlding of the world."[24] She encapsules the difficulty of this task in what she describes as the slow and careful labor of "un-learning our privilege as our loss," of behaving "as if you are part of the margin" but doing so by relinquishing the privileges that attach to choosing marginality as a space of radical resistance from a position

---

[24] Gayatri Chakravorty Spivak, "Subaltern Studies: Deconstructing Historiography," in *Selected Subaltern Studies*, R. Guha and G. C. Spivak, eds, New York: Oxford University Press, 1988: 20. See also Spivak's *In Other Worlds: Essay in Cultural Politics*, New York and London: Routledge, 1988.

in the center.[25] She goes on reworlding marginality in her own personal act of displacement.

> I find the demand on me to be marginal always amusing.... I'm tired of dining out on being an exile because that has been a long tradition and it is not one I want to identify myself with.... In a certain sense, I think there is nothing that is central. The centre is always constituted in terms of its own marginality. However, having said that, in terms of the hegemonic historical narrative, certain peoples have always been asked to cathect the margins so others can be defined as central.... In that kind of situation the only strategic thing to do is to absolutely present oneself at the centre.... [W]hy not take the centre when I'm being asked to be marginal? I'm never defined as marginal in India, I can assure you. (1990: 40–1)

Thinking synchronically, in the precise (spatial) circumstances of the present (postmodern) moment, Spivak repositions herself as a *bricoleur*, a preserver of discontinuities, an interruptive critic of the categorical logic of colonizer–colonized, elite–subaltern, global–local, center–periphery, First World–Third World.

> If I have learned anything it is that one must not go in the direction of a Unification Church, which is too deeply marked by the moment of colonialist influence, creating global solutions that are coherent. On the other hand, it seems to me that one must also avoid as much as possible, in the interests of practical effectiveness, a sort of continuist definition of differences, so that all you get is hostility. (1990: 15)

Spivak's emphasis on discontinuities pushes us away from the simple celebration of pluralism and multiculturalism that has attracted so many contemporary liberal scholars; but at the same time sets clear limits upon her desired preservation of the discourses of feminism and marxism. What must be preserved, Spivak argues, is not the continuist definition of differences embedded in the coherent modernist discourses, the unified parochialism that responds to "others" only with hostility; but rather a "radical acceptance of vulnerability" immersed in a deconstructive *pratique sauvage*, a "wild practice," a "practical politics of the open end" in which choosing marginality becomes an invitingly anarchic, centerless, act of inclusion (1990: 18, 54, 105).

---

[25] Gayatri Chakravorty Spivak, in *The Post-Colonial Critic: Interviews, Strategies, Dialogues*, S. Harasym, ed., New York: Routledge, 1990: 9 and 30.

Spivak's remapping-as-reworlding – she once named her project "Feminism" both to preclude other namings and to recognize the most numerous gendered subjects in the margins of the periphery – disrupts the coherent spatiality of territorial imperialism, Eurocentrism, and spatial science, as well as the spatiality of modernist feminism and Marxism. And she does so with a strategic twist that she describes as bringing "hegemonic historiography to crisis," a critique of Western historicism that becomes a vital turn in her reworlding project and in the progressive spatialization of the new cultural politics of difference. Spivak engages in this counter-hegemonic critique of what Lefebvre metaphilosophically called the "world of representations" primarily via a deconstructive analysis of "texts" rather than of "spaces." But as I hope will become clearer in subsequent chapters, this direct critique of Western historicism and "hegemonic historiography" is another important, if only indirectly spatial, pathway into Thirdspace, one which is given very little critical attention by Lefebvre or hooks.

Spivak's critique of hegemonic historiography is too often sidetracked by the need to defend her neo-Derridean methods against the barrage of attacks, usually from those most to blame for the excesses of historicism, claiming that her project is anti-historical, that it is out to destroy rather than deconstruct critical historical reasoning itself. Although there is an abundance of spatial metaphors and references in her work, the reactive attack of the historicists and more innocently disciplined historians has prevented Spivak from directly confronting the problematic interplay between space and time, geography and history, that is imbricated in the crisis of hegemonic historiography, or what might be termed the historicist "subalterning" of critical spatial thinking. To explore this debate and its relevance to the conceptualization of Thirdspace, one must turn to the far-reaching critique of "Orientalism" contained in the works of Edward Said.

The Imaginative Geographies of Edward Said

Said moves the critique of historicism and hegemonic historiography towards the possibility of constructing a critical geohistory of uneven development at a global scale, a project advanced by Lefebvre and others but often constrained by an insufficiently self-critical Eurocentrism. In "The Palestinian Intellectual and the Liberation of the Academy," Barbara Harlow describes Said's project in explicitly spatial terms.

Said's work, with its insistence on the historical narrative and at the same time – theoretically and biographically – informed by the importance of space, the struggle over the land and against military occupation, engages explicitly in the project of a radical reconstruction, around the issue of geography, of the ascendant linear narrative of history led masterfully from the center. Colonialism and the national liberation struggles waged against its controlling influence articulate not just a temporal sequence, but a critical re-elaboration of geopolitical spatial arrangements and the politics of place, what Soja presents – even if with a residual Eurocentric bias – as the "conjunction between periodization and spatialization" (1989, 64). This politics of place must be critically decentered, further sited and re-read around a rethinking of the history of nationalism, Said maintains, and located institutionally as well as territorially.[26]

Said grounds his deconstruction of the various binaries of Orientalism in the impositioning of "imaginative geographies," dominating conventional representations of space as well as material spatial practices. In these real-and-imagined Orientalist geographies, the center is constructed as powerful, articulate, surveillant, the subject making history; while the periphery is defeated, silenced, subordinated, subjected without a history of its own. Through a critique of the spatial practices of colonialism and its formidable representations of space, knowledge, and power, Said also opens up the postcolonial spaces (and powers) of representation to an alternative geohistorical revisioning. In doing so, he enters Thirdspace through a side door and begins engaging in a critical debate that brings him close to the paths taken by hooks and Lefebvre.

That this is a simultaneously historical *and geographical* critique is brought out very clearly in a recent article by the historical geographer Derek Gregory, amplifying on his extensive treatment of Said in *Geographical Imaginations* (1994). Gregory begins with a quote from Said's *Culture and Imperialism* (1993).

Just as none of us is outside or beyond geography, none of us is completely free from the struggle over geography. That struggle is complex and interesting because it is not only about soldiers and cannons but also about ideas, about forms, about images and imaginings.

[26] Barbara Harlow, "The Palestinian Intellectual and the Liberation of the Academy," in *Edward Said: A Critical Reader*, M. Sprinker, ed, Cambridge, MA, and Oxford, UK: Blackwell, 1992: 173–93. The quote is from pp. 190–1. See also Edward Said, *Orientalism*, New York: Vintage Books, 1979.

"What I find myself doing," Gregory cites Said further, "is rethinking geography ... charting the changing constellations of power,
knowledge, and geography."[27]

In "Orientalism Reconsidered,"[28] Said also demonstrates that he
was at the same time rethinking history.

> So far as Orientalism in particular and European knowledge of
> other societies in general have been concerned, historicism
> meant that one human history uniting humanity either culmi-
> nated in or was observed from the vantage point of Europe, or
> the West. What was neither observed by Europe nor docu-
> mented by it was, therefore, "lost" until, at some later date, it
> too could be incorporated by the new sciences of anthropology,
> political economics and linguistics. It is out of this later recuper-
> ation of what Eric Wolf has called people without history, that a
> still later disciplinary step was taken: the foundation of the sci-
> ence of world history, whose major practitioners include
> Braudel, Wallerstein, Perry Anderson and Wolf himself.
> (1985: 10)

> What has never taken place is an epistemological critique of the
> connection between the development of a historicism which has
> expanded and developed enough to include antithetical atti-
> tudes such as ideologies of western imperialism and critiques of
> imperialism, on the one hand [referring back to the world histo-
> ries of Braudel, Wallerstein, Anderson, and Wolf], and, on the
> other, the actual practice of imperialism by which accumulation
> of territories and population, the control of economies, and the
> incorporation and homogenization of histories are main-
> tained.... In the methodological assumptions and practice of
> world history – which is ideologically anti-imperialist – little or

[27] Derek Gregory, "Imaginitive Geographies," *Progress in Human Geography* 19
(1995), 447–85. In celebrating Said's geographical imagination, however, Gregory,
ever the monitor of the intellectual affinities of others, dims Said's formidable critiques of both historicism and Orientalism's "imaginative geographies" by stripping
Said of his multiplicity of identities. While Gregory insightfully attaches Foucault
and Gramsci to Said's intellectual genealogy, he has more trouble in dealing with
Said's irrepressible both/and also recombinations, if not thirdings, of such binaries
as materialism/idealism, modernism/postmodernism, Marxism/anti-Marxism,
humanism/structuralism, historicism/spatialism, and indeed Orientalism/
Occidentalism. With his categorizing pen and unique brand of geohistoricism,
Gregory comes close to wiping out the radical openness and polyphonic flexibility
of Said's Thirdspatial imagination.
[28] Edward Said, "Orientalism Reconsidered," *Race & Class: A Journal for Black and
Third World Liberation* 27 (1985), 1–15. Slightly different versions of this article were
published in *Europe and Its Others*, F. Barker et al., eds, Colchester: University of
Essex, 1985; and in *Cultural Critique* 1 (1985), 89–107.

no attention is given to those cultural practices, like Orientalism or ethnography [he later adds masculinism and patriarchy], affiliated with imperialism, which in genealogical fact fathered world history itself.   (1985: 11)

To advance his critique of historicism, Said also encourages a certain "wildness" and radical openness, a persistent crossing of boundaries that is tethered only by one's political project. He calls for "fragmenting, dissociating, dislocating and decentring the experiential terrain covered at present by universalising historicism" and a push beyond the "polarities and binary oppositions of Marxist-historicist thought (voluntarisms versus determinism, Asiatic versus western society, change versus stasis) in order to create a new type of analysis of plural, as opposed to single, objects." (1985: 11) Historicism, and especially its most radical variants, must be struggled against, deconstructed, and redeployed into radically different projects that are consciously marginal, oppositional, libertarian, against the grain of history itself. With Said's critique we come full circle again, back to the daunting challenges of exploring Thirdspace-as-Aleph. But we have also broken open new terrain, in the problematic interplay of Historicality and Spatiality, that must be kept in mind as we move on.

The Third Space of Homi Bhabha

One last excursion must be noted before concluding this selective tour of the postcolonial literature. Homi Bhabha's writings, drawing heavily on Said, Frantz Fanon, and Jacques Lacan, perceptively explore the nature of cultural difference or what he calls the "location" of culture.[29] He locates his particular postcolonial project in this way: "With the notion of cultural difference, I try to place myself in that position of liminality, in that productive space of the construction of culture as difference, in the spirit of alterity or otherness" (1990a: 209). This "productive space" is quite distinct from liberal relativist perspectives on "cultural diversity" and

---

[29] The key works that will be discussed here are Homi K. Bhabha, "The Third Space: Interview with Homi Bhabha," in *Identity, Community, Culture, Difference*, J. Rutherford, ed., London: Lawrence and Wishart, 1990a: 207–21; "The Other Question: Difference, Discrimination and the Discourse of Colonialism," in *Out There: Marginalization and Contemporary Cultures*, R. Ferguson et al., eds, Cambridge, MA: MIT Press and New York: New Museum of Contemporary Art, 1990b: 71–88; "Postcolonial Authority and Postmodern Guilt," in *Cultural Studies*, L. Grossberg et al., eds, New York and London, Routledge, 1992: 56–68; and *The Location of Culture*, London and New York: Routledge, 1994. See also Homi K. Bhabha, ed., *Nation and Narration*, London and New York: Routledge, 1990c, especially his own concluding chapter, "DissemiNation: Time, Narrative, and the Margins of the Modern Nation."

"multiculturalism," which form another discursive space. In that space "Western connoisseurship" locates cultures "in a kind of *musée imaginaire*," in a grid of its own choosing, wherein the urge to universalize and historicize readily acknowledges the social and historical diversity of cultures but at the same time transcends them and renders them transparent, illusive. Speaking in particular of Great Britain, Bhabha describes the liberal "entertainment and encouragement of cultural diversity" as a form of control and "containment":

> A transparent norm is constituted, a norm given by the host society or dominant culture, which says that "these other cultures are fine, but we must be able to locate them within our own grid". This is what I mean by a *creation* of cultural diversity and a *containment* of cultural difference.   (1990a: 208)

For Bhabha, the difference of cultures cannot be contained within the universalist framework of liberal democracy, or for that matter Marxist-historicism, for these different cultures are often *incommensurable*, not neatly categorized, a triggering observation for identifying a "third space" of alternative enunciation. This argument connects to his critique of colonial discourse and Orientalism, whose "predominant function is the creation of a space for a `subject peoples' through the production of knowledges in terms of which surveillance is exercised and a complex form of pleasure/unpleasure is incited" (1990b: 75).

Against this "containment of cultural difference," he introduces the notion of *hybridity* and locates it in another example of trialectical thirding-as-Othering. Denying the essentialism that comes from either the genealogical tracing of cultural origins or the representational act of cultural "translation" (a term he borrows from Walter Benjamin), Bhabha. posits a third space that echoes the chosen marginality of bell hooks and, like Spivak and Said, explicitly challenges hegemonic historiography.

> all forms of culture are continually in a process of hybridity. But for me the importance of hybridity is not to be able to trace two original moments from which the third emerges, rather hybridity to me is the "third space" which enables other positions to emerge. This third space displaces the histories that constitute it, and sets up new structures of authority, new political initiatives, which are inadequately understood through received wisdom. . . . The process of cultural hybridity gives rise to something different, something new and unrecognisable, a new area of negotiation of meaning and representation.
>
> (1990a: 211)

As he notes in another context, the Third Space – inspired by his deep involvement with the Third Cinema[30] – is located in the margins, and "it is from the affective experience of social marginality that we must conceive a political strategy of empowerment and articulation" (1992: 56). He also firmly roots the Third Space, assertively capitalized again, in the experience of postcoloniality.

> It is significant that the productive capacities of the Third Space have a colonial or postcolonial provenance. For a willingness to descend into that alien territory – where I have led you – may reveal that the theoretical recognition of the split-space of enunciation may open the way to conceptualizing an *inter*national culture, based not on the exoticism of multiculturalism or the *diversity* of cultures, but on the inscription and articulation of culture's *hybridity*. To that end we should remember that it is the "inter" – the cutting edge of translation and negotiation, the *in-between* space – that carries the burden of the meaning of culture ... And by exploring this Third Space, we may elude the politics of polarity and emerge as others of our selves.   (1994: 38–9)

Along with other key figures such as Stuart Hall and Paul Gilroy, Bhabha adds his voice to a particularly black British tradition of postcolonial debate that connects with, yet is distinct from, the North American tradition exemplified by hooks, West, Said, Spivak, Anzaldúa, Gómez-Peña, and most of the postmodern spatial feminists I have highlighted.[31] Setting these (cultural?) differences aside, the project that is shared across the Black Atlantic (a title of one of Gilroy's recent books) is the exploration of new identities that build alternatives to a "double illusion" created, on the one side, by a cultural essentialism that promotes such polarizing exclusivities as black nationalism; and on the other, by an unlimited cultural relativism that dissolves black subjectivity in a universalist melting pot or a pluralist jumble of equals.

The Third Space of Homi Bhabha is occasionally teasingly on the edge of being a spatially ungrounded literary trope, a floating

---

[30] See *Questions of Third Cinema*, J. Pines and P. Willamen, eds, London: British Film Institute, 1989. In his chapter, "The Commitment to Theory," Bhabha, for the first time I have been able to find, uses the term "Third Space" as an intervention, an alternative space of enunciation, an "alien territory" that "carries the burden of the meaning of culture." A revised version of this chapter appears in Bhabha, 1994: 19–39.

[31] For an interesting extension of this British debate to geography, see Steve Pile, "Masculinism, the Use of Dualistic Epistemologies and Third Spaces," *Antipode* 26 (1994), 255–77. It will be interesting to see if Bhabha's British hybridities will change after his recent appointment to teach and write at the University of Chicago.

metaphor for a critical *historical* consciousness that inadvertently
masks a continued privileging of temporality over spatiality.[32]
Nevertheless, Bhabha effectively consolidates in *The Location of
Culture* a strategic envisioning of the cultural politics of Thirdspace
that helps to dislodge its entrapment in hegemonic historiography
and historicism. Several passages from this book also serve as a tran-
sitional summary, linking where we have been to where we are
going next.

The introductory chapter, "Locations of culture," begins with a
quote from Heidegger: "A boundary is not that at which something
stops but, as the Greeks recognized, the boundary is that from which
something begins its presenting." Bhabha then moves to "Border Lives:
The Art of the Present." The following passages summarize his
synchronic synthesis.

> It is the trope of our times to locate the question of culture in
> the realm of the *beyond*. At the century's edge, we are less exer-
> cised by annihilation – the death of the author – or epiphany –
> the birth of the "subject." Our existence today is marked by a
> tenebrious sense of survival, living on the borderlines of the
> "present," for which there seems to be no proper name other
> than the current and controversial shiftiness of the prefix
> "post": *postmodernism, postcolonialism, postfeminism....*
>
>   The "beyond" ["meta"?] is neither a new horizon, nor a leav-
> ing behind of the past.... [I]n the *fin de siècle*, we find ourselves
> in the moment of transit where space and time cross to produce
> complex figures of difference and identity, past and present,
> inside and outside, inclusion and exclusion. For there is a sense
> of disorientation, a disturbance of direction, in the "beyond": an
> exploratory, restless movement caught so well in the French
> rendition of the words *au-delà* – here and there, on all sides,
> *fort/da*, hither and thither, back and forth.

[32] Bhabha opens *The Location of Culture*, London: Routledge, 1994, with this very
representative quote from Frantz Fanon's *Black Skins/White Masks*: "The architecture
of this work is rooted in the temporal. Every human problem must be considered
from the standpoint of time" (1994: xiv). Although Bhabha tries creatively to spatialize temporality, the Fanonian dictum remains central and Bhabha never directly
confronts the problematic interplay between temporality and spatiality or between
history and geography. At critical moments throughout the text, his assertive spatializations are abruptly diluted by a historical relabeling. What makes Bhabha's
privileging of the historical imagination more ironic are the gruff critiques of his
work as ahistorical, relentlessly timeless, ignorant of postcolonial historicity. One
such critique appears in Robert Young, *White Mythologies: Writing History and the
West*, London and New York: Routledge, 1990: 146. Following Young's blindly
unspatialized historicism, Derek Gregory, who should know better, chooses not to
discuss Bhabha's work in his *Geographical Imaginations* (1994) because of Bhabha's
"apparently relentless setting aside of historicity" (Gregory, 1994: 167–8). Perhaps
Gregory should look *beyond* appearances.

The move away from the singularities of "class" or "gender" as primary conceptual and organizational categories, has resulted in an awareness of the subject positions – of race, gender, generation, institutional location, geopolitical locale, sexual orientation – that inhabit any claim to identity in the modern world. What is theoretically innovative, and politically crucial, is the need to think beyond narratives of originary and initial subjectivities and to focus on those moments or processes that are produced in the articulation of cultural differences. These "in-between" spaces provide the terrain for elaborating strategies of selfhood – singular or communal – that initiate new signs of identity, and innovative sites of collaboration, and contestation, in the act of defining the idea of society itself.

The social articulation of difference, from the minority perspective, is a complex, on-going negotiation that seeks to authorize cultural hybridities that emerge in moments of historical transformation. The "right" to signify from the periphery of authorized power and privilege does not depend on the persistence of tradition; it is resourced by the power of tradition to be reinscribed through the conditions of contingency and contradictoriness that attend upon the lives of those who are "in the minority." (1994: 1–2)

He elaborates further on the notion of "going beyond."

"Beyond" signifies spatial distance, marks progress, promises the future; but our intimations of exceeding the barrier or boundary – the very act of going *beyond* – are unknowable, unrepresentable, without a return to the "present" which, in the process of repetition, becomes disjunct and displaced. The imaginary of spatial distance – to live somehow beyond the border of our times – throws into relief the temporal, social differences that interrupt our collusive sense of cultural contemporaneity. . . . If the jargon of our times – postmodernity, postcoloniality, postfeminism – has any meaning at all, it does not lie in the popular use of the "post" to indicate sequentiality – *after* feminism; or polarity – *anti*modernism. These terms that insistently gesture to the beyond, only embody its restless and revisionary energy if they transform the present into an expanded and excentric site of experience and empowerment. . . . If the interest in postmodernism is limited to a celebration of the fragmentation of the "grand narratives" of postenlightenment rationalism then, for all its intellectual excitement, it remains a profoundly parochial enterprise. (1994: 4)

Being in the "beyond", then, is to inhabit an intervening space, as any dictionary will tell you. But to dwell "in the beyond" is also, as I have shown, to be part of a revisionary time, a return to the present to redescribe our cultural contemporaneity; to reinscribe our human, historic commonality; *to touch the future on its hither side.* In that sense, then, the intervening space "beyond" becomes a space of intervention in the here and now.   (1994: 7)

5

Heterotopologies: Foucault and the Geohistory of Otherness

It is surprising how long the problem of space took to emerge as a historico-political problem.

(Michel Foucault, *Power/Knowledge*, 1980: 149)

Foucault's writing is perfect in that the very movement of the text gives an admirable account of what it proposes: on one hand, a powerful generating spiral that is no longer a despotic architecture but a filiation *en abîme*, a coil and strophe without origin (without catastrophe either), unfolding ever more widely and rigorously; but on the other hand, an interstitial flowing of power (where the relations of power and seduction are inextricably entangled). All this reads *directly* in Foucault's discourse (which is *also* a discourse on power). It flows, it invests and saturates, the entire space it opens.

(Jean Baudrillard, *Forget Foucault*, 1987: 9)

A Second Discovery

Hidden in the underbrush of the Thirdspaces explored in the previous chapters is the body and mind of Michel Foucault. Bringing into full view his spiraling excursions into human spatiality saturates Thirdspace in what Baudrillard described as "an interstitial flowing of power."

Crossing Paths: Lefebvre and Foucault

Little is known and much less written about the relations between Foucault and Henri Lefebvre, although they were clearly aware of each other and Lefebvre, at least, takes an occasional potshot at the more ambiguous politics and the "psychologism" he saw arising from Foucault's writings. Lefebvre argued that Foucault's enraptured individualism failed to explore the "collective subject," that his frequent use of floating spatial metaphors obscured the political concreteness of social spatiality, and that the manysidedness of Foucault's conceptualization of power/knowledge took too little note of "the antagonism between a knowledge [*savoir*] which serves power and a form of knowing [*connaissance*] which refuses to acknowledge power."

These comments, extracted from the first ten pages of *The Production of Space*, refer particularly to Foucault's *The Archeology of Knowledge*, published in French in 1969. Lefebvre's rather flippant critique revolved around what he saw as Foucault's failure to "bridge the gap" between "the theoretical (epistemological) realm and the practical one, between mental and social, between the space of the philosophers and the space of people who deal with material things." In short, Foucault seemed to be distancing himself too far from the Marxian critique of capitalism and from an explicitly politicized history and geography of the present.

Lefebvre's "centered peripheralness" made him particularly concerned with the political implications of what he saw as Foucault's parochial preoccupation with the periphery.

> The issue of prisons, of psychiatric hospitals and antipsychiatry, of various converging repressions, has a considerable importance in the critique of power. And yet this tactic, which concentrates on the peripheries, simply ends up with a lot of pinprick operations which are separated from each other in time and space. It neglects the centers and centrality; it neglects the global. (Lefebvre, 1976: 116)

Lefebvre's views reinforced a similar reaction to Foucault's punctiform micropolitics of disruption that was developing throughout Western Marxism, driven by the fear of fragmentation and localism that had for so long been seen as divisive and diversionary within the larger, revolutionary political project. Within Marxist geography, especially among those most influenced by Lefebvre, such as David Harvey and Neil Smith, these reactions to Foucault seemed to put his rethinking of human spatiality off limits as politically incorrect.

It is becoming increasingly clear, however, that Foucault began developing his own conceptualization of something very much like

Thirdspace in the mid-1960s, at about the same time a spatial problematic was recentering the work of Lefebvre. In the five years leading up to the events in Paris in May 1968, Lefebvre and Foucault crossed paths in many different ways. In addition to three major books, Foucault produced a series of essays on Nietzsche, Freud, Marx, Sartre, the art historian Irwin Panofsky, the writing of history, philosophy and psychology, and, in the journal *Critique*, on "distance, aspect, origin" (1963) and "the language of space" (1964), themes and subjects that directly paralleled Lefebvre's interests.[1]

Foucault also became increasingly engaged with architects and architecture during this period, reflecting in part his growing interest in medicine, clinical psychology, institutional design, urban planning, and the (spatial) order of things. He accepted an invitation from a group of architects to do "a study of space" and on March 14, 1967, presented a lecture in Paris entitled *"Des Espaces autres"* ("Of Other Spaces"), a lecture which I will excavate in great detail later in this chapter.[2] In 1967, interestingly enough, Foucault was under consideration for an appointment in psychology at the University of Nanterre in Paris, where Lefebvre was then professor of Sociology and where the earliest sparks of the "irruptions" of May 1968 would take place. Lefebvre apparently supported the appointment and an offer was made. Foucault, however, turned it down to take up a position at the newly created suburban University of Vincennes as chair of Philosophy.

In contrast to Lefebvre, Foucault never developed his conceptualizations of space in great self-conscious detail and rarely translated his spatial politics into clearly defined programs for social action. It can nevertheless be argued (and Foucault would, when prompted, agree) that a comprehensive and critical understanding of spatiality was at the center of all his writings, from *Folie et déraison: Histoire de la folie à l'âge classique* (1961) to the multivolume work on the history

---

[1] The three books were *Naissance de la clinique. Une archéologie du regard médical*, Paris: Presses Universitaires de France, 1963, tr. A. S. Smith as *The Birth of the Clinic: An Archeology of Medical Perception*, New York: Pantheon, 1973; *Raymond Roussel*, Paris: Gallimard, 1963, tr. C. Ruas as *Death and the Labyrinth: The World of Raymond Roussel*, New York: Doubleday, 1986; and *Les Mots et les choses. Une archéologie des sciences humaines*, Paris: Gallimard, 1966, tr. as *The Order of Things: An Archeology of the Human Sciences*, New York: Pantheon, 1971. For the full references of the journal articles of Foucault, see the excellent bibliography at the back of James Bernauer and David Rasmussen, eds, *The Final Foucault*, Cambridge, MA, and London, UK: MIT Press, 1988. During the same period, 1963–8, Lefebvre published 11 books: *Karl Marx. Oeuvres choisies* (with Norbert Guterman, 2 vols 1963–4); *La Vallée de Campan. Étude de sociologie rurale* (1963); *Marx* (1964); *Métaphilosophie* (1965); *La Proclamation de la Commune* (1965); *Sociologie de Marx* (1966); *Le Langage et la société* (1966); *Position: contre les technocrates* (1967); *Le Droit à la ville* (1968); *L'Irruption de Nanterre au sommet* (1968); and *La Vie quotidienne dans le monde moderne* (1968). Lefebvre rested in 1969, when Foucault's *L'Archéologie de savoir* (Paris: Gallimard) appeared.
[2] The lecture was published in *Architecture-Mouvement-Continuité* 5 (1986), 46–9; and in English as "Of Other Spaces," tr. Jay Miskowiec, in *Diacritics* 16 (1986), 22–7.

of sexuality published in English translation shortly before (and after) his death in 1984.[3] And, even without prompting, he would infuse a spatial politics into these writings that ranged from "the great strategies of geopolitics" played out on a global scale to "the little tactics of the habitat," his own preferred milieu (1980: 149).

Foucault's commitment to a spatial problematic and a spatial praxis can be most easily traced through his interviews with geographers, architects, and anthropologists. In an interview with a group of French geographers in 1976, after a playful evasion ("I must admit I thought you were demanding a place for geography like those teachers who protest when an education reform is proposed, because the number of hours of natural science or music is being cut"), Foucault readily acknowledged that "Geography acted as the support, the condition of possibility for the passage between a series of factors I tried to relate," adding "Geography must indeed lie at the heart of my concerns."[4] In the same interview he commented on his "obsession" with space: "People have often reproached me for these spatial obsessions, which have indeed been obsessions for me. But I think through them I did come to what I had basically been looking for: the relations that are possible between power and knowledge" (1980: 69). The power–knowledge link is acknowledged by every Foucauldian scholar, but for Foucault himself the relationship was embedded in a trialectic of power, knowledge, *and space*. The third term should never be forgotten.

In another interview, with the anthropologist Paul Rabinow, Foucault follows up on this trialectic of space–knowledge–power to specifically embrace two other spatial disciplines, Architecture and Urban Planning.[5] It is worth quoting one of his responses in full, for it provides another epigrammatic introduction to this chapter:

[3] *Folie et déraison: Histoire de la folie à l'âge classique,* Paris: Plon, 1961; in English as *Madness and Civilization: A History of Insanity in the Age of Reason,* New York: Pantheon, 1965. Foucault's three-volume *Histoire de la sexualité* (*The History of Sexuality*) has been translated by R. Hurley as *I: An Introduction,* New York: Pantheon, 1978; *II: The Use of Pleasure,* New York: Pantheon, 1985; and *III: The Care of the Self,* New York: Pantheon, 1985.

[4] "Questions on Geography," in *Power/Knowledge: Selected Interviews and Other Writings, 1972–1977,* tr. C. Gordon, New York: Pantheon, 1980: 63–77. The quote cited appears on the last page. The original interview was published as *"Questions à Michel Foucault sur la géographie,"* in the first issue of the radical geography journal, *Hérodote* (a name claiming Herodotus, usually considered one of the first critical historians, as the first radical geographer), 1 (January–March 1976), 71–85. Subsequent issues of *Hérodote* contained questions from Foucault to the geographers, *"Des questions de Michel Foucault à Hérodote"* (July–September 1976), 9–10; and the responses to them (April–June 1977), 7–30.

[5] Paul Rabinow, "Space, Knowledge, and Power. Interview: Michel Foucault," *Skyline* (March 1982), 16–20; also in Rabinow, ed., *The Foucault Reader,* New York: Pantheon, 1984, 239–56. The interview in *Skyline* was preceded by a short essay by Gwendolyn Wright and Paul Rabinow, "Spatialization of Power: A Discussion of the Work of Michel Foucault," 1982: 14–15.

Yes. Space is fundamental in any form of communal life; space is fundamental in any exercise of power. To make a parenthetical remark, I recall having been invited, in 1966, by a group of architects to do a study of space, of something that I called at that time "heterotopias," those singular spaces to be found in some given social spaces whose functions are different or even the opposite of others. The architects worked on this, and at the end of the study someone spoke up – a Sartrean psychologist – who firebombed me, saying that *space* is reactionary and capitalist, but *history* and *becoming* are revolutionary. This absurd discourse was not at all unusual at the time. Today everyone would be convulsed with laughter at such a pronouncement, but not then.   (1982: 20)

I reprieve and emphasize this "firebombing" of Foucault, for it centers our attention around not just the problematic of space but, in particular, the thick ideological layers that have built up over the past century concerning the relationships between Spatiality and Historicality. I will return again and again to Foucault's creative recasting of this relationship in this and the next chapter.

Other Crossings

The spatial feminist and postcolonial critiques discussed earlier have been deeply influenced by Foucault, much more so than by Lefebvre, who is rarely even mentioned except in some of the early modernist feminist urban critiques, and then primarily for his critique of everyday life in the modern world. For most of the present generation of postmodern cultural critics who take space seriously, and, for that matter, most of those who do not, Foucault is a primary catalyst. As such, he draws particularly intense critical attention, both positive and negative, generating around his writings an ever widening swirl of publications and treatises ranging from inspirational puffery to aggressive dismissal. Whatever one's position on his achievement, however, every contemporary discourse and debate on postmodern culture, and I will add on the conceptualization of Thirdspace, must at one time or another pass through Foucault.

In his *Geographical Imaginations* (1994), Derek Gregory provides an extraordinarily comprehensive tour of Foucault's filiations into almost every nook and cranny of contemporary critical discourse. Although there is no special section of the book devoted to Foucault, he appears as a reference point on nearly every page of Gregory's labyrinthine "Chinese Encyclopedia" of geographical thinking. Foucault is introduced centrally, for example, to weave together the first chapter, "Geography and the World-as-exhibition," a heady gambol through the powers of visualization and the disciplinary borderlands between geography and sociology, anthropology, and economics.

> Foucault seems to suggest that a dispersed and anonymous system of spatial sciences emerged in the eighteenth century – quite apart from the "bibliographic dinosaurs" that haunt the pages of the usual histories of geography – as part of a generalized medico-administrative system of knowledge that was deeply implicated in the formation of a disciplinary society of surveillance, regulation, and control. Its characteristic figure is the Panopticon, from which Foucault observes:
>
>   "Our society is not one of spectacle, but of surveillance; under the surface of images, one invests bodies in depth; behind the great abstraction of exchange, there continues the meticulous, concrete training of useful forces; the circuits of communication are the supports of an accumulation and centralization of knowledge; the play of signs defines the anchorages of power; it is not that the beautiful totality of the individual is amputated, repressed, altered by our social order, it is rather that the individual is carefully fabricated in it, according to a whole technique of forces and bodies." (1994: 62–3)

Gregory sums up the Foucauldian view of geography's disciplinary history – and, I would add, Foucault's particular envisioning of Thirdspace – in a brief passage that cracks open the old stories of Geography's past to the more contemporary spatial critiques of Mark Poster, Donna Haraway, William Gibson, Giovanni Vattimo, and Jean Baudrillard.

> The construction of this discursive triangle between power, knowledge, and spatiality effected a colonization of the life-world in which "space" was given both metaphorical and material resonance: as Foucault declared, "Space is fundamental in any exercise of power." (1994: 63)

Gregory then moves with Foucault along "a trans-disciplinary voyage into `deep space'" that takes us into the next chapter, "Geography and the Cartographic Anxiety," a stunning 135-page tour de force into the post-prefixation of the geographical imagination. After moving through the landscapes of the new cultural Geography (represented by the Kandy of James Duncan) and the thick ethnographies of Clifford Geertz's Balinese cockfights, which aim to make the strange and exotic familiar and textually interpretable, Gregory asks us to "think for a moment of Foucault," and

his "quite contrary ability to deliberately make strange our otherwise familiar fictions of the past, and in so doing to raise urgent and
unsettling questions about the construction of subjectivities and differences that escape these textualities" (1994: 150). Foucault's heterotopias – as "marginal sites of modernity, constantly threatening to
disrupt its closures and certainties" (1994: 151) – are then used to
introduce the next section, on "Postmodern cartographies and spatialities," a trip that ultimately takes us to Fredric Jameson and the
Bonaventure Hotel (sites to which we will return, with Gregory, in
chapter 7).

Foucault is foregrounded again in Gregory's exceptional excursion
into the postcolonial literatures, especially in conjunction with his
multi-sided presentation of the work of Gayatri Spivak. I condense
his treatment of Spivak but try to retain the shifting quality of the
Gregorian critique in the following excerpts:

> In her later essays Spivak shows an acute sensitivity to the *terri-
> toriality* (or more generally, the spatiality) of subject-construc-
> tion. ... Following Foucault ... one might see this as charting
> what he called "the insurrection of subjugated knowledges."
> But Spivak is plainly interested in mapping more than the con-
> tours of so many local knowledges – and here the term is again
> Foucault's ... – because she wants to identify what could be
> called (and perhaps should be called) a "global localism," the
> universalizing claim of Eurocentrism. ... [I]t is, in fact,
> Foucault's problematic that is a special concern, and when she
> urges the realization of "space-specific subject-production" –
> and the phrase *is* hers, not mine – she is drawing attention to
> the European provenance of poststructuralism. ...
>
> She explains what she means in a brief but revealing critique
> of Foucault. .....One of Foucault's most imaginative contribu-
> tions has been his critique of the repressive hypothesis, and his
> (contrary) claim that power is productive, constitutively
> involved in the double process of *assujettissement*: subjection
> and "subjectification." But Foucault concedes that his attempts
> to elucidate the productive capacities of power have had
> recourse to "the metaphor of the point which progressively irra-
> diates its surroundings." Spivak's concern is that, without great
> care, "that radiating point, animating an affectively heliocentric
> discourse, fills the empty place of the agent with the historical
> sun of theory, *the Subject of Europe* (my emphasis). In her view,
> such a possibility is ever-present in the architecture of
> Foucault's project. "Foucault is a brilliant thinker of power-in-
> spacing," she declares, but he tacitly confines those spaces to
> particular maps. "The awareness of the topographical

inscription of imperialism does not inform his presuppositions," and in consequence his "self-contained version of the
West ... [ignores] its production by the imperial project."

Revealed here, with great clarity, is Spivak's ability to position herself in an ambivalent relation to the theoretical method
that is being deployed and to use that position, as Young says,
"so as to disconcert and disorient the reader from the familiar
politico-theoretical structures which it seems to promise"...
[But], in a close reading of Foucault's *The Order of Things*
Bhabha ... argued that the colonial and postcolonial are indeed
present in the margins of the closing essay, and he concludes
that this implies and imposes the need to reinscribe Foucault's
perspective ... "in the supplementary spaces of the colonial and
slave world." (1994: 190–3)

One final quote from Gregory on Foucault is particularly useful in
moving us forward. It pops up in chapter 6, a lengthy critical review
of my *Postmodern Geographies*, along with a companion quote from
Donald Lowe's *History of Bourgeois Perception*.[6] From Foucault:

"Metaphorizing the transformation of discourse in a vocabulary
of time necessarily leads to the utilization of the model of individual   consciousness   with   its   intrinsic   temporality.
Endeavoring on the other hand to decipher discourse through
the use of spatial, strategic metaphors enables one to grasp precisely the points at which discourses are transformed in,
through and on the basis of relations of power."
                              (1994: 261; from Foucault, 1980: 69–70)

From Lowe:

"In bourgeois society, development-in-time despatialized historiography. With the extension of the historical landscape, time
now possessed a depth and diversity which it previously
lacked.... [B]ourgeois society discovered the concept of historicism." (1994: 262; from Lowe, 1982: 43)

There are many more passages to mine from *Geographical
Imaginations* on Foucault's infiltration of contemporary cultural criticism and critical theory. At this point, however, I will venture briefly
into one corner of the Foucauldian world of influence that Gregory

[6] Donald Lowe, *History of Bourgeois Perception*, Chicago: University of Chicago
Press, 1982.

seems to have overlooked. This connection moves us back again to the African-American cultural critique and helps to guide our upcoming journeys to Los Angeles and other real-and-imagined places.

To begin, consider Cornel West's comments on Foucault in "The Dilemma of the Black Intellectual," another chapter in his and bell hooks's *Breaking Bread*.[7] West passes through three particularly attractive "models" for the activist black intellectual: as Humanist (the Bourgeois Model); as Revolutionary (the Marxist Model); and as Postmodern Skeptic (the Foucauldian Model). On the latter he writes:

> The Foucaultian [sic] model and project are attractive to Black intellectuals primarily because they speak to the Black postmodern predicament, defined by the rampant xenophobia of bourgeois humanism predominant in the whole academy, the waning attraction to orthodox reductionist and scientific versions of Marxism, and the need for reconceptualization regarding the specificity and complexity of Afro-American oppression. Foucault's deep anti-bourgeois sentiments, explicit post-Marxist convictions, and profound preoccupations with those viewed as radically "other" by dominant discourses and traditions are quite seductive for politicized Black intellectuals wary of antiquated panaceas for Black liberation.... The Foucaultian model promotes a leftist form of postmodern skepticism; that is, it encourages an intense and incessant interrogation of power-laden discourses in the service of neither restoration, reformation, nor revolution, but rather of revolt.   (hooks and West, 1991: 142–3)

West states that black intellectuals can learn much from each of these models but, learning from the Postmodern Skeptic, he argues that no one of them should be uncritically adopted, especially since each fails to speak to the "uniqueness" of the Black cultural predicament. Instead he argues for an Insurgency Model, for the black intellectual as Critical Organic Catalyst, drawing upon "the Black Christian tradition of preachment," the "Black musical tradition of performance," and the "kinetic orality and emotional physicality, the rhythmic syncopation, the protean improvisation, and the religious, rhetorical, and antiphonal elements of Afro-American life" (1991: 136, 144).

I cite West's observations both as an invitation to learn from Foucault and as a warning against the uncritical adaptation of any

---

[7] Cornel West, "The Dilemma of the Black Intellectual," in hooks and West, *Breaking Bread*, Boston: South End Press, 1991: 131–46.

single conceptualization of Thirdspace, whether from Foucault or Lefebvre, Spivak or Said, bell hooks or Cornel West. I must also include my own appropriations and representations of their work and ideas, for I too cannot speak to the uniquenesses of your subject position; nor can I simply mimic its specific cultural and political identity as if it were my own. But I can attempt to keep Thirdspace open to limitless reinterpretation by following Foucault and his post-modern skepticism; and to seek, with West, the opportunities that might arise from this reinterpretation to mobilize and stimulate a radical and postmodern politics of (spatial) resistance that redraws the boundaries of identity and struggle.

In Thirdspace with Michel Foucault

Just before his death in June 1984, an old batch of lecture notes pre-pared by Foucault in 1967 was released into the public domain for an exhibition in Berlin. These notes were never reviewed for publication by Foucault and were not recognized as part of the official body of his works until their appearance in 1984, under the title *"Des Espaces autres,"* in the French journal *Architecture-Mouvement-Continuité*. They were subsequently published under the title "Of Other Spaces" in the Texts/Contexts section of *Diacritics* (Spring, 1986) and in the architectural journal *Lotus* in the same year. Still ignored by most Foucauldian scholars, "Of Other Spaces" provides a rough and patchy picture of Foucault's conception of a new approach to space and spatial thinking that he called *heterotopology* and described in ways that resemble what is being described here as Thirdspace.

One can derive a richer and more detailed picture of Foucault's version of Thirdspace from the larger body of his works and in a few interviews where he speaks specifically about space and geography. But this is not only beyond what I will try to do here but is a some-what risky venture, for Foucault only rarely addressed the "spatial problematic" directly and explicitly in his major writings. Only in "Of Other Spaces" does he, with some fleeting attempt at systematic directness, present his particular views about space and, as more than an aside, about the relations between space and time.

There is, of course, another risk in focusing on "Of Other Spaces." It was never published by Foucault and may be seen as just an early, preliminary sketch that was forgotten and discarded as he moved on to other projects. I am willing to take that risk and to extract from "Of Other Spaces" a collection of insights that add significantly to a practical and theoretical understanding of Thirdspace – and also to what might be called the geohistory of otherness.

Utopias and Heterotopias

"Of Other Spaces" begins with a brief history of space that sets space against time and against history itself, culminating in an assertion that "the anxiety of our era has to do fundamentally with space, no doubt a great deal more than with time" (1986: 23).[8] I used Foucault's opening paragraph in the first chapter of *Postmodern Geographies* (1989) as a springboard for a critique of a prevailing historicism in modernist critical thinking, a blinkering of perspective that has persistently constrained our ability to think critically about space and the spatiality of human life in the same ways we have learned to think critically about time and the "making of history." I will return with Foucault to this critique of historicism in the next chapter, but it is useful now to reflect again the opening words to "Of Other Spaces."

> The great obsession of the nineteenth century was, as we know, history: with its themes of development and of suspension, of crisis and cycle, themes of the ever-accumulating past, with its great preponderance of dead men and the menacing glaciation of the world. The nineteenth century found its essential mythological resources in the second principle of thermodynamics. The present epoch will perhaps be above all the epoch of space. We are in the epoch of simultaneity: we are in the epoch of juxtaposition, the epoch of the near and far, of the side-by-side, of the dispersed. We are at a moment, I believe, when our experience of the world is less that of a long life developing through time than that of a network that connects points and intersects with its own skein. One could perhaps say that certain ideological conflicts animating present-day polemics oppose the pious descendants of time and the determined inhabitants of space. Structuralism, or at least that which is grouped under this slightly too general name, is the effort to establish, between elements that could have been connected on a temporal axis, an ensemble of relations that makes them appear as juxtaposed, set off against one another, implicated by each other – that makes them appear, in short, as a sort of configuration. Actually, structuralism does not entail a denial of time; it does involve a certain manner of dealing with what we call time and what we call history. (1986: 22)

This epochal foregrounding of space over time, spatiality over historicality, sets the stage and tone for Foucault's ruminations on

---

[8] All quotes and page references are from *Diacritics*, 1986.

"other spaces." It is immediately followed by a "very rough" retracing of the "history of space" from the Middle Ages (dominated by "the space of emplacement," a hierarchic ensemble of places: sacred and profane, protected and open, urban and rural); then, starting with Galileo and the 17th century, an "opening up" of this space of emplacement into an infinitely open space of "extension" rather than "localization;" and finally to a contemporary moment when "the site" as a new form of localization has been substituted for extension.

This is an especially crucial point in the lecture, for it swings our attention to a spatiality of *sites*. "The site is defined by relations of proximity between points or elements," formally as series, networks, or grids (as in computers or traffic systems) or "more concrete" in terms of "demography" ("the human site or living space"). Today, he concludes, *"space takes for us the form of relations among sites."* The centering of contemporary spatiality on sites and the spatial relations among sites (or what might alternatively be called the "situatedness" of sites) is a distinctive feature of Foucault's conceptualization of heterotopology. Although less infused with allusions to the production process, the sites and situations of Foucault take on insights that reflect Lefebvre's critique of everyday life in the modern world and his trialectic of the perceived, the conceived, and the lived. Following up on this in-sitedness, Foucault takes us through various pathways to understanding contemporary spatiality that also strikingly resemble the trialectics and thirdings of Lefebvre, although again without the assertive foregrounding of an explicit political project.

Foucault first notes that our lives are still governed by "sanctified" (modernist?) oppositions (e.g. between private and public space, family and social space, cultural and useful space, leisure and work space) "nurtured by the hidden presence of the sacred." Whereas history, he interjects, was detached from the sacred in the 19th century and a certain theoretical desanctification of space occurred after Galileo, "we may still not have reached the point of a practical desanctification of space." Much remains hidden in our lived spaces, buried in these "oppositions" that "our institutions and practices have not yet dared to break down," that we continue to regard as "simple givens." Capitalism is not mentioned, but its intonations are there in the post-Galileo "space of extension" and in that "menacing glaciation of the world" Foucault associates with the 19th century's "great obsession" with history.

For guidance into the hidden worlds of what Lefebvre would call the spaces of representation, Foucault turns to the "monumental work" of Gaston Bachelard (author of *The Poetics of Space*) and the "descriptions" of phenomenologists. They help us to know that we do not live in an empty or homogeneous space, but one filled with quantities and qualities and "perhaps thoroughly fantasmic as well." But these analyses of "the space of our primary perception, the space of our dreams and that of our passions," though unquestionably important and fundamental, are concerned primarily with "internal space" rather than what Foucault defines as "external space."

> The space in which we live, which draws us out of ourselves, in which the erosion of our lives, our time and our history occurs, the space that claws and gnaws at us, is also, in itself, a hetero-geneous space. In other words, we do not live inside a void, inside which we could place individuals and things. We do not live inside a void that could be colored with diverse shades of light, we live inside a set of relations that delineates sites which are irreducible to one another and absolutely not superimpos-able on one another.   (1986: 23)

In his own way, Foucault is setting aside, with appropriate apprecia-tion, both the "luminous" illusion of transparency that Lefebvre associated with "conceived space," with representations of space that tend to see spatiality entirely as a dematerialized mental space; and the "realistic" illusion of opacity and oversubstantiation, which reduces spatial reality to empirically definable spatial practices, material or natural objects, to the geometry of things in themselves. In the context of a sympathetic critique of this "double illusion," as Lefebvre termed it, Foucault opens his search for "other spaces" and "other sites," especially those that "have the curious property of being in relation with all the other sites, but in such a way as to sus-pect, neutralize, or invert the set of relations that they happen to des-ignate, mirror, or reflect." Here is another example of what I have called a critical thirding-as-Othering.

The first of these "curious" sites, linked to all others but contradict-ing them as well, are *utopias*, "sites with no real place ... [that] pre-sent society itself in a perfected form, or else society turned upside down...fundamentally unreal places." More "real" are what Foucault defines as *heterotopias* in a conceptualization that resonates with what might be called the micro- or site geography of Thirdspace.

> There are also, probably in every culture, in every civilization, real places – places that do exist and that are formed in the very founding of society – which are something like *counter-sites*, a kind of effectively enacted utopia in which the real sites, all the other real sites that can be found in the culture, are simultane-ously represented, contested, and inverted. Places of this kind are outside of all places, even though it may be possible to indicate their location in reality.   (1986: 24; emphasis added)

Real places that contain all other places, represented, contested, inverted in all their lived simultaneities and juxtapositions, each standing clear, "absolutely different from all the sites they reflect and speak about" – Foucault too has seen the Aleph of Thirdspace.[9]

Reflecting on the relations between utopias and heterotopias, Foucault raises that favorite device of the surrealists, Lacanian psychoanalysts, and Althusserian structuralists: the mirror.

> The mirror is, after all, a utopia, since it is a placeless place. In the mirror, I see myself there where I am not, in an unreal, virtual space that opens up behind the surface; I am over there, there where I am not, a sort of shadow that gives my own visibility to myself, that enables me to see myself there where I am absent: such is the utopia of the mirror. . . .

He then adds to this virtual space a (more?) material reality:

> But it is also a heterotopia in so far as the mirror does exist in reality, where it exerts a sort of counteraction to the position that I occupy. From the standpoint of the mirror I discover my absence from the place where I am since I see myself over there. Starting from this gaze that is, as it were, directed toward me, from the ground of this virtual space that is on the other side of the glass, I come back toward myself; I begin again to direct my eyes toward myself and to reconstitute myself there where I am. The mirror functions as a heterotopia in this respect: it makes this place that I occupy at the moment when I look at myself in the glass at once absolutely real, connected with all the space that surrounds it, and absolutely unreal, since in order to be perceived it has to pass through this virtual point which is over there.   (1986: 24)

These brief ontological musings on subjectification, objectification, and emplacement open up interesting avenues for discussion of the existential spatiality of Being and Becoming, presence and absence, the inside and the outside. But I shall leave these ontological excursions aside to move on with Foucault's nomadic wanderings through his other spaces.

[9] In *The Order of Things: An Archeology of the Human Sciences*, New York: Pantheon, 1971: xv, published soon after the preparation of *"Des Espaces autres,"* Foucault looks to another of Borges's Aleph-like creations, the "Chinese encyclopedia," in which "strange categorizations" of knowledge seemingly shatter "familiar landmarks of thought." Elsewhere, Foucault refers to Borges's description of the "Library of Babel," that impossible-to-comprehend "infinite Hexagon" of language and literature (Borges, in *Ficciones*, New York: Grove Press, 1962). The intent in all these Borgesian allusions is to *disorder* the presumed orderliness of knowledge, "to break up all the ordered surfaces and all the planes with which we are accustomed to tame the wild profusion of existing things." See also Borges's temptingly titled collection, *Labyrinths*, D. A. Yates and J. E. Irby, eds, New York: Modern Library, 1983, first edn 1964.

The Principles of Heterotopology

In the remainder of his lecture, Foucault attempts "a sort of systematic description" of heterotopias, or what he says, with some amusement at its faddishness, might be called a symptomatic "reading" of these other, different, spaces. Didactic and anti-didactic at the same time, Foucault romps through the "principles" of heterotopology with unsystematic autobiographical enjoyment and disorderly irresponsibility. I will let him speak for himself, with little comment or critique other than to warn the reader against seeking in these principles any "ordered surfaces" or axiomatic neatness.

First: heterotopias are found in all cultures, every human group, although they take varied forms and have no absolutely universal model. However, two main categories of heterotopias are identified, one of "crisis" and the other of "deviation." Whether these cover all heterotopias is left unclear.

The crisis heterotopias are associated with "so-called primitive societies" and are reserved for individuals who, with respect to the society and human environment in which they live, are in a state or stage of personal crisis: "adolescents, menstruating women, pregnant women, the elderly, etc." These are privileged or sacred or forbidden places and sites that have been persistently disappearing in "our society," although a few remnants can still be found, usually in association with a preferred location, outside the home, for, say, an individual's first sexual experience. The examples given for young men are the 19th-century boarding school and military service, and for young women the "honeymoon trip" in trains and hotels – anywhere other than the still-sanctified homeplace.

The crisis heterotopia, Foucault claims, is today being rapidly replaced by the more modern heterotopia of deviation, spaces in which those whose behavior is deviant from "required" norms are placed. These include rest homes and psychiatric hospitals, "and of course prisons." Tracing the transition from heterotopias of crisis to those of deviation and exploring the social and personal meaning of the multiplication of deviant or "carceral" spaces has, of course, been central to nearly all of Foucault's life work.[10]

---

[10] In *Madness and Civilization* (1967) – note the date – Foucault speaks of the "geography of haunted places," those institutional sites beyond the pale, outside the spaces of civil society, where the "deviant" are emplaced, incarcerated. It is also interesting to note that *The Birth of the Clinic: An Archeology of Medical Perception* (1973: ix; original French publication in 1963) begins with the statement "this book is about space"; while *Discipline and Punish: The Birth of the Prison* (1977: 141; in French, 1975) expounds upon the central thesis that "discipline proceeds from the distribution of individuals in space." For a good overview of Foucault's persistent and insightful engagement with space, see Chris Philo, "Foucault's Geography," *Environment and Planning D: Society and Space*, 10 (1992), 137–61.

Second: heterotopias can change in function and meaning over time, according to the particular "synchrony" of the culture in which they are found. The only example given is the cemetery, a "strange heterotopia" that inspires two paragraphs of locational analysis and synchronic genealogy. Until the late 18th century, Foucault notes, cemeteries (with charnel houses, tombstones, mausoleums defining a hierarchy of sites) were placed at the "heart" of the city in or next to the church, and were still deeply associated with sacred resurrection and the immortality of the soul. Later, cemeteries were removed to the suburbs in a "bourgeois appropriation" aimed at improved health, with death becoming closely associated with "illness;" and the individualization of the dead, with each family possessing its dark resting place in "the other city," located in sites and spaces specifically designed and designated for the socially Othered.[11]

Third: the heterotopia is capable of juxtaposing in one real place several different spaces, "several sites that are in themselves incompatible" or foreign to one another. Here Foucault looks at places where many different spaces converge and become entangled, jumbled together, using as his model the rectangular stage of the theatre or the cinema screen as well as the oriental garden, the smallest parcel of the world that has, since antiquity, been designed to represent the terrestrial totality. Foucault compares the oriental garden with the carpet or rug: "the garden is a rug onto which the whole world comes to enact its symbolic perfection, and the rug is a sort of garden that can move across space." These little Alephs were modernized first into sprawling zoological gardens and later into "other spaces" that contain all places and spaces (with unstated allusions ranging from Parisian arcades to world's fairs and exhibitions and ultimately to Disney World).

Fourth: heterotopias are typically linked to slices of time, which "for the sake of symmetry" Foucault calls *heterochronies*. This intersection and phasing of space and time (Thirdspace and Thirdtime?) allows the heterotopia "to function at full capacity" based on an ability to arrive at an "absolute break" with traditional experiences of time and temporality. In the modern world, many specialized sites exist to record these crossroads of space and time. "First of all, there are heterotopias of indefinitely accumulating time," such as museums and libraries, where "time never stops building up" in an attempt to establish a "general archive," a "place of all times that is itself outside of time and inaccessible to its ravages."

In opposition to these heterotopias are those more fleeting, transitory, precarious spaces of time. Noted are the festival site, the

---

[11] Can we read into this the emergence of more contemporary forms of urban spatial segregation and Othering – the ghetto, the barrio, the gang turf? I think so.

fairgrounds, the vacation or leisure village. In a foresighting of a more disneyed world, Foucault sees both forms increasingly converging in compressed, packaged, "invented" environments that seem both to abolish and preserve time and culture, that appear somehow to be both temporary and permanent.

Fifth: heterotopias always presuppose a system of opening and closing that simultaneously makes them both isolated and penetrable, different from what is usually conceived of as more freely accessible public space. Entry and exit are regulated in many different ways: through rituals of religious purification (the Muslim hammam) or hygienic cleansing (the Scandinavian sauna); but also through more subtle enactments, such as the illusions of freedom and openness that hide closure and isolation (the "famous" bedrooms that once existed on "the great farms of Brazil" to welcome the uninvited traveler for a night's sleep or the equally "famous" American motel rooms "where illicit sex is both absolutely sheltered and absolutely hidden") and other forms of boundary maintenance and discipline unmentioned by Foucault.

Through such forms of spatial regulation the heterotopia takes on the qualities of human territoriality, with its conscious and subconscious surveillance of presence and absence, entry and exit; its demarcation of behaviors and boundaries; its protective yet selectively enabling definition of what is the inside and the outside and who may partake of the inherent pleasures. Although not mentioned explicitly in "Of Other Spaces," implicit in this heterotopian regulation of opening and closing are the workings of power, of what Foucault would later describe as "disciplinary technologies" that operate through the social control of space, time, and otherness to produce a certain kind of "normalization."[12]

Last: heterotopias have an even more comprehensive function, in relation to all the space that remains outside of them. This "external," almost wraparound function is described as unfolding between two extreme poles:

> Either their role is to create a space of illusion that exposes every real space, all the sites inside of which human life is partitioned, as still more illusory (perhaps that is the role that was played by those famous brothels of which we are now deprived). Or else, on the contrary, their role is to create a space that is other, another real space, as perfect, as meticulous, as well arranged as ours is messy, ill constructed, and jumbled.

[12] In *Discipline and Punish* (1977) and elsewhere, Foucault explores the consummate heterotopia of surveillance and disciplinary power that is embodied in Bentham's Panopticon prison to illustrate his provocative point, noted earlier, that "discipline proceeds from the distribution of individuals in space" (1977: 141).

> This latter type would be the heterotopia, not of illusion, but of
> compensation, and I wonder if certain colonies have not func-
> tioned in this manner. In certain cases, they have played, on the
> level of the general organization of terrestrial space, the role of
> heterotopias.   (1986: 27)

The colonies mentioned as heterotopias of compensation are first,
the Puritan societies of 17th-century America ("absolutely perfect
other places"); and then, in a more elaborate exultation, the "marvelous, absolutely regulated colonies in which human perfection
was effectively achieved:" the Jesuit villages of Paraguay and elsewhere in South America. Foucault ends his lecture with an exemplary tribute to the boat, the "heterotopia *par excellence*." The boat is
a place without a place, a floating piece of space that exists by itself
yet is the greatest reserve of the footloose imagination: "given over
to the infinity of the sea," floating from port to port, tack to tack,
brothel to brothel, "as far as the colonies in search of the most pre-
cious treasures they conceal in their gardens." "In civilizations without boats," Foucault concludes, "dreams dry up, espionage takes the
place of adventure, and the police take the place of pirates."

The Heterotopology of Thirdspace

Foucault's heterotopologies are frustratingly incomplete, inconsis-
tent, incoherent. They seem narrowly focused on peculiar microgeographies, nearsighted and near-sited, deviant and deviously
apolitical. Yet they are also the marvelous incunabula of another
fruitful journey into Thirdspace, into the spaces that difference
makes, into the geohistories of otherness. Are they similar or are
they different from the Thirdspace of Lefebvre, bell hooks or Homi
Bhabha? The answer, to both questions, is yes. And it is this intentional ambiguity that keeps Thirdspace open and inclusive rather
than confined and securely bounded by authoritative protocols.
    In "Of Other Spaces" is encapsuled the sites of Foucault's unend-
ing engagement with spatiality, with a fundamentally spatial problematic of knowledge and power. In so many ways, space was as
central to Foucault as it was to Lefebvre, the former inflecting primarily through the nexus of power what the latter persistently
parsed through the meanings of social production. They both also
shared another parallel positioning that has been obscured in the
interpretations of even their most fervent followers, and it is to this
peculiar obfuscation that I wish to devote a few closing (and opening) remarks.
    In recent years, many in the most spatial disciplines (Geography,
Architecture, Urban Studies) and others (social theorists, historians,

anthropologists, sociologists, feminists, postcolonial critics) have looked to Foucault and/or Lefebvre for intellectual, philosophical, and political legitimization of either their distinctive and longstanding spatial perspectives or their newfound discoveries of the importance of space. Such celebratory attachments have played a major role in the contemporary reassertion of a critical spatial perspective and geographical imagination throughout all of the human sciences. As a further elaboration on the meanings of Thirdspace, however, I want to suggest that these celebrations have missed the central point that Lefebvre and Foucault were making in their different yet similar conceptualizations of spatiality: *that the assertion of an alternative envisioning of spatiality (as illustrated in the heterotopologies of Foucault, the trialectics and thirdings of Lefebvre, the marginality and radical openness of bell hooks, the hybridities of Homi Bhabha) directly challenges (and is intended to challengingly deconstruct) all conventional modes of spatial thinking.* They are not just "other spaces" to be added on to the geographical imagination, they are also "other than" the established ways of thinking spatially. They are meant to detonate, to deconstruct, not to be comfortably poured back into old containers.

This intentional disordering and disruption of the geographical imaginations and spatial sciences and philosophies that have evolved to the present is a necessary first step en route to understanding Thirdspace. Without this recognition of the constraints and illusions of conventional spatial discourses, the meanings of Thirdspace cannot be comprehended and the works of Foucault and the others become little more than a self-nourishing smorgasbord for spatial thinkers who see nothing wrong with their taken-for-granted perspectives. Because they have been so deeply socialized and disciplined to think spatially in particular ways, such fields as Geography and Architecture – often singled out as "privileged" viewpoints in the postmodern literature – may find it more difficult to enter and explore Thirdspace than others less traditionally constrained and spatially focused.

6

Re-Presenting the Spatial
Critique of Historicism

Historicism works on the new and the different to diminish newness and mitigate difference.... The new is made comfortable by being made familiar.

> (Rosalind Krause, "Sculpture in the Expanded Field," in
> Foster, *The Anti-Aesthetic*, 1983: 31)

A growing skepticism concerning older explanatory models based in history has led to renewed interest in the relatively neglected, "under-theorized" dimension of space. ... It has become less and less common in social and cultural theory for space to be represented as neutral, continuous, transparent or for critics to oppose "dead ... fixed ... undialectical ... immobile" space against the "richness, fecundity, life, dialectics" of time, conceived as the privileged medium for the transmission of the "messages" of history. [viz. Foucault, 1980] Instead spatial relations are seen to be no less complex and contradictory than historical processes, and space itself refigured as inhabited and heterogeneous, as a moving cluster of points of intersection for manifold axes of power which cannot be reduced to a unified plane or organized into a single narrative.

> (Dick Hebdige, "Subjects in Space," in *New Formations*,
> 11, 1990: vi–vii)

If this distinction [between Representations of Space and Spaces of Representation] were generally applied, we should have to look at history itself in a new light. We should have to study not only the history of space, but also the history of representations, along with that of their relationships – with each other, with practice, and with ideology. History would have to take in not only the genesis of these spaces but also, and especially, their interconnections, distortions, displacements, mutual interactions, and their links with the spatial practice of the particular society or mode of production under consideration.

(Henri Lefebvre, *The Production of Space*, 1991a: 42)

A New Introduction

A general critique of "hegemonic historiography" and Eurocentric historicism has been central to the postcolonial discourse, and with a few postcolonial critics such as Spivak and Said the critique became specifically attached to unsilencing the hidden importance of real-and-imagined geographies and spatialities. As noted by Dick Hebdige, a similar space-opening critique occurred within the broader realms of cultural studies, creating an often explicitly post-modern discourse that committed itself to the belief that, at least in the contemporary world, it is space more than time that "hides con-sequences" from us, that provides the most revealing critical per-spective for making practical and theoretical sense of the present.

This interpretive privileging of space over time in contemporary cultural criticism has many sources. The phrase about hiding conse-quences, for example, comes from the English radical critic of art and visual representation, John Berger, whose work, as much as anyone's since Walter Benjamin, deserves to be described as infused with the "trialectics of seeing."[1] With great constancy over the past three decades, Berger has infiltrated his cultural criticism, novel-writing, poetry, photography, and other "ways of seeing" with the triadic interplay of spatiality, historicality, and sociality. In politicizing the "look of things," however, Berger strategically tilts his trialectics of seeing to focus on the particular importance of space and spatiality.

In an essay on modern portrait painting, Berger took a sideward glance to speak about the crisis of the modern novel, a crisis which he associated with a dramatic change in the "mode of narration." It is no longer possible, he argued, to tell a straight story sequentially unfolding in time, for today we are too aware of what continually cuts across the storyline laterally, of the "simultaneity and extension of events and possibilities." He then states:

There are many reasons why this should be so: the range of modern means of communications: the scale of modern power:

[1] John Berger, *Ways of Seeing*, London and Harmondsworth: BBC and Penguin Books, 1972; *The Look of Things*, New York: Viking, 1974; *About Looking*, New York: Pantheon, 1980; *And Our Faces, My Heart, as Brief as Photos*, New York: Pantheon, 1984. My description of Berger enlarges on the excellent book on Walter Benjamin by Susan Buck-Morss, *The Dialectics of Seeing: Walter Benjamin and the Arcades Project*, Cambridge, MA: MIT Press, 1989.

the degree of personal political responsibility that must be accepted for events all over the world: the fact that the world has become indivisible; the unevenness of economic development within that world; the scale of exploitation. All these play a part. Prophecy now involves a geographical rather than historical projection; it is space not time that hides consequences from us.... Any contemporary narrative which ignores the urgency of this dimension is incomplete and acquires the oversimplified character of a fable. (1974: 40)

Within the American tradition of critical cultural studies, the most powerful voice proclaiming a similar ascendency of the spatial has been that of Fredric Jameson. Influenced by Lefebvre's "conception of space as the fundamental category of politics and of the dialectic itself," Jameson inflected his self-acknowledged historicism to recognize the postmodern "weakening of historicity" and the concomitant rise of an effulgent spatial problematic in his well-known article, "Postmodernism, or the Cultural Logic of Late Capitalism" (1984). He stated emphatically that "we cannot return to aesthetic practices elaborated on the basis of historical situations and dilemmas which are no longer ours," to which he added that "a model of political culture appropriate to our own situation will necessarily have to raise spatial issues as its fundamental organizing concern."[2] As he would later elaborate in a book carrying the same title, "our daily life, our psychic experience, our cultural languages are today dominated by categories of space rather than by categories of time."

What is missing from the works of Berger and Jameson, and also from the assertive spatiality of Lefebvre and so many others, is a deeper critique of historicism's inhibiting hold on the critical spatial imagination. It may be true that, in the postmodern world of today, space more than time hides consequences from us and geographical issues are more fundamental in contemporary politics and daily life than historical ones. But cloaking these observations and assertions is a residual historicism that continues to constitute and historicize the new spatial critique along traditional lines, smothering it with unselfconscious historical (mis)understanding.

For the most forceful and direct critique of the genealogy and contemporary persistence of this particular form of space-blinkering historicism, one must turn again to Foucault. The outlines of his explicitly spatial critique of historicism have already been drawn. It began with Foucault's recognition, in "Of Other Spaces," of the

---

[2] Fredric Jameson, "Postmodernism, or the Cultural Logic of Late Capitalism," *New Left Review* 146 (1984), 88–9. The same title was used for a subsequent book by Jameson, Durham: Duke University Press, 1991.

19th-century obsession with history: "with its themes of development and of suspension, of crisis and cycle ... of the ever-accumulating past ... and the menacing glaciation of the world." Out of this infatuation with evolution, revolution, development, and the "great transformation" from traditional to modern societies unfolding in the spread of industrial capitalism came a deep historicization of critical consciousness that permeated all the "grand houses" of Western social and cultural theory and philosophy by the end of the "long" 19th century (1789–1914).

Of particular importance was the emergence over this long century of a new division of labor in the practice of modern historiography, built around a split between "idiographic" and "nomothetic" approaches. A specialized discipline of "History proper" defined itself increasingly around the accurate and detailed description and interpretation of particular events, backed by a distinctive historiographical research method designed above all for textual objectivity and empirical accuracy. Within this now capitalized and institutionalized modern discipline of "History," a particular form of historicism was embraced, internalized, and almost intrinsically taken for granted. After all, wielding the sophisticated objectivity and interpretive power of the modernized historical imagination was the designated (and unquestionable) task of the proper professional historian.

In contrast, a more theoretically ambitious, generalization-seeking, and transdisciplinary "philosophy of history" also took shape at the same time around various grand figures of continental European critical philosophy. This more nomothetic approach to history, presumed to be following ancient traditions, deeply affected two other 19th-century developments: the formation of the specialized social sciences (Sociology, Political Science, Economics, Psychology, Anthropology, etc.) and the emergence of scientific socialism, primarily around the assertively historical materialism of Marx and Engels. This enabled the critical historical imagination, and along with it an almost presuppositional leaning toward many different forms of historicism, to diffuse throughout the human sciences and into most critical and theoretical interpretations of society and sociality.

Prior to this 19th-century division, history and geography were almost inextricably connected. I would even go as far as to suggest that there were no "pure" historians or geographers prior to 1800. The labelings of Herodotus or Eratosthenes, or even Kant and Hegel, as true historian or geographer are figments of 20th-century disciplinary and disciplined imaginations seeking historical roots and legitimacy. If any of these eminent figures were to be asked, "are you a historian or a geographer?", I strongly suspect the answer would be "both/and also." After 1900, however, History and Geography occupied very separate intellectual niches. The seeds of this separation may have been planted by Kant in his identification of time and space, history and geography, as generic ways of thinking, all-encompassing and synthesizing schemata for understanding everything in the empirical world. The historical and geographical imaginations together thus filled, as Kant put it, "the entire circumference of our perception." In contrast, the "substantive" and more analytical disciplines were set in more specialized compartments of knowledge formation.

Something happened in the late 19th century to reconstitute the more spatio-temporally balanced Kantian inheritance. In each of the three major realms of socialist, idealist, and empiricist thought (often attached most closely with, respectively, French, German, and British intellectual traditions), putting phenomena in a temporal sequence (Kant's *nacheinander*) somehow came to be seen as more significant and critically revealing than putting them beside or next to each other in a spatial configuration (Kant's *nebeneinander*). Time and History thus absconded with the dynamics of human and societal development – agency, evolution, revolution, change, modernization, biography, the entire ontological storyline of the "becoming" of being and sociality – while the empirical dead weight of space and geography was shuttled into the background as extra-social environment, a stage for the real action of making history.

A new field of Modern Geography also took shape during the same period. Like History, it was deeply influenced by the mainly German debates about idiographic and nomothetic approaches to knowledge. But there were several significant differences. The discipline of "Geography proper," paralleling History proper, defined itself primarily around idiographic descriptions of areal or spatial variations on the surface of the earth (geo-graphy, literally earth-writing). Geography, however, maintained both a physical and a human-social perspective, while History jettisoned nearly all of its direct concerns with physical and biological processes except as their outcomes might possibly affect the course of human events or, more broadly, human "society."

But it was what happened outside Geography proper that was most revealing. With very few exceptions, a complementary nomothetic and critical "philosophy of geography" or of human spatiality, comparable to the philosophy of history, did not develop until the late 20th century. With this much weaker transdisciplinary diffusion of the geographical imagination, what often passed for a philosophy of geography or a history of geographic thought was little more than an introverted institutional and disciplinary history of interest only to "proper" geographers. While every discipline, including the physical sciences, had and continues to have its own respected historians, very few have recognized a need for resident geographers to explore the substantive and conceptual significance of spatiality and/or nature within the disciplinary fold.

In a closely related development, the city and urbanism became peripheralized as important (nomothetic) subjects in the philosophy of history, in the mainstreams of historical materialism, and in the evolution of Western social theory. The city and the specificities of urban life became increasingly conceptualized as mere background or container for the dynamics of human and societal development, happenstantial settings that had some influence on history and society but only by coincidence and/or incidental reflection, not by any inherent social or historical qualities that made the spatialities of social life important subjects for study in themselves. Much more important were the general and generalizable social processes, social relations, and social subjectivities that could be abstracted from the particularities of specific times and places.[3]

The recent development of just such a transdisciplinary philosophy (or metaphilosophy) of geography, spatiality, and critical urbanism is a recurrent theme in *Thirdspace* and may, in retrospect, be one of the most important philosophical and intellectual developments of the 20th century. But however this spatial turn is eventually interpreted, it is useful to go back again with Foucault to that critical turning-point in the 19th century and to see it with him as a key element in the emergence of a new *modernism* that would take hold in the *fin de siècle* and persist, with only minor modifications, up to the present.

Foucault noticed a perplexing side effect rooted in the disciplinary origins of History proper and, especially, of the new philosophy of history. "Did it start with Bergson or before?" he asked, referring to the French philosopher who perhaps more than any other infused history with transdisciplinary *élan*. Why is it that while *time* was treated as richly filled with life, with agency and collective action and social will, with the dynamics of societal development, with contradictions and crises that carried all human beings along the rhythmic paths of an "ever-accumulating past," *space* was treated as something fixed, lifeless, immobile, a mere background or stage for the human drama, an external and eternal complication not of our own choosing? History was socially produced. Geography was naively given.

---

[3] This is what made Lefebvre's emphasis on the urban and especially his argument that social relations become concrete only in their spatialization so bewildering – and unacceptable – to urbanists, Marxists, historians, sociologists, philosophers, and others.

This peculiar ontological, epistemological, and theoretical back-grounding of space seemed absurd to Foucault, who was always committed to disordering and disrupting disciplinarity wherever he found it. He reminisced about the "Sartrean psychologist" who "fire-bombed" him in the 1960s. "Space is reactionary and capitalist," the critic exclaimed; "history" and "becoming" are revolutionary. Making this radical privileging of time over space, historicality over spatiality, even more absurd for Foucault was his conviction that "the present epoch" was becoming "above all the epoch of space." Those who stubbornly adhered to the 19th-century obsession were described as the "pious descendants of time," and their pieties, if expressed today, he thought, would convulse us all with laughter.

It is important to emphasize, however, that Foucault did not see space as replacing time in some new hegemonic spatiology. Instead, he saw as necessary the breakdown/deconstruction of a space-blink-ered historicism and the reconstitution of an-Other history and histo-riography that revolved around the trialectic of space/knowledge/power, with some variant of structuralism playing a key role in rethinking "what we call time and what we call history." To affirm this trialectic, he proclaimed that "a whole history remains to be written of *spaces* – which would at the same time be the history of *powers* (both these terms in the plural) – from the great strategies of geopolitics to the little tactics of the habitat" (1980: 149). For Foucault, the "other" history was *geohistory*, an inseparable combina-tion of heterotopologies and heterochronies that explicitly focused on the spatio-temporal interpretation of the power–knowledge rela-tion.

The spatialization of time and history did not occur as Foucault (and Lefebvre) presumed it would. A vigorous and significantly feminist postmodernist, poststructuralist, and postcolonial critique was waged against "hegemonic historiography" to expose all its silencings, including the critical importance of space. But the explic-itly spatial critique of historicism was blunted in several different ways. Historians sympathetic to the cultural critiques of historiogra-phy embraced attractive ideas from cultural studies and postmod-ernism but either, like Hayden White, ignored the spatial critique or worked to construct a "new historicism" that was equally oblivious to the contemporary retheorization of space.[4] As Rosalind Krause notes in the epigraph to this chapter, the constant aim of the acade-mic historicist is to work on "the new and the different" to make them fit the familiarly historical and hence to avoid any necessity to think differently about the past, the present, and the future.

[4] For a particularly glaring example, see Brook Thomas, *The New Historicism*, Princeton: Princeton University Press, 1991; especially chapter 2, "The New Historicism in a Postmodern Age."

Many of the post-prefixed critics (Jameson comes immediately to mind) enthusiastically added a spatial emphasis to their writings but protected their radical historicizing from the most disruptive effects of the spatial critique. And a few critical human geographers and social theorists, especially those with established historical and/or cultural leanings (Derek Gregory and Anthony Giddens, for example), espoused a "historico-geographical materialism," a "new cultural geography," or a "spatio-temporal sociology" that propounded a balanced integration of spatiality and historicality but either failed to achieve this balance, or dialectic, in their writings or, more significantly, failed to see the fundamental ontological and epistemological problems involved in combining the historical and spatial imaginations. Space and geography may have been foregrounded, but an unquestioned historicism lurked offstage to confuse and constrain the spatialization of historical discourse.

Before turning to a few illustrations of how the spatial critique of historicism was either dissipated in confusion or disregarded entirely, it may be useful to state clearly what the spatial critique is *not*. First of all, it is not an anti-history, an intemperate rejection of critical historiography or the emancipatory powers of the historical imagination. Rather than anti-history, it can best be described as an attempt to restore the ontological trialectic of sociality–historicality–spatiality, with all three operating together at full throttle at every level of knowledge formation. Every history that is not merely a chronicle or a fable must presume to be intrinsically spatial, to be about spatiality, in much the same way that history is presumed to be intrinsically social, about the sociality of human life. But while the parenthetical (social) is normally taken for granted by historians, the parenthetical (spatial) must be insistently asserted – as geohistory or spatio-temporality, never history or temporality alone – as a consciousness-raising *aide-mémoire*.

The spatial critique should also not be construed as an assertion of a deterministic spatialism or a hegemonic new form of critical thinking, even if we accept the epochal arguments made by Foucault, Berger, and others about the history and geography of the present. Any privileging of spatiality – or of Thirdspace – has to be understood as temporary, a strategic foregrounding of the weakest part of the ontological triad designed to restore a more balanced trialectic. It must be remembered, therefore, that every geography, every journey into real-and-imagined Thirdspaces, is also filled with historicality and sociality, with historical and social as well as spatial "determinations." To say that space today provides a more revealing critical perspective than time is thus not a statement of eternal hegemony but of strategic sensibility. History and critical historiography continue to be of central importance in making sense of the contemporary world, especially when they are radically open to "other" critical perspectives.

I must also add that all the problems associated with historicism are not the fault of historians or philosophers of history alone. They can be found in nearly every department of the intellectual division of labor, including those labeled postmodernist or those occupied by the traditional spatial disciplines. They can probably be found in my own discourse and, if Derek Gregory is correct, they weave their way through all of Lefebvre's writings. A historicism that limits the scope of critical spatial thinking has not disappeared, Foucault's expectations notwithstanding. It remains even today stubbornly transdisciplinary.

Finally, I must explain why the spatialization of history or of discourse or of critical thinking is not simply a declarative matter, nor is it enough just to add spatial metaphors, geographical descriptions, and a few good maps while intoning that "geography matters" and "space makes a difference." Spatialization requires more than such incidental recognition and declarative commitment. And it must also go beyond the "additive" mode that characterizes, for example, the effusively geographical historiography of Fernand Braudel and the *Annales* school, often perceived as being about as geographical as historians can get.

The Braudelian model is much better than space-blinkered historicism, but such "spatialization by adjacency"[5] renders spatiality relatively unproblematic and merely supplemental. Spatiality and human geographies represent no major challenge to the historian's task other than the accumulation of more and more geographical information and supplementary insights. Moreover, the spatiality that is conventionally addressed is either that of an externalized physical environment frozen into the background of the *longue durée* or else an incidental Firstspace tableau of geographical sites and situations that impinge upon, or perhaps at times deeply influence, historical *évènements* and *mentalités*. The complex and problematic spatialities of Lefebvre or Foucault, not to mention hooks and Bhabha, Anzaldúa and Olalquiaga, Spivak and Said, Rose and Gregory (despite some residual historicism), are rendered essentially invisible.

Another level of spatialization must be reached for and explored, one that critically problematizes the interplay of spatiality and historicality and sees in this problematic geohistory or spatio-temporality the necessity to rethink them together as co-equal modes of representation, empirical inquiry, and (social) theorization. This was the project Lefebvre and Foucault initiated in the 1960s, and there

---

[5] I want to thank Beverley Pitman for suggesting this phrase to me.

were significant, earlier attempts at something similar in the writings of Nietzsche, Hegel, Heidegger, Sartre, Baudelaire, the surrealists, Kracauer, Simmel, and especially Walter Benjamin.[6] To move forward the present narrative, I will briefly address only the last-mentioned of these significant historical sources.

Benjamin was particularly concerned with the alluring and illusive "narcotic of the narrative," the discursive and non-discursive constraints imposed by narrativity on the writing as well as the making of history. He saw this addiction to the narrative form, with its compulsion toward linear, sequential, progressive, homogenizing conceptions of history, as requiring a withdrawal strategy that involved "blasting" the embedded historical subject out of its temporal matrix and into a more visual, imagistic, and spatial contextualization.

Derek Gregory, in *Geographical Imaginations*, incisively captures Benjamin's spatializing blast.

> Benjamin sought to interrupt this process [of homogenizing history through narrativity] by calling into question its endless suppression of difference beneath repetition. What was distinctive about his attempt to do so was that it went beyond disclosure of the logic of capital *to assault the modalities of representation*. A concern with what Wolin [1982: 100] calls the "image-character" of truth became a vital moment in Benjamin's work ... "for in this way he sought to confer equal rank to the spatial aspect of truth and thereby do justice to the moment of representation that is obscured once truth is viewed solely as a logical phenomenon." In other words, Benjamin effectively "spatialized" time, supplanting the narrative encoding of history through a textual practice that disrupted the historiographical chain in which moments were clipped together like magnets.   (1994: 234)

With Benjamin's blasting in mind, I turn next to Hayden White, one of the contemporary world's leading critical historians and a

---

[6] If I were more of a traditional historian, I would feel compelled to trace back each of these historical sources to unravel their distinctive approaches to the space–time, geography–history dialogue, a project which, once begun, would no doubt push me back further in time, through Kant, Leibniz, Spinoza, Galileo, Copernicus, Vitruvius, Aristotle, Plato, Eratosthenes, and so on. What would severely hinder such an archeology, however, is *the almost complete absence of a secondary literature that explicitly and perceptively addresses the problematic relation between historicality and spatiality* presumed to be embedded in these sources. Before Lefebvre and Foucault and especially after the mid-19th century, such questions were rarely raised; and up to the present there are still very few attempts to excavate the past debates on this arcane problematic. This represents an enormous challenge to contemporary trialectical scholarship, for it will entail a radical rewriting and reinterpetation of the history of philosophy and the philosophy of history, geography, and all the human sciences.

stunning exemplar of the persistence of an "underspatialized" historicism.

The Persistence of Historicism

Why Loving Maps is Not Enough

In such books as *Metahistory* (1973), *The Tropics of Discourse* (1978), and especially *The Content of the Form* (1987), Hayden White develops an argument about historicality and narrative discourse that so closely parallels my earlier writings on spatiality and critical human geography that I am tempted to recompose his writings under the mimetic title *Postmodern Historiographies: The Reassertion of Historicality in Contemporary Critical Theory*.[7] As a "metahistorian," White brilliantly exposes both the lasting strengths and the persistent weaknesses of the historical imagination through a deconstruction and reconstitution of the narrative form and a rethinking of "the value of narrativity in the representation of reality" and "the politics of historical interpretation" (the titles of two essays in *The Content of the Form*, henceforth COF).

For White, history is about "the real world as it evolves in time," but as he also notes "it does not matter whether the world is conceived to be real or only imagined; the manner of making sense of it is the same." (1978: 98) In this way, White opens historiography and the narrative discourse to "fictionalization," to a poetics of interpretation that draws from literature and literary criticism to represent a real world that is always simultaneously real-and-imagined. In his retheorization of narrativity, White dissolves the old dichotomy between "realist" and "imagined" or "fictional" narratives, between "scientific" and "poetic" historiography, and seeks instead a recombinant third way of looking at history that returns to narrativity as one of its "enabling assumptions," a commitment that he laterally attaches in the preface to COF to "cultural critics, Marxist and non-Marxist alike" and to "a whole cultural movement in the arts, generally gathered under the name postmodernism" (1987: xi).

White also speaks of "lived time" in ways that echo Lefebvre's lived space of representations, suggesting for historicality something very much like the trialectics of spatiality. Lived time (Thirdtime?) transcends the "scientific" historiography of atomistic "events" and the poetic and metaphorical representations of historicality to

[7] The full titles are *Metahistory: The Historical Imagination in Nineteenth-Century Europe* (1973); *Tropics of Discourse: Essays in Cultural Criticism* (1978); and *The Content of the Form: Narrative Discourse and Historical Representation* (1987). All three are published in Baltimore by Johns Hopkins University Press.

become something more than their mere combination. It creates a metahistoricality that is filled with multiplicities, many different planes of social time, as well as with hidden experiences, undecipherable codings, unexplainable events.

Yet, for all these parallels and echoes, there is a peculiar blockage in White's work to exploring the dialectics of spatiality–historicality and the implications of the spatial critique of historicism for critical historiography and narrative discourse. Take, for example, White's extensive use of Paul Ricoeur's hermeneutic philosophy of time and the narrative.[8] Although Ricoeur's ideas about space and geography barely budged from the Bergsonian, he filled his approach to narrativity with subtly double-coded terms and concepts which, in French and English, resound with ambivalent spatial and temporal meaning: plot, emplotment, configuration, context, world, trope, trajectory, *peripeteia*, time-span, story-line. Whereas I would like to believe that Ricoeur was aware of the pronounced spatiality of time that rings in these terms and concepts, there is almost no spatial content to White's narrative form or tropics of discourse (to use additional double-coded terms).

In his calculated embrace of the contemporary poststructuralist, postmodernist, and postcolonial literatures, White devotes a chapter each in COF to Fredric Jameson and his "redemption of the narrative" and to Michel Foucault's "historiography of anti-humanism." There are also long discussions here and in *Tropics of Discourse* of the varied structuralisms of Lévi-Strauss, Lacan, Piaget, Althusser, Derrida, and others. In all this, there is almost no mention of the forceful spatializations promoted in these literatures. This omission is especially pronounced in the discussions of Jameson and Foucault.

Foucault's agonistic commentaries on the "intensification of historical consciousness" in the 19th century are seen by White as fundamentally mistaken. Far from a cause for alarm, this emerging historicism was for White what Foucault stated it certainly was *not*: "an advance in learning, a progressive movement in the history of thought caused by the realization of the 'error' contained in the earlier conception of knowledge" (1987: 124–5). What was this "error"? Hinting that it had something to do with a privileging of space, White argues that "the Classical age had no place for time in its *episteme* … that it had purchased its certitude at the expense of any awareness of the reality of time, of the finitude of existence." White thus uses Foucault to explain why time came to be assertively treated as richness, fecundity, life, dialectic, in the 19th century, but ignores him to accept implicitly the accompanying de-intensification

---

[8] See Paul Ricoeur, *Time and Narrative*, 3 vols, tr. K. McLaughlin and D. Pellauer (Chicago and London: University of Chicago Press, 1984, 1985, 1988).

of spatial consciousness, the relegation of space into the background of history: fixed, dead, undialectical, immobile, quantifiably homogeneous: what Marx, ever the historical materialist, would trenchantly describe as the annihilation of space by time.

Jameson's spatializations are similarly ignored if not misconstrued. In a sentence which offers abundant opportunities to spatialize, White sees only historicization and the celebration of historicism:

> Jameson's tack is to grant that Marxism can never be a finished creed, but always a system in evolution, the vitality of which consists in its capacity to "narrativize" its own development, to "situate" its successive incarnations within the context of their formulations, and to uncover the "plot" in which they play their parts and contribute to the articulation of their unifying "theme." (1987: 155)

"His rule," White adds, "good historicist that he is, is 'contextualize, always contextualize.'"[9] White's spatially unconscious narrative reaches its peak shortly after this distorted claim in a comment on "the fact that the artwork [referring to the Marxist notion of the paradox of art] 'reflects' the conditions of the time and place of its production," but at the same time becomes generalizable beyond its specific contextualization. The first part of the paradox, White casually adds, means that the production of art and knowledge are therefore to be regarded as "timebound," ignoring what the postcolonial critics would latch on to with regard to hegemonic historiography, that the production of art and knowledge is also *placebound*, rooted in the "where" as well as the "when" (1987: 156).

This reflexively conditioned collapsing of space into time is the hallmark of historicism's stranglehold on the historical imagination. It has pervaded even the best of critical historiography for more than a century. Seek it out when you read histories, for it is nearly always there. The problematic spatiality of human life is either swallowed whole into history, without a trace, or it is metaphorized into something else and then subsumed as context, situation, content, culture, community, milieu, environment. So pervasive is this subordination of space to time that time's conquest infiltrates the dictionary. Looking up "space" in the Shorter Oxford English Dictionary, one finds the primary definition to be "Denoting time or duration," exemplified further as follows: 1. Without article: Lapse or extent of

---

[9] Ironically, the injunction which Jameson actually uses at the beginning of *The Political Unconscious: Narrative as a Socially Symbolic Act* (Ithaca: Cornell University Press, 1981, 9) is "Always *historicize*" rather than "contextualize." That White cannot quite tell the difference is telling indeed.

time between two definite points, events, etc.... 2. Time, leisure, or opportunity for doing something.... 3a. The amount or extent of time comprised or contained in a specified period.... b. The amount of time already specified or indicated, or otherwise determined.... 4. A period or interval of time.

Moving on, why is it that White in particular is so oblivious to the spatial critique or even to the addable geography that is contextualized in history? The easiest answer is that White is a historian and, after all, every cobbler prefers his or her own leather. White's subject was narrative discourse and historical representation, certainly not postmodern geographies or Thirdspace. Geography and spatiality are the subjects of other specialized discourses and disciplines, not the job of the proper historian. But White's approach to time, history, and the narrative is intentionally much more inclusive than such "property" reasoning implies. His purpose has been to break open historiography to other perspectives and discourses, to erase the distinction between disciplinary history and the philosophy of history by developing a metahistory that makes historicality and the narrative form truly transdisciplinary. In such a radically open historicality, there must be other reasons for White's exclusion of the spatial critique.

A second possible explanation, not entirely divorced from the first, has to do with White's not uncommon envisioning of contemporary historiography and the narrative as being in crisis and under siege from nearly all quarters. It must be defended to survive. In such essays as "The Burden of History" and "The Absurdist Moment in Contemporary Literary Theory," from *The Tropics of Discourse*, and "The Question of Narrative in Contemporary Historical Theory" (chapter 2 in COF), White seems almost paranoically to line up most critics of the narrative form as "time bandits" (my term, *pace* the Monty Python film) determined to push history and historiography deeper into crisis with their absurdist literary critiques or structuralist and poststructuralist narrative-bashing.

"Certain semiologically oriented literary theorists and philosophers," including Barthes, Foucault, Derrida, Kristeva, and Eco (Said is also peripherally mentioned), are presented as not just viewing the narrative as only one of many more or less useful discursive codes but as attacking the narrative as "an instrument of ideology ... the very paradigm of ideologizing discourse in general" (1987: 33). Braudel and the *Annalistes*, described as "certain social scientifically oriented historians," are seen as regarding "narrative historiography as a nonscientific, even ideological representational strategy, the extirpation of which was necessary for the transformation of historical studies into a genuine science" (1987: 31). A special category is set aside for "historians who can be said to belong to no particular philosophical or methodological persuasion but speak from the standpoint of the *doxa* of the profession, as defenders of the craft of historical studies" (1987: 31). They care little about theory or theoretical reflection and absolve themselves from the debates in eclectic, but dangerous, isolation.

As for *historicism*, White conflates the differences between "a properly historical and a historicist approach to history" with the conventional distinctions between idiographic historiography and the more nomothetic philosophy of history. Both these longstanding distinctions "obscure more than they illuminate of the true nature of historical representation." Every historical representation, White goes on to note, contains most of the elements of what conventional theory calls "historicism" (1978: 101–2). Historicism thus becomes as impregnable to attack as the narrative form itself.

White valiantly defends his more flexible and open narrative historiography against any and all presumed attackers. Amidst this *tour de force*, it is no wonder that the spatial critique of historicism, if it was recognized at all by White, was relegated to a minor position of peripheral peskiness, merely another form of anti-history and anti-narrativity. Spatiality always remains in its proper place, subservient to the historicality of social life. This subservient positioning brings up still another way of explaining White's geographical myopia. Every good historian, White certainly included, usually considers him/herself to be a good geographer as well (just as every good geographer is also, to some degree, a good historian). With little prodding, the good historian will proudly state "I love maps" and easily agree that "geography matters" in history, even that it is somehow becoming more important in contemporary historiography. But this is a geography and a spatiality defined on the historian's own terms and turf, oblivious not only to the new spatial critique of historicism but to the lively recent developments and restructurings in all the spatial disciplines.

White's writings are symptomatic of the persistence and power of a fundamentally underspatialized historicism not only in history proper but in all branches of contemporary critical theory. That even the most far-reaching, open-minded, and imaginative theoretician of history should be blind to the need to spatialize historicality, especially given his presentist attachment to cultural self-understanding, to creative responses to the precise circumstances of our present moment (remembering West's phrase), and to the configurative real-and-imagined tropics of discourse, is the strongest argument possible for continuing to press the spatial and spatializing critique of historicism.

Hayden White Meets Henri Lefebvre

A new opportunity to illustrate further the implications of White's occluded spatial vision serendipitously presented itself in a recent review by White of Henri Lefebvre's *The Production of Space*, published in the History and Theory section of a special issue of *Design Book Review*.[10] Most of the issue was devoted to the theme of "Orientalism," featuring articles by Edward Said, Barbara Harlow, Janet Abu-Lughod, Michael Watts, and others. The richly visual editors of this excellent journal of architecture, urbanism, landscape, and design announced White's review on the front page next to an Escher drawing and playfully enlivened the actual text with three pictures captioned: (1) "Anomalous space in William Hogarth's print, *Perspectival Absurdities*, for a 1754 textbook (From *Art and Illusion*)"; (2) "In Saul Silverberg's 1954 drawing for *The New Yorker*, a single line changes function and meaning in a series of spaces (From *Art and Illusion*)"; and (3) "Worm's-eye axonometric of the Nofamily House; Lars Lerup, 1978–82 (From *Planned Assaults*)."

The review itself is a sincere and occasionally insightful attempt to capture what White calls Lefebvre's "revisionist-Marxist theory of postindustrial space." But there is a fascinating subtext which devours Lefebvre's spatiality into the bowels of time and history. The challenges to historiography presented by Lefebvre are brought to the surface of the text, but they are left dangling in the air, with no real substance or grounding, to be easily swatted away, when the time comes, by any good historian. The review demands its own critical re-review.

The first paragraph carefully locates Lefebvre in time, as "one of the last representatives of the fading social class of 'intellectuals'" who have witnessed "most of the historically significant events of this century." He is said to write on such "arcane" subjects as metaphilosophy and "the end of history." His birthdate is noted, but not his birthplace. That he is French seems enough to contextualize him, to put him in his place.

Lefebvre is described in the second paragraph as one of the great French interpreters of Marxist thought whose "relentlessly dialectical" social criticism allows Marx to be revised "in the light of changing historical circumstances" and not just "limited to consideration of the experience of his time." White next fixes on Lefebvre's "periodizations," tracing over time the rise of what White calls, following *La Révolution urbaine* (1970) rather than *La Production de l'espace* (care-

---

[10] Hayden White, review of *The Production of Space*, by Henri Lefebvre, *Design Book Review* 29/30 (Summer/Fall 1993), 90–3.

fully noted as being originally published in 1974), an "urban" mode of production (a phrase nowhere to be found in the book being reviewed).[11]

The third paragraph begins promisingly. "Lefebvre views space as a dimension of human practice, which is experienced in so many different ways that it resists analysis." At this point the subplot thickens. Such a view of space, White notes, "resembles that other dimension of human existence – time – similarly difficult to contemplate without anomaly." He then proceeds to read Lefebvre as a philosopher of history:

> Yet, according to Lefebvre, space is ultimately more important than time, because it is spatial arrangements that determine the rhythms and periodicities of time. Indeed, he argues, space rather than time provides a basis for the understanding of human history. The notion that the "content" or deep meaning of history is time or temporality (as Paul Ricoeur, author of *Time and Narrative*, has recently argued) is, in Lefebvre's view, a typically idealist mystification. The fundamental subject matter of history is space . . . the secret of history's meaning.

Rather than refuting this "Marxist ... but also post-Marxist" position, White kindly but clumsily goes on in the next three paragraphs to assure the reader that Lefebvre is speaking about a very special kind of "social" space, "the space created by human groups in specific locales, using specific modes of production, and engaging in collective exertions of an economic, social, and political kind, to achieve purposes more or less human." Is that a snigger I hear? A "succession" of these "local" spaces, from "ancient Egypt, Greece, Rome, the European Middle Ages, the Renaissance, the 18th century, and so on," lead up to the "modern era" where "the historical significance of our time" is captured in Lefebvre's critique of a "new hegemonic social space" that has emerged in the wake of the "urban revolution," a now global space "produced . . . as a commodity," the only space "we humans can experience or know."

In paragraph after paragraph White continues to historicize Lefebvre's spatiality, not as part of a dialectic but as a subordinate duality, thereby always keeping Lefebvre's challenges at a comfort-

---

[11] White describes Lefebvre's dating of this epochal transition, "as disruptive as that wrought by the transition from agrarian to industrial society during the 19th century," to the period after World War II. Lefebvre himself usually dated the beginning of this radical change in the relations between society and space to around 1910.

able distance. Lefebvre's lengthy discussion of the transition from "absolute" to "abstract" space, which is both a brilliant exposition on the emerging specific geography of capitalism and one of Lefebvre's most historicist moments, is glossed over by White as a shift from "natural" space to that of mathematicians and philosophers, another example of how "history displays the human capacity to substitute 'culture' – products of its own labor power – for 'nature.'" Lefebvre's "conceptualization of history" as a "sedimentation of different kinds of socially organized space" is immediately compared with Mikhail Bakhtin's conception of the "chronotope" – Bakhtin being one of the most attractive quasi-spatializers of history and the narrative form. There is no mention of Benjamin.

Everything builds toward what might seem to be the ultimate denouement of Lefebvre's postindustrial-urban-global theory of history: the increasing irrelevance of "older social systems" and the historical past itself. Quoting Lefebvre, White writes:

> So irrelevant are the older (even early 20th-century) rules of political maneuver to the real forces shaping the new social space that the older notions of history, historicity, and the determinisms associated with these *temporal notions* lose their meaning.

This is indeed a very powerful statement, with significant ramifications not only in the spatial critique of historicism but also in every one of the post-prefixed critiques of modernism. It returns us to the beginning of this chapter, to Rosalind Krause, Dick Hebdige and John Berger, to the arguments that challenge history's lasting hegemony in interpreting the "real forces" that shape our lives. How does White respond? Not with a vigorous counterargument – why bother? – but with a series of asides politely addressed to his presumed audience of architects, designers, and urbanists.

First, White rightly criticizes Lefebvre for his romanticizing of the medieval and Renaissance town, whose organicism and good life have always been exaggerated by Lefebvre. But White also exaggerates Lefebvre's urbanism, imposing on him the proclamation of a new "urban" mode of production, with "urbanization" as "its process" and an "urbanized" social space "its product." This leads logically to a summary depiction of the book as "not a history of the city or of urbanization" but rather "a philosophy of the history of the city and urbanization," comparable to the key works of Lewis Mumford, which, White notes, are not mentioned. Historicizing the urbanization of Lefebvre thus seems complete.

But in the same paragraph there is another opening, a larger project defined by Lefebvre for which White says "A historical materialist consideration of the history of space will provide an answer." In
the only lengthy quote from the book, White almost innocently presents one of Lefebvre's most explosive and challenging assertions,
which I have quoted in chapter 1, that "the social relations of production have a social existence to the extent that they have a spatial
existence; they project themselves into space, becoming inscribed
there, and in the process producing the space itself." This provocative challenge to all traditional modes of theorizing historicality and
sociality, and historical materialism in particular, is reduced by
White to a tired whimper from a fading and now anachronistic intellectual.

"As one can see from a passage such as this," the short penulti-
mate paragraph begins, "Lefebvre's book is an example of what our
British colleagues disparagingly call 'grand theory.'" With one swipe
of his pen, White absolves himself from directly dismissing
Lefebvre's work, leaving it all to "our British colleagues," who
I might add probably include the most notoriously historicist
and francophobic critics writing in English. Protected from
having his reading of Lefebvre have any effect whatsoever on
his own thinking, White meekly concludes by obviously straining to
discover something, anything, that might interest those reading his
review.

"I kept asking, what are the *practical* implications of this revisionist-Marxist theory of postindustrial space for anyone having to *work*
with what are essentially spatial forms and relationships?" What he
finds is simply, but not insignificantly, a weakness in "architectural
and urbanistic criticism" having to do with an improper identification of "their object of study, which is space rather than the material
objects and the relations between them that appear to 'inhabit'
space." Lefebvre can help these critics, White offers, "to imagine
'social space' as the possible artwork it *might* yet become." And so
ends the review.

I have chosen to deconstruct White's historicist tropics of discourse not because he is wicked or ignorant, but because he is so
good at what he does. That the editors of the *Design Book Review*
chose an eminent historian to review this important critical and theoretical book about space, rather than an architect, urbanist, or geographer, is significant in itself, for it illustrates again the subtle
hegemony of critical historiography and historicism even within the
spatial disciplines. It is time for an-Other History to be created, one
that is comfortable with a decentering of the historical imagination
and, to paraphrase Spivak and hooks, works to un-learn its privilege
and behave as if it were part of the margin, in a space-time or geohis-

tory of radical openness. To make this work, a simple exercise is helpful: whenever you read or write a sentence that empowers history, historicality, or the historical narrative, substitute space, spatiality, or geography and think about the consequences.

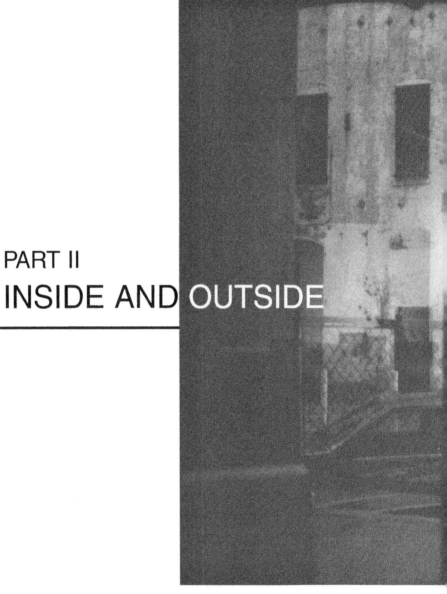

PART II
INSIDE AND OUTSIDE

LOS ANGELES

Remembrances:
A Heterotopology of the
Citadel-LA

Geography is nothing but history in space.... Geography is not an immutable thing. It is made, it is remade every day; at each instant, it is modified by men's [and women's] actions.

(Elisée Reclus, 1905–8; quoted in Kristin Ross,
The Emergence of Social Space, 1988: 91)

The city exists as a series of doubles; it has official and hidden cultures, it is a real place and a site of imagination. Its elaborate network of streets, housing, public buildings, transport systems, parks, and shops is paralleled by a complex of attitudes, habits, customs, expectancies, and hopes that reside in us as urban subjects. We discover that urban "reality" is not single but multiple, that inside the city there is always another city.

(Iain Chambers, *Popular Culture:
The Metropolitan Experience*, 1986: 183)

In the Spring of 1989, I helped to organize an exhibition within the spaces of Perloff Hall, then the home of the Graduate School of Architecture and Urban Planning at UCLA, where I have been teaching for more than two decades. The exhibition was part of a multi-year celebration on campus of the bicentennial of the French Revolution. Paris followed many of UCLA's organizational initiatives and major events were simultaneously videoscreened at Royce Hall, on campus, and at the Arc de Triomphe and other symbolic sites in the French capital. For our part in this multi-media and cross-disciplinary extravaganza of global simultaneities, we chose the theme *1789/1989 – Paris/Los Angeles – The City and Historical Change* and tried to compress within the exhibition space-time a 200-year visual geohistory of the present urban scene.

Remembering the events and the particular sites and sights of the exhibition provides an opportunity to begin grounding Thirdspace in the specificities of the urban, at least as these specificities reverberate through the space–time of Paris and Los Angeles, 1789 to 1989. In the absence of a written catalogue for the exhibition, I will try to capture its sites and sights in a present-tensed stroll through my personal remembrances of things passed ... and things still to come, adding a few more twists and turns to my original memories.[1]

Pictured at an Exhibition

The Preliminary Discourse

Leading up to the exhibition's opening was a series of public lectures that centered on urban evolution and revolution in Paris.

- James Leith (Queen's University, Canada), "La Bastille and Paris: 1789 to 1989"
- Dora Weiner (UCLA), "Sacred and Secular Space: Transformations of Religious into Medical Buildings, 1789 to 1820"
- Jean Baudrillard (University of Paris IX), "Revolution and the End of Utopia"
- Josef W. Konvitz (Michigan State University), "Spatial Change and the Centralization of Power: Paris Before and After the Revolution"

This was followed by a Colloquium that introduced the exhibition's central theme.

- Richard Lehan (UCLA), "The City and Literature: Pre-modern Paris and Post-modern Los Angeles, Cities at the End of Time"
- James Leith, "Planning for the Louvre: A Case for the Longue Durée"
- Dolores Hayden (UCLA), "The Power of Place Project: Planning for the Preservation of the Urban History of Los Angeles"

---

[1] What follows is a substantially expanded and revised version of Edward W. Soja, "Heterotopologies: A Remembrance of Other Spaces in the Citadel-LA," *Strategies: A Journal of Theory, Culture and Politics* 3 (1990), Special Issue: *In the City*, 6–39. The journal is a publication of the Strategies Collective of the University of California, Los Angeles. The first part of this article dealt with Foucault's concept of heterotopology and formed the basis for another expansion and revision that appears as chapter 5 above.

- Edward Soja (UCLA), "Taking Los Angeles Apart: Fragments of a Postmodern Geography"
- Jean Baudrillard, "From Beaubourg to the Arche de la Défense: Architecture, Urban Space, and the Power of Simulacra: A Commentary"

Some memories of these discursive events linger on as we move, along the preferred route, into the exhibition itself.

Getting to Biddy Mason's Place

We enter the building from the courtyard and garden. To the right is a hallway wall embedded with a great rectangular chunk of Louis Sullivan's masterful ornamental facade of the old Gage building in Chicago, donated to the School by Sullivan's family. It is now a permanent part of Perloff Hall. Opposite the facade chunk to the left is the first entry to the more temporary exhibition, a glassed-in display cabinet containing a mini-exhibit arranged by "The Power of Place," the name and restorative aim of a non-profit, design-oriented, memory preserving organization that was initiated by Dolores Hayden, one of the speakers at the Colloquium and a leading architectural historian and spatial feminist.[2]

Through photos, maps, and texts, the visual display, curated by Donna Graves (then assistant director for the project) is a recreation of Biddy Mason's symbolic Place in the history and geography of Los Angeles. It is a tribute to the work of Hayden, Graves, artist and graphic designer Sheila Levrant de Bretteville, artist Susan King, and sculptor Betye Saar, who together reshaped an asphalt parking lot near the famous Bradbury Building into a cruciform pedestrian park off Broadway with an opening to Spring. It is also a launching pad for remembering the evolution of Black power and presence in the urban landscape, and the submergence of this heritage under the veils of popular imagery and hegemonic urban historiography.

Biddy Mason (1818–91) arrived in Los Angeles as a slave owned by a Mormon master who had trekked from Mississippi to Utah and on to Southern California in the turbulent years following the American conquest and California statehood. With the assistance of several local Black residents, Biddy and her family were freed in

---

[2] For an introduction to the aims of this organization, see Dolores Hayden, "The Power of Place: A Proposal for Los Angeles," *The Public Historian* 10 (1988), 15–18. More recently, the Biddy Mason project served as the centerpiece for Hayden's new book, *The Power of Place: Urban Landscapes as Public History*, Cambridge, MA and London, UK: MIT Press, 1995. See chapters 5 and 6, "The View from Grandma Mason's Place" and "Rediscovering an African American Homestead," for elaborated stories of Grandma Mason's life and its artistic simulation in downtown Los Angeles.

Third Street

BRADBURY BUILDING

1965
1992

Broadway

Spring Street

BIDDY MASON PARK

Fourth Street

January 1856. Ten years later, the first African-American woman to own substantial property in the city established her home at 331 South Spring Street, on the North–South street that would later become the city's prime financial corridor, the "Wall Street of the West." From her home, Grandma Mason – given the name Grandma as a token of respect and admiration – practiced as a midwife and nurse, educator, nurturer, and entrepreneur, a forceful figure engraining an assertive Black presence in Los Angeles.

Among the many emanations from Biddy Mason's homeplace was the founding (in 1872) of that most famous of local Black religious institutions, the First African Methodist Episcopal (FAME) church, originally located at Eighth and Towne, where a plaque today honors Biddy Mason as founder. Over the years, the church would become a nurturing ground for many future generations of Black leaders, including longtime mayor, Tom Bradley. Most recently, the church's power of place was felt around the world during the events of April–May 1992 as the televised site for converging countercurrents of political opinion and debates about the future of the burnt out city. One particularly memorable televised moment presented the stunned face and stumbling voice of newscaster Ted Koppel as he unexpectedly listened to gang leaders from the Crips and Bloods present their sophisticated plans for rebuilding Los Angeles. "You speak very well," he mumbled in clumsy surprise, evoking many other memories.

There is a long history of placemaking at and around the site of Biddy Mason's home. In 1905, the Tuskegee Institute (forerunner of the Urban League) proposed to commemorate Mason with a six-storey building at the home site designed to house a community center to train Black youths to find employment. Various other proposals, such as for a Biddy Mason Museum and Community Center, were made by the Black community for the original site of the First African Methodist Episcopalian church at Eighth and Towne. All these sites were celebrated in 1976 as part of the wider celebrations of the bicentennial of the American Revolution;[3] and in 1988 a spiritual filled ceremony recommemorated Mason's grave at the Evergreen Cemetery in Boyle Heights.

Through the efforts of The Power of Place, Grandma Mason's positioning in time and space has been rethought as a designated urban mnemonic, remembered in 1989 in two places at once. The first is concretely close to the original homesite on Spring Street in a

---

[3] For this event, see Donna Mungen, *The Life and Times of Biddy Mason: From Slavery to Wealthy California Landowner*, Los Angeles: no publisher, 1976. Also see the pamphlet displayed at the exhibition, *The Power of Place: Los Angeles*, by Dolores Hayden, Gail Dubrow, and Carolyn Flynn, Graduate School of Architecture and Urban Planning, UCLA; and Hayden, 1995.

walk-through mews tucked away from the busy streetscape of downtown Los Angeles; the second is in the boxed-in mementos and inscriptions of the display case in Perloff Hall. In both sites and sights, the successful intentions of The Power of Place to affect the real-and-imagined urban scene are recreated and made clear. The actual and the simulated sitings are similarly provocative, not just of memories of Black Los Angeles and the forgotten role of African-American women, but also of broader questions having to do with the intentional preservation of the historicality and spatiality of the city.

The restoration of Grandma Mason's Place lies at the crossroads of several movements shaping the looks of downtown Los Angeles as well as other American cities. The Art in Public Places Program, an organization linked to the Community Redevelopment Agency (CRA)[4] and subsidized from public and private redevelopment sources, has contributed significantly to the dense sprinkling of public art in downtown LA and provided much of the funding for Grandma Mason's Place. The now well-established movements to preserve the historical and cultural heritage of the city, especially the Los Angeles Conservancy, helped to empower the project. And the wider struggle for more visible representation of the under-represented (mainly workers, women, minorities) has given the project its force and direction.

In contemplating The Power of Place and its projects today, two strings of questions come to mind. First: Whose history is to be preserved in these designs for commemoration? In what places is this heritage most appropriately encased? What forms shall the memories take? How can choosing a past to preserve help us to construct a better future? Does choosing sites also mean choosing sides? Through preserving urban history, can we improve our understanding of the current geography and political economy of the city – and work better to change it in significantly beneficial ways? And second: By redesigning the built environment can we, must we, reinterpret the past and not take it for granted? Given what Foucault called the synchrony of culture, is what we construct today destined to be a misleading representation of history, a simulation that accrues to itself only its own immediate contemporary meaning? Can we ever recapture and preserve a historical site when its set of relations to

---

[4] The CRA is a state agency established in 1949 to help redevelop "blighted" areas. After declaring most of downtown Los Angeles blighted, it became the central city's master planner and heritage preserver-destroyer, the primary referee deciding on what is remembered from the past and what is obliterated. In an ironic juxtaposition of sites, the CRA headquarters is presently located in the Banco Popular building, at 354 South Spring Street, just across the way from the Grandma Mason memorial. The building, a Beaux Arts tower constructed in 1903, was for many years considered the city's finest office building.

other real sites has been erased by time? Would not a deeper understanding of the contemporary dynamics and political economies of
urban design and development serve us better than recovering the
past in constructing a better future? Is it space more so than time that
hides consequences from us?

At first sight, it would seem that The Power of Place has appropriately responded to both sets of questions. Yet I have a nagging feeling that something is being missed in this well-intentioned attempt
to re-orient urban histories to cultural geographies and popular spatial struggles over gender, race, and ethnicity. Is it the relative
absence of a more radical perspective on class and the political economy of the cityscape? Or is it a blindspot for the countervailing powers of place to co-opt and oppress rather than reveal or emancipate?
Or perhaps it is the unquestioned acceptance that the binary division
of space into public and private still exists as a "contested terrain"
for contemporary struggles in Los Angeles? Each of these points can
be argued against The Power of Place, especially as represented in
Hayden (1995). But something else is more troublesome to me. It has
to do with different ways of looking at places and spaces, at history
and geography, at the past and the present. And it makes me return
to the two strings of questions raised in the paragraph above.

Answers to the first string of questions are typically informed by
an emancipatory sense of the power of history and an emplotment
within a meaningful historical narrative, a sequential archeology or
genealogy, or, as Hayden (1995) describes it, in "storytelling with the
shapes of time." The past is searched through as a potential means
of empowering the present. As Hayden quotes Kevin Lynch:
"Choosing a past helps us to construct a future" (1995: 226). By
recovering and preserving the history of places and spaces, we can
recover and preserve our collective selves much better than if we forget the past and repeat our mistakes and injustices. Here history
defines the power of place.

Answers to the second string lead in a different direction. They are
derived from an emancipatory sense of the cartography and heterotopology of power, from an emplotment within a meaningful interpretive geography. Here, in a kind of role reversal, the present is
researched both for its insights in reinterpreting the past and for the
sites from which to act to recontextualize the future. Both can be
described as spatio-temporal perspectives, but the first (and most
familiar) works primarily to historicize and preserve fixed geographies, while the second (more difficult to grasp) works in the
opposite direction, to spatialize history and historiography. Too
often, the first perspective's power and confidence renders the
second invisible. Recalling the discussion in chapter 6, I continue to
think about this spatial reductionism and how memory and

historical preservation can diminish the real-and-imagined power of lived spaces as I move through the rest of the exhibition.

Remembering the Bastille: 1789–1989

We turn the corner from the glassed case into the building's central hallway. Here a long illuminated wall guides the viewer along a chronological corridor picturing the richly heterotopic site of the Place de la Bastille, from its revolutionary storming in 1789 through the many attempts over the years to reconfigure its symbolic meaning (including pictorial recreations of the papier-mâché elephant of gargantuan proportions that was once set into the Place) to a picture of the newly completed Opera House that set off its own stormy controversy and symbolic dissonance at the time of the revolutionary bicentennial. The visual and aural stimulation is intense.

The photo-narrated walkthrough peaks at a spectacular video-taped presentation *auteur*-ed by Robert Manaquis, the director of the UCLA Bicentennial celebrations and co-organizer with me of the 1789/1989 urban exhibition. As we look at the Opera House and the video screen, behind us is a huge hanging satellite reproduction of "Los Angeles – From Space," a semi-permanent fixture in the central open enclosure of the building, floating in the so-called "two-storey space" where architects hold their juries and crits. It is a heady walk along this bridge of sights, from that translucent moment when the 14th-century fortress of the Bastille disappeared brick by brick in the fervor of revolution, through its many different preservational reappearances and simulations, each selectively trying in various ways to recapture the past in the confines of the insistent present-day power of place.

In this present-day, all the memories boggle as a freshly commodified fortress of culture replaces and reconstitutes the historical site of the Bastille as one of several spectacular bicentennial emplacements into the space of Paris: the intrusively imposing Arche de la Défense that Jean Baudrillard examined in his colloquium presentation; the wackily postmodern Parc de la Villette with its planned garden designed jointly by Peter Eisenman and Jacques Derrida amidst red archi-follies by Bernard Tschoumi; and the Pei-emplaced pyramid puncturing the heart of the Louvre, another subject of symposium discussion.[5] Will Paris – or the French Revolution – ever be the same? The question hangs in the air.

On the video screen, the images provide both immediacy and transition, juxtaposing in one place several sights and sites that are

---

[5] One also recalls American bicentennial-like intrusions on the urban landscape: the Statue of Liberty spectaculars in New York City, the 1984 Olympics in Los Angeles (three years after its official 200th birthday), and more.

themselves incompatible. Nostalgic music plays as the just-seen photo-narrative blends into old cinematic representations of the revolutionary Bastille, complete with a hell-hag Madame Defarge mouthing her words as the fortress burns. What we see is a transformation of historical memory into heroic (and anti-heroic) modernist (mainly Hollywood) imagery. Then, as we listen to the fulsome contemporary debates over the construction of the Opera House and the power plays among its promoters – and see what is happening elsewhere around the Place – we suddenly begin to realize that even the familiar modernist images are themselves being displaced by an entirely new set of time-eroding simulacra, forcing the past, against its will, into the heterochrony of the present.

As the videotape ends and prepares itself to begin again, I am reminded once more of Foucault. "The heterotopia," he wrote, "begins to function at full capacity when men arrive at a sort of absolute break with their traditional time." Perhaps this is what has been happening to the Place de la Bastille and to "other spaces" in the historical city. The power of place is being first neutralized and then inverted. New places of power emerge, writ larger, as the history of modernity is forcefully collapsed into contemporary conservative postmodern geographies. To continue interpreting them from modernist perspectives is to miss the point.

The Main Event: Symbolizing the Civic Center[6]

This is the end of art "as we know it." It is the end of the art of art history. It is the end of urban art with its dialectical struggles. Today this simulated art takes place in cities that are also doubles of themselves, cities that only exist as nostalgic references to the idea of the city and to the ideas of communication and social intercourse. These simulated cities are placed around the globe more or less exactly where the old cities were, but they no longer fulfill the function of the old cities. They are no longer centers; they only serve to simulate the phenomenon of the center. And within these simulated centers, usually exactly at their very heart, is where this simulated art activity takes place, an activity itself nostalgic for the reality of activity in art.
(Peter Halley, "Notes on Nostalgia," in *Collected Essays*, 1987)

---

[6] Many people helped to put together this portion of the exhibition, but there from the beginning to the end were Taina Rikala and Iain Borden.

We arrive now at the most central place of the exhibition, the center of centers, site of many sites, a small gallery located just beyond the video monitor showing films of the French Revolution. At the entrance is a spotlit bird of paradise, the City of Los Angeles's official flower, in a vase atop a gray wooden box. The flower is doubly appropriate, since the time of the exhibition coincided almost exactly with *Floreal*, the eighth month of the Revolutionary (Republican) Calendar, extending from April 20 to May 19. From the temporality of the previous sites we enter present-day Los Angeles and, at this point, the narrative breaks down into a cluster of revealing emplotments imitating the splintered labyrinth of the city before us.

Entrancement

Entrancement comes first. To tell you where you have been as well as to introduce you to where you are going is a massive sculptural form that dominates the gallery space and powerfully catches your eye. Half of it rises earthily from a billowy base of matted brown butcher paper to the crenellated turrets of a simulated Bastille. The other half sits on top of a slightly cracked bunker of grey concrete upholding the gleaming bronzed-glass towers of a simulated Bonaventure Hotel, the symbolic microcosmos of postmodernized Los Angeles. The two sides of the soaring sculpture blend into one another in brilliantly executed adjacencies that highlight the epochal transition between crumbling old fortress and resurgent new citadel, and other transitions as well. With its twinned peaks, it is called the "Bastaventure."[7]

---

[7] The artful architects of the Bastaventure are Ali Barar and James Kaylor.

The twinning is intentional. The Bonaventure Hotel, until recently owned by the Japanese and struggling with bankruptcy, symbolizes and simulates the *geographical* experience of postmodernity just as the Bastille and its memorializations symbolize and simulate the *historical* experience of the French Revolution. Despite the quibbling of experts on the "true" qualities of the Bonaventure/Bastille, their intentional twinning is made clearer when we follow Foucault and view them together as contemporary heterotopias, evocative "counter-sites" in which all other (and absolutely different) real sites within the synchronous culture are "simultaneously represented, contested, and inverted."

These specialized Aleph-sites function at full capacity within a specifically periodized slice of time – and after a break with deeply entrenched historical traditions. Idiographic historians might not think the Bastille was that much of a prison for revolutionaries, noting its use as a restful and relatively luxurious place of incarceration for the wealthy. And the architectural stylists will always categorize the Bonaventure as, at the very best, Late Modern and insist that it must be read as such. But seen and read as real-and-imagined heterotopias, lived spaces of representation, (partially) decodable social and cultural hieroglyphs, they take on new meanings.

Fredric Jameson, in a now memorable article published in the same year (1984) as the Los Angeles Olympics and Foucault's death, and just after Henri Lefebvre's most extended stay in California, was the first to begin reading the Bonaventure Hotel (which he originally called the Bonaventura) heterotopologically.[8] For Jameson, this "populist insertion into the city fabric" has become a "hyperspace" of both illusion and compensation, a new kind of cultural colony and brothel (my combining of Foucault's separate allusions) that exposes many of the archetypal "performative" conditions of contemporary postmodernity: depthlessness, fragmentation, the reduction of history to nostalgia, and, underlying it all, the programmatic decentering of the subordinated subject and the rattling awareness that the individual human body has been losing the capacity "to locate itself, to organize its immediate surroundings perceptually, and cognitively map its position within a mappable external world" (1984: 83).

While Jameson presents these unsettling conditions as indicative

---

[8] Fredric Jameson, "Postmodernism, or the Cultural Logic of Late Capitalism," *New Left Review* 146 (1984), 53–92. In 1984, while the article was in press, Jameson, Lefebvre, and I wandered through the Bonaventure, rode its glass-encased elevators, and had some refreshments in the rooftop revolving restaurant overlooking downtown. In 1989, I took much the same trip with Robert Manaquis and Jean Baudrillard, when Baudrillard was participating in the revolutionary bicentennial.

of a new *schizophrenia*, Celeste Olalquiaga, in *Megalopolis*, describes
this postmodern syndrome more aptly, as a special kind of urban
psychasthenia:[9]

> Defined as a disturbance in the relation between self and sur-
> rounding territory, psychasthenia is a state in which the space
> defined by the coordinates of the organism's own body is con-
> fused with *represented space*. Incapable of demarcating the limits
> of its own body, lost in the immense area that circumscribes it,
> the psychasthenic organism proceeds to abandon its own iden-
> tity to embrace the space beyond. It does so by camouflaging
> itself into the milieu. This simulation effects a double usurpa-
> tion: while the organism successfully reproduces those ele-
> ments it could not otherwise apprehend, in the process it is
> swallowed up by them, vanishing as a differentiated
> entity.   (1992: 1–2; emphasis added)

Olalquiaga's psychasthenia helps us situate the Bonaventure more
clearly in the trialectics of spatiality, at a place where the power of
Secondspace representations is opening up new ways to reproduce
its dominance, not just over the perceived space of daily practices
but over the whole of lived space and its primary sites, especially in
the increasingly "camouflaged" human body.

Directly experiencing the Bonaventure, with all its very real plea-
sures, is like entering a mutated cityscape of seductive simulations,
a Blade Runner-built environment that uncannily makes an esthetic
= anesthetic, enticing both the subject and object of history and
geography beneath a numbing amnesiac blanket of exact copies for
which no original ever really existed. In the bunkered and beaconed
fortress of the Bonaventure – if you will excuse my playful reference
to the site of the hotel on Bunker Hill, which Jameson originally
incorrectly called Beacon Hill – entrance (entrancement?) and exit
ways are curiously unmarked at many different levels, as if,
Jameson says, "some new category of closure [was] governing the
inner space of the hotel itself." Inside and out, one is lost in a
"placeless dissociation," an "alarming disjunction between the body
and the built environment" that Jameson compares (and attempts to
link) to the experience of Los Angeles itself and, even more point-

[9] Celeste Olalquiaga, *Megalopolis: Contemporary Cultural Sensibilities*, Minneapolis:
University of Minnesota Press, 1992. This use of the term *psychasthenia* originates
with Roger Caillois, an important figure (along with the better known Georges
Bataille) in the dissident surrealist movement that emerged in Paris just before
World War II and influenced the spatial thinking of both Walter Benjamin and
Henri Lefebvre. Caillois called it "legendary psychasthenia"; see "Mimicry and
Legendary Psychasthenia," tr. John Shepley, *October* 31 (1984), 16–32. Olalquiaga
gives this historical reference an explicitly spatial twist. Note: in Modern Greek, *psy-
chasthenia* is the general term used for mental disorder.

edly, beyond to the growing incapacity of our minds to cognitively map not just the city but also "the great global multinational and decentered communicational network in which we find ourselves caught as individual subjects." As John Berger saw, space seems more so than time to hide consequences from us at every scale of experience.

I covered some similar tracks in depicting the Bonaventure:

Like many other Portman-teaus which dot the eyes of urban citadels in New York and San Francisco, Atlanta and Detroit, the Bonaventure has become a concentrated representation of the restructured spatiality of the late capitalist city [what I now prefer to call the postmetropolis]: fragmented and fragmenting, homogeneous and homogenizing, divertingly packaged yet curiously incomprehensible, seemingly open in presenting itself to view but constantly pressing to enclose, to compartmentalize, to circumscribe, to incarcerate. Everything imaginable appears to be available in the micro-urb but real places are difficult to find, its spaces confuse an effective cognitive mapping, its pastiche of superficial reflections bewilder coordination and encourage submission instead.... Once inside ... it becomes daunting to get out again without bureaucratic assistance.

(Soja, 1989: 243–4)

Jameson's intentionally spatializing interpretation of the Bonaventure Hotel (and, I might add, of postmodern historicality) has occupied the eye of a still-unsettled storm of reaction and criticism that vividly illustrates those present-day conflicts Foucault predicted would arise between "the pious descendants of time and the determined inhabitants of space," or, remembering my earlier reference, between the historicizers of geography and the spatializers of history. Given the continuing role of hegemonic historiography in contemporary critical studies, even sympathetic critics of Jameson's first significant attempt to spatialize his own well-established historicism have been decidedly perplexed by Jameson's Bonaventure-ous moment. For one group of critics, Jameson's new agenda for history and criticism, revolving as it does around his slippery notion of cognitive mapping, seems to slide too blithely into ahistorical or, worse, anti-historical discourse. Characteristically, the response of these critics is an almost nostalgic defense of the 19th-century privileges of historicism.

Thus Mike Davis, in an otherwise illuminating interpretation of the "decadent tropes" of modernism to be found in Los Angeles, berates Jameson for presenting a geography "deprived of historical coordinates."[10] To find these coordinates, Davis seeks what he calls the "true temporalities" that can be read from the spatial text of Bunker Hill and the Bonaventure Hotel, and excavates them in a sequenced saga of urban renewal as urban removal of the poor to make way for the shining monuments of global capitalism. Jameson's explicit attempt at establishing historical coordinates from the Marxist heterochronies of Ernest Mandel's long waves is dutifully (and accurately, I think) criticized by Davis for faulty periodization and getting its dates wrong.[11] What Davis was primarily concerned with was not that Jameson was being confined by his Marxism but that he, Jameson, was misapprehending the politically correct meaning of the urban "renaissance" of downtown Los Angeles as seen from an avowedly historical materialist perspective.

Davis's alternative reading is brilliantly executed and much more prescriptively, informatively, and confidently political. But it remains straitjacketed none the less (or should I say all the more) by a lingering *diachronomania* that draws too uncritically from that most historicist of epistemological fallacies: *post hoc, ergo propter hoc*, after this, therefore because of this. No lateral or synchronic connections are allowed except for the insidious impact of global capitalism. For

---

[10] Mike Davis, "Urban Renaissance and the Spirit of Postmodernism," *New Left Review* 151 (1985), 53–92. See also his "Chinatown, Part Two? The 'Internationalization' of Downtown Los Angeles," *New Left Review* 164 (1987), 65–86.

[11] But not, it should be noted, for wrapping his critique too tightly with Marxism, a position fiercely taken by Derek Gregory in his attempt at "Mapping Jameson" (using Mike Davis) in his *Geographical Imaginations*, 1994: 278–82. The chapter in which Gregory's dismissal of Jameson as a manipulative Marxist historicist appears is entitled "Chinatown, Part Three? Uncovering Postmodern Geographies," a title that references the original film about Los Angeles as well as Davis's earlier article and my book, which was the primary focus for this lengthy chapter-as-book review. Gregory seems blind to tarring the equally deserving Davis with the same anti-Marxist anti-historicist brush he applies to Jameson. This blindness blunts and distorts his critique. By foregrounding and romancing Davis's historical materialism, Gregory obscures the possibility of interpreting Jameson's commentary on the Bonaventure as a discursive turning-point, a crude beginning in a personal attempt to spatialize his well-established historicism in the face of the contemporary moment. Rather than moving into Jameson's more recent writings (from which an argument can be made that he has certainly not advanced his spatializing project very far), Gregory looks backwards to Jameson's *The Political Unconscious: Narrative as a Socially Symbolic Act*, which begins with an injunction to "Always historicize!" and is filled with the same space-blinkered historicism I criticized earlier regarding Hayden White. The implication here appears to be: "once a historicist, always a historicist," a logic that reeks with the worst of historicism's overdeterminations. Perhaps Gregory was uncomfortable himself with these tactics. An otherwise careful reference bibliographer, Gregory leaves out the date of publication for *The Political Unconscious* (1981) in the appropriate footnote (1994: 278) and omits the book entirely from the Select Bibliography (which contains seven other works by Jameson). But I must cut short this footnote, lest it become *Chinatown*, Part Four.

Davis, the Bonaventure of the present must be properly narrativized in a politicizable sequential skein of past events unfolding over time at the site in question: a typical form of historicizing geographies. With the present so unproblematically a product of the past, the possibility that postmodernity poses new challenges to radical discourse and politics, that it marks in some way a break, or at least a significant deflection, of that "menacing glaciation" of the past, virtually disappears.[12]

Donald Preziosi, an art historian who, like Davis, is a long-time resident in Los Angeles, approaches Jameson's Bonaventure from an entirely different point of view in "La Vi(ll)e en Rose: Reading Jameson Mapping Space."[13] Turning Davis's critique upside down and inside out, he criticizes Jameson for not escaping far enough from historicism, for being "supremely historical," for never moving outside "that grand master narrative plot," that "commonplace, totalizing historicism central to art historical discourse since the nineteenth century institutionalizations on both sides of the Atlantic." These observations emanate from an important critical stream in the field of art history that has been deeply concerned with the stranglehold of historicism on the vision of the discipline since the 1920s. Preziosi, representing this critical stream, sees the same visionary dangers in Jameson's narrative tour of the Bonaventure.

> For Jameson, his version of Marxism is a place coextensive with the space of History. To arrive in that space, it is necessary to "pass through" texts, and above all the texts and hyperspaces of postmodernism, in order to grasp the latter's "absent causes": their History. (1988: 91)

Captivated, perhaps inadvertently, by the phrase "the space of History," Preziosi unfortunately ventures too far beyond his insightful critique of historicism to demand that we "position ourselves *out-*

---

[12] Davis's more recent work, *City of Quartz: Excavating the Future in Los Angeles*, London: Verso, 1990, is more sensitive to the spatial critique of historicism and to the need to critically spatialize the historical (materialist) narrative. The growing tension between his residual historicism and his emerging spatializations is one of the most interesting features of this remarkable book.

[13] In the inaugural issue of the UCLA students' journal *Strategies: A Journal of Theory, Culture and Politics* 1 (1988), 82–99. The third issue of *Strategies* (1990), under the title "In the City," contained not only another brief statement by Preziosi ("Oublier La Citta") and a reprinting of Rosalyn Deutsche's "Men in Space," but also the extended critique of *Postmodern Geographies* by Derek Gregory ("Chinatown, Part Three?") that would appear revised as chapter 4 in his *Geographical Imaginations* (1994), and my "Heterotopologies: A Remembrance of Other Spaces in the Citadel-LA" (1990). See also Donald Preziosi, *Rethinking Art History: Meditation on a Coy Science*, New Haven and London: Yale University Press, 1989.

side or beyond not simply 'postmodernism' itself, but outside of
time, space, and *history*." Rather than Jameson's "rhetorical over-
complications of the relatively simplistic Bonaventure Hotel," we are
ultimately guided toward a "metacommentary on architectonic representation itself," wherein all the devilish faults of the "historical
canon(s)" have already been sublimely exorcised. History, the primary target, is thus stripped of its problematics and becomes little
more than what Baudrillard once called it: "an immense toy."
Unwittingly perhaps, it is made to appear that spatiality is also dis-
pensed with in the process, except in the form of an ungrounded and
unlived "metacommentary" on architectonic representation.

In his more recent outsider's grand tour of the Jameson–
Bonaventure debates, Derek Gregory displays a critical style that is
characteristic of his *Geographical Imaginations*. He begins by presenting Jameson's views from the Bonaventure in a sympathetic light,
but then proceeds to dim the lights by undermining Jameson in an
eclectic barrage-collage of critiques that leaves the reader a little
more confused than enlightened. In the ten pages of text in which
Jameson's Bonaventure is the fulcrum (1994: 150–60), Gregory
amasses an army of attack that draws upon Davis, Preziosi,
Olalquiaga, Lefebvre, Soja, Jacques Lacan, Deleuze and Guattari, the
architectural critics Peter Eisenman and Michael Sorkin, the geographer Michael Dear (whose sharp photograph of the Bonaventure is
included), Jean Baudrillard, Walter Benjamin, the philosopher
Gillian Rose, the sociologist Sharon Zukin, and ultimately Donna
Haraway.

As footnoted earlier (note 11), a prime target for Gregory is
Jameson's (and indirectly Soja's and Zukin's but, significantly, not
Davis's) categorically Marxist-historicist armor. Only Gregory and a
few chosen others are apparently able to move freely into and out of
Marxism and historicism without being caught in their sticky web
of overdeterminisms. Even Lefebvre, constructed by Gregory as an
unwavering Hegelian Marxist and therefore ineluctably historicist,
is eventually thrown into the same basket as Jameson. In a much
more cogent critique, however, Gregory builds on Olalquiaga and
towards Haraway, thereby effectively exposing the more crucial
weaknesses in Jameson's work with respect to gender, sexuality,
postcolonial strategies of resistance, and that most originary spatiality, the body. The spatial psychasthenia induced by such postmodern places as the Bonaventure thus becomes a more powerful
personal-and-political concept than even Jameson imagined. As
Olalquiaga states:

Bodies are becoming like cities, their temporal coordinates
transformed into spatial ones. In a poetic condensation, history

has been replaced by geography, stories by maps, memories by scenarios. We no longer perceive ourselves as continuity but as location.... It is no longer possible to be rooted in history. Instead, we are connected to the topography of computer screens and video monitors. These give us the language and images that we require to reach others and see ourselves. (1992: 93)

Following this quote, Gregory adds:

In such circumstances, if we even approximately recognize ourselves in the description, a passionate appeal to history is unlikely to be enough. Olalquiaga presents two options: either contemporary identity "can opt for a psychasthenic dissolution into space," affixing itself to any scenario by a change of costume, or "it can profit from the crossing of boundaries, turning the psychasthenic process around before its final thrust into emptiness, *benefitting from its expanded boundaries.*"
(Gregory, 1994: 160; emphasis added to quote from Olalquiaga, 1992: 17)

Moving in completely opposite directions, Davis and Preziosi missed the budding Thirdspace perspective that can be excavated from Jameson's cognitive remapping of the Bonaventure. By recentering his critique of Jameson's accomplishments around the avowedly radical and assertively spatial postmodernism of Olalquiaga, and in following this with effective excursions through the feminist and postcolonial literatures, Gregory begins building his own version of Thirdspace. The trouble with the Gregorian version, and the reason why I have not excavated it at greater depth in *Thirdspace*, is that it is buried under such a thick mesh of criss-crossing omni-directional critiques, points and counterpoints moving in so many different directions at once, that it is difficult to tell where Gregory stands – and whether there is anyone else left standing with him.

In the end, what is missing from Davis, Preziosi, and Gregory is a critically balanced sense of the ontological trialectic of spatiality–historicality–sociality and a deeper appreciation of Lefebvre's trialectic of perceived–conceived–lived space. In their very different unhingings of this trialectical balance, spatiality and postmodernity are represented in ways that can too easily be totalized as unequivocally evil forces infecting contemporary lifeworlds. This was not necessarily the intent of any of these scholars, but the impressions they leave behind in their treatments of the Bonaventure are susceptible to just such an interpretation, and especially to reversions to one or another of the trialectical extremes: an over-idealized and over-empowered historicism, spatialism, or socialism.

To conclude this lengthy sidetrack from the exhibition, I return again to Jameson's original (bon)adventure. With the benefit of hindsight, it seems now that in entering, and indeed stimulating, the debate on "Postmodernism and the City," Jameson stumbled into a new and different project that he may not have been fully aware he had entered. Neither a total denial of the usefulness of critical historiography nor a fulsome celebration of historicism, the project opened up in the windowed reflections of the Bonaventure pointed toward a more balanced and critically problematic trialectic of spatiality, historicality, and sociality rooted experientially rather than just materially or metaphorically in Lefebvre's real-and-imagined lived spaces of representation.

Jameson's brief exploration of Thirdspace was tentative and clumsy. It was constrained by a historicism and a historical materialism that filled postmodern spatiality too deterministically with the menacing glaciation of capitalism. It was thus too little informed by the emerging postmodern feminist, post-Marxist, and postcolonial critiques. More specifically, it also depended too much on Kevin Lynch's limiting conception of cognitive mapping and urban design rather than on the more profound, and more familiar, inspiration of Henri Lefebvre, who was with Jameson at the time he was preparing the 1984 article for publication. After all, it was Lefebvre more than anyone else who made the author of *The Political Unconscious* (1981) begin considering the need to change his demand to "always historicize" as the "one absolute and ... 'transhistorical' imperative of all dialectical thought," to an-Other version: *always spatialize as you historicize sociality*! Jameson may not have continued to practice this strategic spatialization in all his subsequent writings, but his brief encounter with Thirdspace should not be buried under the hypercriticism – or the hyperpraise – his work has received. But we have lingered long enough. It is time to move on to other spaces.

CITADEL-LA

Returning to our tour we can look beyond the entrancement with the Bastaventure to the heterotopological symbols that are concentrated in and around the CITADEL-LA, the "little city" that defines the power-filled "civic center" of the polynucleated Los Angeles region. The CITADEL-LA is colorfully mapped right before you, on the gridded main wall of the gallery. Amidst the dense clustering of building footprints – gold for government sites, bright red orange for the sites of high culture, beaux arts colors I am told – a boldly written placard announces another reading of the spatial text of downtown Los Angeles.

The first cities appeared with the simultaneous concentration of commanding symbolic forms, CIVIC CENTERS designed to announce, ceremonialize, administer, acculturate, discipline, and control. In and around the institutionalized locale of the CITADEL (literally, a "little city") adhered people and their spatially focussed social relations, creating a CIVIL SOCIETY and an accordingly built environment.

The city continues to be organized through two interactive processes, surveillance and adherence, looking out from and in towards the citadel and its panoptic eye of POWER. To be urbanized means to adhere, to be made an adherent, a believer in a collective ideology and culture rooted in the extensions of *polis* (politics, policy, polity, police) and *civitas* (civil, civic, civilian, citizen, civilization)

This representation of the CITADEL-LA seems rather presumptuous, especially for those who are familiar with the actual sites depicted on the cartographic wall. The Civic Center of Los Angeles is, at first glance, rather unremarkable: a rectangular band of more-than-usually dull buildings, with a few annexed extensions, stretching across the top of Bunker Hill, a few minutes away from the Elysian fields to the north, where the city was born;[14] and a stone's throw away from the glassy Bonaventure, nestled in the skyscrapered financial center to the south, where the city, it seems, is being born again. Surely the commanding powers described in the proclamations above no longer apply here, of all places, in the world's

---

[14] At the North Broadway entrance to Elysian Park, a monument marks the first white siting of Los Angeles, the place where Gaspar de Portola and Father Juan Crespi made camp on August 2, 1769, during the first European expedition through California. By the time of the French Revolution, the indigenous Yang-Na had virtually disappeared without a trace of commemoration in what were once their primary hunting grounds. Nearly two centuries later, the nearby Chavez Ravine was also cleared of its inhabitants – many of whom were rebellious descendants of the old Californios and other poor "natives" – to make room for Dodger Stadium, one of the most profitable sports sites in the world. Elysian Park today also contains California's first botanic garden, filled with exotics from every continent, and the much more domestic Los Angeles Police Academy, where young recruits are taught more contemporary hunting and gathering skills.

most symbolic space of urban decenteredness, of dissociated neighborhoods, of the ex-centric idiocy of urban life.[15]

Yet, the power-filled centrality of the citadel is astounding in Los Angeles. Even as things fall apart, the center holds firmly, and, lest we forget, it has done so for more than 200 years, in part as a historical residual from the first siting but also as an imposing contemporary accretion of the commanding powers associated with newer modes of urban surveillance and adherence, discipline and control, what Foucault specifically calls governmentality. Nowhere else outside the federal citadel in Washington, D.C., is there a larger concentration of government offices, employees, and authority. Looking at the map represented in Illustration 7, the commanding sites of power cascade from west to east, occasionally lapping over the formal borders defined by Temple and 1st Streets from the Harbor Freeway to Alameda Street. Let us walk through the sites.[16]

(1) At the western edge, the County Health Building and the County Health Department Headquarters sit adjacent to the enormous Harbor-Pasadena and Hollywood-Santa Ana Freeway interchange, one of the earliest and most dazzling of "spaghetti junctions," where many feel the ultimate test of healthy freeway driving can be found in crossing eight lanes of traffic from right to left to exit the Harbor and join the Hollywood. The view is both frightening and exhilarating.

(2) The Department of Water and Power Building comes next, the headquarters for the largest public utility in the US. It glows at night as an electrified beacon and is surrounded by a moat of gushing water (drought permitting) that also serves the building's air conditioning system. One can write an entire history of Southern California around the terms Water and Power. . . .

(3) The next block, between Hope and Grand, houses the Ahmanson Theater, the Mark Taper Forum, and the Dorothy Chandler Pavilion, which together comprise the Music Center, a white marble acropolis for the performing arts, LA's version of NY's Lincoln Center. Just to

---

[15] Such idiocy derives from the Greek root *idios*, one's own, private, separate, apart, as in idiosyncratic, describing someone who acts in a way peculiar to him- or herself; and also idiographic, literally writing in just such a way, emphasizing the peculiarity and particularity of things and events. The original Greek meaning of "idiocy" was unlearned in the way of the traditional *polis*, displaying rural versus urban traits and behavior; hence Marx's use of the term in "the idiocy of rural life." That "idiot" (from *idiotes*, private person) has come to mean someone who is "so deficient mentally as to be incapable of ordinary reasoning or rational conduct; a blockhead, and utter fool" is testimony to the peculiar hold of the *polis*, the political city, on Western consciousness, ideology, and language. I have been helped in my understanding of Greek by Antonis Ricos and Costis Hadjimichalis.

[16] The numbers below are keyed to the map.

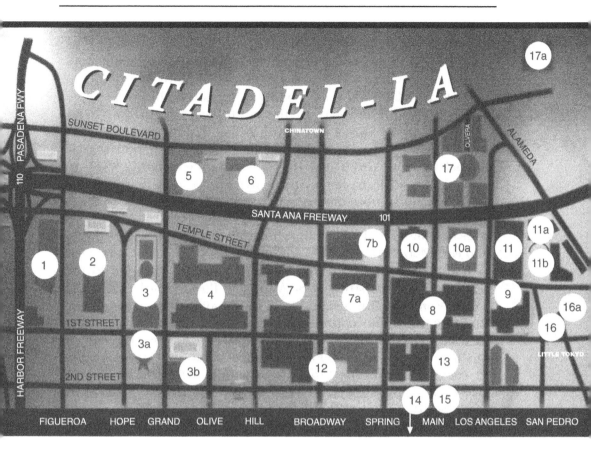

the south is the site of the still-being-built, Frank Gehry-designed Disney Concert Hall (3a), and just across from that space sits the low-rising, pyramid-topped, red sandstone-clad Museum of Contemporary Art (MOCA), designed by Arata Isozaki as a massing of solids and voids, with the main gallery underground (3b). Above ground, from the surface of MOCA's plaza, the view eastward is blocked by a wall of high-rise condominiums that aggressively separates the old downtown below from the cultural acropolis and the citadel.

(4) The big block stretching between Grand and Hill (and terminating Olive Street in its coverage) contains the administrative headquarters of the County of Los Angeles, the County Offices and Hall of Administration in one building, the massive County Courthouse in another. The County Board of Supervisors deliberates over a constituency of more than 9 million people, by far the largest local government unit in the country. A powerful fiefdom of five white males for most of its existence, the Board of Supervisors has, grudgingly,

begun to change its colors and gender in recent years. The County Courthouse and the nearby Criminal Courts Building are reputed to be the busiest anywhere, churning incessantly, day and night, to feed the country's largest urban prison system, with more inmates than New York City and Cook County (Chicago) combined.

(5) Across the 101 Freeway to the north, near where the Grand Avenue axis meets Sunset Boulevard, is the headquarters of the LA City's Board of Education, serving a student population that is approaching New York City's in size. Once called Charity Street to follow up on adjacent Hope (there never was a Faith, it seems), Grand Avenue has become the main cultural axis cutting through Bunker Hill.

(6) East of Grand and still just north of the Freeway, the land slopes up to Fort Moore Hill, an almost forgotten historical site where what has been called "the most spectacular man-made waterfall in the US," built by the Department of Water and Power, marks the white Pioneer Memorial. An inscription in the ceramic glaze reads, "May those who live in our naturally arid land be thankful for the vision and good works of the pioneer leaders of Los Angeles, and may all in their time ever provide for its citizens water and power for life and energy." Downslope from the waterfall, which rarely if ever flows these days, is the residential Chinatown and, stretching out everywhere else, *Chinatown* the movie.

(7) Back inside the citadel proper we encounter more of the county's expanding power. Continuing what was originally designed as a Beaux Arts, City Beautiful plaza and promenade, the rather seedy hidden core of the grand County complex extends over Hill Street (in what is called the Paseo de los Pobladores) to append the Hall of Records (neatly designed by Richard Neutra), the Law Library, and, across Broadway, the Criminal Courts Building (7a), home to the Municipal Court of the Los Angeles Judicial District, County of Los Angeles and State of California, built in 1925. Across Temple from the Courts is the fourteen-storey Hall of Justice, now abandoned and ready for demolition (7b). During the infamous O. J. Simpson trial, a makeshift media city occupied one of the courtly parking lots to project the local scene around the world.

(8) City Hall, with the City Hall East and City Hall South buildings just next door, is in many ways the capstone of the governmental citadel. Administering the country's second largest city, its all-American pastiche of Greek, Roman, Byzantine, and Italianate designs is symbolically LA Gotham, projecting the city's Dragnet-noir imagery televisually all over the world. Topped by a local interpretation of the Mausoleum of Halicarnassus, its 27 to 28 storeys was the only exception allowed to break the 13–storey height limit that was maintained by the city until 1957. If one looks hard enough,

revealing inscriptions can be read from the facade of City Hall. One reads: THE CITY CAME INTO BEING TO PRESERVE LIFE, IT EXISTS FOR THE GOOD LIFE. In the hopes that such promises can be delivered today, bands of homeless people regularly squat on the south-facing grass apron, waiting.

(9) Beyond the extensions of City Hall to the east lies Parker Center, headquarters of the Los Angeles Police Department (LAPD), named after the police chief who so insensitively bungled the police response to the Watts riots of 1965 with his racism. Teeming with black-and-whites, the Parker Center makes up for the relatively limited man-power of the LAPD with an arsenal of crack-house ramming tanks, a large air force of sight-trap-and-shoot helicopters, and a barrage of other specialized armaments that builds on the region's vast high-technology defensive and offensive war industries. Not surprisingly, but often missed in the media blitz and glitz, the Parker Center was the first target of attack for the public demonstrations that began the April 1992 uprisings. Other buildings in the citadel followed.

(10) Directly across Temple from City Hall is the Federal Courthouse, abutting on to the freeway channel that separates the contemporary citadel from its ancestral home now contained in El Pueblo de Los Angeles Historical Park. But before the main Federal Complex is entered, there is an entertaining buffer consisting of the touch and playful Children's Museum and the multi-level Los Angeles Mall (10a). Hall and Mall meet in the Triforium, a reinforced concrete, glass, and steel sculpture electronically programmed to synchronize lights and sounds and to serve as the site of noontime concerts for children, office workers, and the homeless of downtown.

(11) The federal climax of the citadel is found in its easternmost wing. Here a new center is rising around Roybal Plaza, out of sight for nearly all residents of Los Angeles (even downtown office work-ers) yet already screened in glimpses shown around the world as the site of the second trial of the police officers responsible for the Rodney King beatings. An older building, housing the US Federal Courthouse, now faces the new Metropolitan Detention Center (11a) that is featured, from a very different perspective, on the cover of Mike Davis's *City of Quartz*. Next to it are the gleaming Roybal Center Office Building and the just completed Veterans' Administration Outpatient Clinic. In the midst of all this is a rather bizarre plaza with a rotunda designed by the artist, Tom Otterness, and an entranceway, within sight of Parker Center, punctuated by Jonathon Borofsky's 30–foot-high *Molecule Men*, a burnished steel sculpture of four interconnected male silhouettes covered from head to toe with a swiss cheese of (bullet?) holes (11b). The sculpture has already been dubbed "the drive-by." We will have to return here for a closer look later on.

(12) The citadel proper laps over southward, across First Street, in several additional extensions. Two huge building blocks straddle Broadway just before it enters the citadel. On one side is the Junipero Serra Building and just south of it the Civic Center Law Building. On the other side of Broadway is the headquarters of the Fourth Estate, Times-Mirror Square, where our times are mirrored daily by one of the most regionally dominant and influential urban newspapers in the world. As its self-advertisement once proclaimed, "It all comes together in the *Times*." Whether it is an annex of the Civic Center, or vice versa, is an interesting question.

(13) An extended governmental axis exists along Main Street. First, there are the old offices of Caltrans, the California Department of Transportation, now absorbed into the Metropolitan Transit Authority (MTA). It serves as the electronic nerve center for controlling traffic operations along the county's 527 miles of freeways. In a windowless room, an advanced computer system connected to a huge illuminated freeway map provides sophisticated surveillance of millions of freeway drivers every day in what one wag called the Orwellian Ministry of Traffic. Seeing everywhere without ever being seen, it reminds you again of Bentham's Panopticon as seen by Foucault.

(14) Further south on Main is the Ronald Reagan State Office Building (no comment) and the Southern California Rapid Transit District Headquarters. The SCRTD has recently been absorbed, with the LACTC (Los Angeles County Transportation Commission), into a new MTA (Metropolitan Transportation Authority) charged with spending as much as $180 in public funds (mainly federal and almost surely not forthcoming) over the next 30 years for mass transit, the latest round of federal subsidization of Los Angeles's development, hopefully filling the voids left by the post-Cold War decay of the Defense industry.

(15) Off to the east side of Main is St Vibiana's Cathedral, a commanding power over more than 4 million adherents in what is reputed to be the world's largest Catholic archdiocese. This modest bastion of the churchly First Estate presses up against Skid Row, the absolute inversion of citadellian empowerment, home to what many say is the largest homeless encampment in the US. When the pope visited St Vibiana's, the streets were cleared and cleansed to make Skid Row invisible. Plans are currently under way to rebuild St Vibiana to increase this invisibility and further obliterate the unsightly surroundings.

(16) A few more outposts might be noted. Inside the rectangle of the citadel, in the extreme southeast corner, is the small Japanese-American National Museum, commemorating the erosion of the old Little Tokyo and the rise of the Big Tokyo-induced and -designed

Little Tokyo just to the south of First Street (16a). In a seedy parking lot to the north, across from the Parker Center, sits the Frank Gehry-designed warehouses of the Temporary Contemporary Museum closed after the permanent establishment of the new Museum of Contemporary Art but now re-opened. And just outside the Alameda boundary to the east, within sight of the Federal Detention Center, is a new Central District Headquarters for the Department of Water and Power, matching its bookending of the citadel to the west. Water and Power are everywhere inside and outside the citadel.

(17) Sitting above all this is the old citadel of El Pueblo de Nuestra Señora la Reina de Los Angeles, centered around a celebratory gazebo and the adjacent La Placita church and Olvera Street tourist arcade. Nearby is Union Station and, above and beyond that, Los Angeles County Jail (17a), one of the largest incarceration centers in the country.

The tour of the CITADEL-LA is thus completed. But its heterotopological exploration has just begun.

Cultural Crown

We move back to the Cultural Crown, symbolically displayed for us in what the architects call a massing model for the newest crown jewel being added metonymically to the CITADEL-LA: the Walt Disney Concert Hall competitively designed by the current king of California architects, Frank Gehry. In the background is posted a collection of working drawings and promotional text dreaming the tomorrows of the contemporary Acropolis, the "topmost city" of the CITADEL-LA, the fortress of fine arts that has put the crest of cut-down-to-size Bunker Hill squarely in the center of the mental map of American urban imagineering.

Compared to the intrusive new Opera House in the Place de la Bastille, the Disney Hall (future headquarters of the Los Angeles Philharmonic) was originally designed to be primordially indigenous, Cali-casually informal, a kind of vegetative climax perfectly adapted to local conditions, a celebration of home-grown inhabitants and heroes. The glassed-in foyer is described in front of you as "the city's living room" open to the mixed use of the masses. The sprawling plazas are seen as the "culmination" of the city's public space. The plants and trees celebrate the locale: no exotics here. Gehry's Pritzker-prized competitors, the outsiders Bohm of Cologne, Hollein of Vienna, and Stirling of London, didn't stand a chance.

A more symbolic pairing than Gehry and Disney is scarcely imaginable. Together, their dreamworlds will aptly complete the deconstruction and reconstitution of Bunker Hill, its transformation from the "old town, lost town, crook town" described by Raymond Chandler to what the archly anarchic architect Arthur Erickson predicts will be the "center of centers of the western world." A brief aside is useful here to commemorate what has been truncated from Bunker Hill. I will draw primarily from a recent essay by Anastasia Loukaitou-Sideris and Gail Sansbury (1996) on the "Lost Streets of Bunker Hill." Both are based in the Department of Urban Planning at UCLA and they call their very interesting work "didactic street biographies."[17]

Bunker Hill is the southernmost extension of a string of hills (Elysian Park, Fort Moore, Bunker) paralleling the Los Angeles River as it cuts through a pass between the Santa Monica and San Gabriel Mountains. The hills were an area of great significance to the indigenous Gabrilieno Indians and in the lower-lying wedge between the hills and the river, El Pueblo de Los Angeles was born, burned, and recycled southward from Elysian Park to its present memorial site around La Plaza. The first significant Anglo occupation of Bunker Hill began in the 1870s and over the next several decades it became a highly fashionable suburb, with "carpenter Gothic" and "Queen Anne-Victorian" buildings (still visible today only in the artfully preserved Carroll Street and adjacent areas in Angelino Heights nearby). In the 1920s and 1930s, the wealthy moved further away from the center and the building stock on Bunker Hill was subdivided into rooming houses for rent, mainly to single, elderly men. Some tenements were also built, as the area became a densely populated pocket of working class life and liveliness, perhaps the closest any neighborhood in Los Angeles came to Manhattan's Lower East Side.

Between 1904 and 1944, the funicular Angel's Flight, above and parallel to the Third Street Tunnel, carried hundreds of people daily up the slope from Hill Street at its base. Old films, photographs, novels, and memoirs depict a vibrant streetlife and cityscape on Bunker Hill throughout this period. John Fante writes about it in his 1939 novel, *Ask the Dust*, and in *Dreams from Bunker Hill*. Life on the streets of Bunker Hill in the 1930s also features strategically in the spatial biography of Harry Hay, founder of the Mattachine Society and a key figure in the development of the gay movement in the US. Hay is described by his biographer, Stuart Timmons, as he attempts to escape from the police, with the help of Bunker Hill residents, after participating in a "Milk Strike," called in 1933 in the shadow of the newly built City Hall by "mothers of the poor and unemployed

---

[17] Anastasia Loukaitou-Sideris and Gail Sansbury, "Lost Streets of Bunker Hill," *California History* 74, 1996: 146–62. I particularly want to thank Gail Sansbury for commenting on chapter 7 and letting me read some of her own writings on "reading/working" in downtown Los Angeles.

to make the government stop allowing surplus milk to be poured down the storm drains to keep the price up."[18]

> [Hay remembered] "This was a dramatic scene! Women were grabbing and shielding their children, and every so often you would see someone go down with a bloody head. The police were being absolutely brutal, without provocation. I think they wanted to incite a riot so they could clear the crowd."
>
> With his hands behind him, Harry backed up to the open door of a bookstore, where stacks of newspapers were held down by bricks. He found his hand resting on a brick. "I made no conscious decision, I just found myself heaving it and catching a policeman right in the temple. He slid off his horse and a hundred faces turned to me in amazement. No one was more amazed than I. Always before, I had been the one who threw the ball like a sissy. This 'bull' was my first bull's-eye ever!"
>
> Sympathizers murmuring in Yiddish, Portuguese, and English grabbed him. He heard, "We've got to hide this kid before the cops get him." Hands led him backward through a building connected to other buildings – a network of 1880s tenements that formed an interconnected *casbah* on the slopes of the sprawling old Bunker Hill quarter. He was pushed through the rooms that immigrant women and children rarely left, across catwalks and planks, up, up, hearing occasional reassurance, "Everything's fine. Just don't look down." Once out of the structure, near the top of the hill, he was hustled to a large Victorian house where he found himself standing, dizzy and disoriented, in a living room full of men drinking coffee. In the center, cutting a cake, was a soft-featured man in woman's attire. The man gestured theatrically when he spoke, and everyone addressed him as Clarabelle and referred to him as her. . . .
>
> Harry had heard of Clarabelle as one of the most powerful of the "Queen Mothers" who traditionally oversaw the temperamental comings and goings in the districts of town where they lived; Harry felt that such figures formed a regional network of salons among some pre-Stonewall gays. "Clarabelle controlled Bunker Hill and had at least a dozen 'lieutenants' covering stations, one called the Fruit Tank – that was our nickname for the jail cell for queers. . . ." She ordered Harry to be hidden in her

---

[18] Stuart Timmons, *The Trouble with Harry Hay: Founder of the Modern Gay Movement – A Biography*, Boston: Alyson Publications, 1990. This book was brought to the attention of Loukaitou-Sideris and Sansbury by Moira Kenney. See Kenney's far-ranging analysis of the spatial struggles of gay and lesbian activists, *Strategic Visibility: Gay and Lesbian Place-Claiming in Los Angeles, 1970–1994*, unpublished doctoral dissertation, Urban Planning, UCLA, 1994.

basement, and a lieutenant led him down five flights of stairs to a room storing Persian rugs, concealed him in a carpeted cave, and promised him coffee and a fried-egg sandwich in due time. (Timmons, 1990: 65–7)

In the 1940s and 1950s, the "blightedness" of Bunker Hill became a matter of widespread civic scorn and frenzied public campaigns, largely because it was in the way of downtown development. Led by the Community Redevelopment Agency (CRA), which began operations in 1949, and the *Los Angeles Times*, a public-private partnership was formed to declare Bunker Hill a blot on the landscape of the burgeoning modern city center, an unhealthy canker that must be removed before it spreads. And removed it was. By the end of the 1960s, all 396 buildings were destroyed, 11,000 residents were displaced, local resistance was defeated, the hill was flattened out, the maze of property ownership was packaged into large lots by the CRA and sold for a song to the superdevelopers. The cut-off hilltop was crowned by the Music Center, some expensive condominiums were added later with some cheaper apartments for the elderly (although not cheap enough for most of the elderly displaced), and a New Downtown began to grow rapidly further southward.

In the sheer slope down from Temple Street to about Eighth Street, between Hill Street and the Harbor Freeway, the New Downtown Central Business District erected its skyline of office buildings: Atlantic Richfield (Arco) Plaza, Bank of America Tower, Bank of California Headquarters, California Plaza, Citicorp Plaza, City National Bank, the Figueroa at Wilshire Building, First Interstate Bank Building, First Interstate World Center (the tallest building west of the Mississippi), Los Angeles Hilton Hotel and Tower, Los Angeles World Trade Center, Manulife Plaza, MCI Center, One Wilshire Building, the Pacific Stock Exchange, Seventh Market Place, Sheraton Grande Hotel, Union Bank Building, Unocal Center, Wells Fargo Center, 444 Building on Hope, 777 Tower on Figueroa, and, of course, in its full and proper name, the Westin-Bonaventure Hotel and Gallery.

As the CRA was funded by tax-increment financing, which meant that the rise in property values on land that they controlled could be tapped for their continued operations, the so-called Central Business District Project promised to be an almost unimaginable windfall. At one time in the 1970s, before the property tax revolt and the passage of Proposition 13, a bill temporarily passed in the City Council that would have potentially made the redevelopment of downtown Los Angeles under the CRA the largest public works project in human history, rivaling, in the scale of public resources invested, the building of the pyramids. This huge moneygusher was eventually turned off and a cap put on the use of tax increment funds, but the CRA still functions by milking cash from the cow of the earliest redevelopments on Bunker Hill. And they continue to promise that Angel's Flight will be reconstructed to commemorate ... what? I think they have forgotten.

Back in the exhibition, a series of placards attached to the future site of Disney Hall on the citadel wall traces the transformation of Bunker Hill in other ways. First there is the already noted vision of Raymond Chandler from *The High Window* (1942):

> Bunker Hill is old town, lost town, crook town.... In the tall rooms haggard landladies bicker with shifty tenants. On the wide cool front porches, reaching their cracked shadows into the sun, and staring at nothing, sit the old men with faces like lost battles.... Out of the apartment houses come women who should be young but have faces like stale beer, men with pulled-down hats and quick eyes that look the street over behind the cupped hand that shields the match flame; who look like nothing in particular and know it, and once in a while even men that actually go to work. But they come out early, when the wide cracked sidewalks are empty and still have dew on them.

Erasing this past, literally shearing it off, has been an avalanche of specifically cultural redevelopment rolling down Grand Avenue (née Charity Street), from the Music Center complex that first injected culture into the rectangular CITADEL-LA in the 1960s; to the vast mixed-use and still-being-completed California Plaza project that now contains the celebrated MOCA, skyscraping almost trompe l'oeil office towers, pock-ettes of good and bad (appropriately capitalized) Public Art, and the site of the future Disney-Gehry extravaganza of performance spaces.

The largest "performance plaza" imagined in the original plans was to span the last block of Olive Street before it ended in the Civic Center. A giant water fountain was to be designed by W.E.T., the company responsible for the water fantasies of both Disneyland and Walt Disney World. It would "dialogue" with the public, at times flooding the plaza with its watery tricks, at other times parching it to permit spectacular displays of local ethnic artistry. Today, it seems, such W.E.T. dreams have dried up, although they may rise again in the future.

The longer-term dreams of the Cultural Crown continue, however, as seen in a placarded extract from a popular pictorial cognitive map of downtown Los Angeles produced by the firm Unique Media Incorporated.

> The Music Center is the cultural crown of Southern California, reigning over orchestral music, vocal performance, opera, theatre and dance.... It tops Bunker Hill like a contemporary Acropolis, one which has dominated civil-cultural life since it was inaugurated in 1964.

Gehry designed Disney Hall to capture the axis of the old Music Center building across First Street and then to bend it directly towards MOCA, architect Arata Isozaki's understated, low-ridered, Hollywood bonsai.

A third placard reads:

> [MOCA is] a temple ... in which the gods cavort, amuse, and delight, even as they inspire.... [Its] ambience invokes the power and wonder of the Roman Pantheon, though Isozaki replaced cylinder and dome with cube and pyramid. The result is an equally uplifting space whose awesome splendor invites contemplation and joy ... Isozaki's design embodies the exquisite shape and proportion of Marilyn Monroe – classic, voluptuous, and sensuously draped to enhance and tantalize ... [just as] Frank Gehry's Temporary Contemporary proclaims the inimitable bone structure of Katherine Hepburn, magnificently lean and rugged yet indisputably regal, even in work clothes.... Two goddesses of a very different sort – yet both cast an aura that transcends mere physical presence and hovers somewhere closer to the realm of pure energy.[19]

The last one returns to the words of the architect, Arthur Erickson.

> California Plaza on Bunker Hill will become the *center* for all these other centers [of culture, government, commerce, ethnic life] ... Los Angeles [can thus] express what is unique about itself and at the same time begin to fulfill its future role as a center of centers of the western world.

Looking behind you, after squinting at the placards on the citadel wall, you can see again the original massing model and drawings of Gehry's Disney Hall, now being called "the Eiffel Tower of Los

[19] This fulsome quote from Sherry Geldin, MOCA's Associate Director at the time, is taken from an unpublished doctoral dissertation by the critical art historian, Jo-Anne Berelowitz, and from her "A New Jerusalem: Utopias, MOCA, and the Redevelopment of Downtown Los Angeles," *Strategies* 3 (1990), 202–26, a cutting critique of the utopian phallocentrism of LA's cultural mandarinate. I have not read the original Geldin article, "One Very Lucky Museum," *Individuals: A Selected History of Contemporary Art*, H. Singerman, ed., New York: Abbeville Press, 1987.

Angeles."[20] In subsequent deconstructive design readjustments, Gehry has soared in his unifying eclecticism to produce a vision of Los Angeles that simultaneously appears as a "galleon in full sail," "a pile of pottery shards from an archeological dig," "a stack of broken crockery," "the cracked open pieces of a fortune cookie," and, increasingly to my eyes, an artform abstractly referenced – intentionally or not, it is difficult to tell – to the cardboard boxes and other life paraphernalia of the homeless on Skid Row.

Looking up, a floating sky of chicken wire can be seen linking up the entire exhibition space. From it dangles a collection of volunteered symbols of Los Angeles 1989: surfboards, palm tree silhouettes, bumper stickers, simulated plates of sushi, movie cameras and clickboards, the Hollywood sign, traffic signals, insignia from past Olympics, signposts saying "Trespassers Will Be Shot," and other signatures on the contemporary landscape donated for commemoration by architecture and urban planning faculty and students. The cloud of symbols infects the space with a tensely hovering sense of ersatz unity that is at the same time temporary and contemporary, revealing and concealing, up-to-the-moment and hopelessly out-of-date.

Palimpsest

Another wall in the exhibition space has been created to carry the burden of history on its surface. Hidden behind it is an area that has yet to be explored, for the wall is actually a partitioning of the gallery constructed purposefully to conceal (like history itself). On its surface, facing what we have already seen, is painted a giant replica of the French tricolor, flag of the revolution, tall columns of blue, white, and red, on which a thickening black time line is drawn to move Los Angeles eventfully along a path defined by nine turning-point dates in the history and geography of Paris: 1789 – 1830 – 1848 – 1871 – 1889 – 1914 – 1940 – 1968 – 1989.

The historical display runs from right to left, synchronically and diachronically reconnecting Paris and Los Angeles via the particular story of El Pueblo de Nuestra Señora la Reina de Los Angeles, the UR-ban birthplace of the regional metropolis that has taken (part of) its name. At each of these crucial dates for Paris, little spatial histories are told in words and pictures for the specific site of El Pueblo. Let me try with hindsight to recapture these moments.

---

[20] See Diane Haithman, "$50-Million Cost Hike May Delay Disney Hall Opening," *Los Angeles Times*, August 23, 1994. The article describes the rising costs of construction for the Hall and the county-financed parking garage, especially after Gehry's redesigns and the earthquake that caused other rumblings earlier in the year. Costs are now estimated at around $260 million and may force postponement of the planned opening well beyond 1997.

<u>1789:</u> **Paris** The French Revolution, storming of the Bastille, abolition of feudal system, Declaration of the Rights of Man, nationalization of church property (also: first US Congress meets in New York, Washington inaugurated as President, Belgium declares independence from Austria).

El Pueblo Located somewhat to the north of the present-day site, El Pueblo is a successful agricultural settlement and trading post, with an appointed mayor (a Christianized Indian) and a powerful *Zanjero*, keeper of the irrigation ditch constructed to tap the Porciuncula (Los Angeles) River. Enclosed by a rough earthen wall are almost 30 adobe residences clustered around an open plaza. A small chapel had recently been constructed and the *rancho* system had begun, with soldiers being given grazing permits for the lands outside the wall.

The first Spanish explorers had come 20 years earlier, to be greeted by the local inhabitants of Yang-Na village (near present-day City Hall) and by several small earthquakes, one of which Father Crespi described as "half as long as an ave maria." In 1781, 44 settlers arrived, more than half with some African blood: 11 men, 11 women, and 22 children, mainly from Sinaloa and Sonoma. Of the men, 2 were white Spaniards; the others were Indian (4), mestizo (1), African (2), and mulatto (2). All the women were either mulatto or Indian. By 1789, the population had reached nearly 140. The pueblo here would be destroyed by a flood in 1815, relocated, destroyed again by flood in 1824–5, and relocated again to the present-day site.

<u>1830:</u> **Paris** Revolution again on the streets of Paris, Charles X abdicates as king of France, Louis Philippe – the Citizen King – becomes King of the French, France captures Algeria, the anarchist geographer Elisée Reclus is born (also: social unrest in Brussels and Warsaw, Ecuador secedes from Gran Colombia, religious society of Mormons established by Joseph Smith).

El Pueblo Excluding Indians, the population reaches about 770. In the period since 1789, the Indian population of California drops from 300,000 to 150,000. Locally, a major venereal disease epidemic decimated the Gabrilieno in the period around 1805, the date of the first American visitor to El Pueblo. In 1830, the era of the *Californios*, resolutely Mexican Los Angeles, is in its full glory, dominated by those with such family names as Pico, Carillo, Lugo, and Sepulveda. The settlement is the center of sprawling ranchos and a major springboard for the secularization of the missions, such as those at San Gabriel, established in 1771, and San Fernando, built in 1797. Secularization had many advantages. It opened new land for cattle ranching and provided a growing supply of cheap, quasi-slave labor – as long as the supply of Indians would last.

Twelve years after the French Revolution-inspired Mexican

Revolution of 1810, all of California had become part of the Republic of Mexico. In the same year (1822) a new Catholic Church was built next to what would become the central Plaza a few years later. The oldest building still standing today is the Avila Adobe, built in 1818. El Pueblo and the Mexican-mulatto-mestizo Californios were thriving in 1830. The pueblo would be officially declared a city (*ciudad*) five years later by the Mexican Congress, and in the census of 1836, the city was shown to have grown to 2,230: 605 men, 421 women, 651 children, 553 Indians. Of this total, 50 were described as "foreigners."

1848: Paris Another revolutionary period, worker uprisings, Louis Philippe abdicates, Louis Napoleon elected President of the French Republic, *The Communist Manifesto* of Marx and Engels circulates on the streets, as do the more "positivist" ideas of Auguste Comte (also: revolutions and uprisings in Vienna, Venice, Berlin, Milan, Parma, Prague, Rome, Budapest; the Treaty of Guadalupe Hidalgo ends US–Mexico War, ceding California to the US, along with present-day Texas, New Mexico, Utah, Nevada, Arizona, and parts of Colorado and Wyoming).

El Pueblo For the first time, El Pueblo resonates with the radical changes taking place in Paris. Los Angeles had been made the capital of Mexican California three years earlier and ever since the Californios had become embroiled in major power struggles – among themselves, with Mexican thugs (*cholos*) and bandits, with Yankee looters and thieves, and with the US army. The cattle ranches boomed in 1848, supplying beef to the adventurers in the north drawn by the discovery of gold at Sutter's Fort early in the year. As El Pueblo churned with invasion, rebellion, and new commercial wealth, the era of the Californios had clearly reached a crucial turning-point.

The US declared war on Mexico in 1846 and resistance fighting immediately began in and around El Pueblo. After some significant initial victories, the Californios were defeated in the "Battle of Los Angeles" in early 1847 and Commodore Stockton led his marching troops triumphantly down present-day Alameda Street, Aliso Street, and Los Angeles Street directly into La Plaza, the heart of the Californio citadel. Stockton occupied the Avila Adobe and made it his war headquarters, while "Military Commandant" John C. Fremont and his freebooting troops settled into the area near present-day Universal Studios to negotiate the Treaty of Cahuenga, which effectively ended the war in Los Angeles.

By the February 1848 signing of the Treaty of Guadalupe Hidalgo, El Pueblo had already begun to adjust to the Yankee conquest. The Californios were promised equal rights with US citizens; Andres

Pico, his brother, former Governor Pio Pico, and most of the other leaders of the Californio rebellion were pardoned; and for this and the next few years what would become Los Angeles County experienced only modest changes, especially when compared with the flood of 40,000 Yankees that turned the "miserable village" of Yerba Buena into the new metropolis of San Francisco to the north. By 1850, the new county counted a total population of 3,530, many of them "domesticated Indians." The city of Los Angeles, with a little over 1,600 inhabitants, was predominantly Californio and would remain so for the next two decades.

1871: **Paris** The Paris Commune, two months of popular rule in the citadel after France was defeated in the Franco-Prussian War and William I, King of Prussia was declared German Emperor in Versailles; two French workers compose "L'Internationale" (also: labor unions legalized in Britain, Carroll publishes *Through the Looking Glass*, Darwin's *The Descent of Man* appears, Stanley meets Livingstone at Ujiji).

El Pueblo Los Angeles enters the world's consciousness for the first time in the Chinese Massacre. In October, a vigilante band of Yankees (with a few Mexicans and several police officers) go on a vicious racial rampage in El Pueblo, killing at least 20 Chinese (out of a total resident population of 200) and looting Chinese stores and homes in response to the accidental death of an Anglo man making a citizen's arrest of someone he assumed was involved in a tong feud along the notorious *Calle de los Negros* (Nigger Alley).

The event culminated two decades of extraordinary violence, lawlessness, and interracial conflict in "Hell Town," as Los Angeles was called during the period following the American conquest. During most of these years, a murder occurred every day on the average, as Mexican bandits, revolutionaries, and cholos, Yankee hustlers, cattle thieves, and desperadoes, and a few remaining Indian rebels fought against lynch mobs, hispanophobic vigilante groups, the armed Los Angeles "Rangers," and the official American government. The Wild West flavor of these times would be remembered forever. That it was also a period of such intense "Anglo-Americanization" that it could be labeled "ethnic cleansing" has been intentionally forgotten.

The Chinese Massacre was an international embarrassment. New public and private coalitions were formed in 1871 to erase the images of racism and violence, and to imaginatively "rebuild LA" along different lines. A railroad had already been constructed from the city center to the port of San Pedro; a small real estate boom had begun the subdivision of the rancho lands and consequent urban sprawl; and in a few years a rail link would be completed between Los Angeles and San Francisco, opening up the region to mass immigration and rapid urban growth. For the historic El Pueblo, 1871 was a major turning-point, the beginning of a long period of decay and neglect, as the American city turned its back on the old Plaza and its embarrassing memories to construct a new center on its southern flank. Ironically, Paris too turns away from the sites and sights of the Paris Commune to rebuild the city in very different directions.

1889: **Paris** Celebrations of the centennial of the French Revolution, the Eiffel Tower (the world's tallest structure) is built for the Paris World Exhibition, May Day celebrated for the first time, *la belle époque* is in its fullest glory (also: the Austrian Crown Prince commits suicide at Mayerling, the Dakotas, Montana, and Washington become states, London dock strike, Brazil declared a republic, Hitler and Heidegger born).

El Pueblo The City population reaches nearly 50,000, but the *belle époque* of El Pueblo is clearly over. None the less, the Plaza remains an important gathering-point and political rostrum for Mexican migrant workers, political exiles, and revolutionaries, as well as for Mexican and American labor organizers, who had successfully held the first strike in Los Angeles in 1883, against Colonel Otis and the *Times* (which began publishing two years earlier during the centennial of the founding of El Pueblo). A new Chinatown, housing the city's main vegetable growers and sellers, is taking shape in the Garnier Block near the Plaza; and the first well-established Black neighborhood has developed along Los Angeles Street (a southern extension of the former *Calle de los Negros*), with Biddy Mason's Place close by on Spring.

The earlier era of the missions and the "domestication" of Indian life is experiencing a romantic revival and reconstruction after the enormous popular success of Helen Hunt Jackson's idyllic *Ramona* (published in 1884). A new urban mythology for Los Angeles is in full bloom, sifting through the past to erase discomforting racial memories and to recreate Los Angeles as the first truly (white) American city, freed from the ties to both Californio and European urban traditions. In 1889, while the French were celebrating their revolution's centennial and a new county of Orange was carved out in Southern California, Mexican Los Angeles was historically recreated as a reliquary museum, with the encasing of El Pueblo as the city's first "historical park" by the newly established Parks and Recreation (sic!) Department.

1914: **Paris** The beginning of World War I, the pacifist socialist Jean Jaurès murdered, British troops land in France, Germans occupy Lille and Rheims (also: Panama Canal opened, more than 10 million immigrants enter the US from southern and eastern Europe over the preceding decade, US Marines occupy Vera Cruz, forcing Mexican President to resign).

El Pueblo   The population of booming Los Angeles is fast approaching half a million and it will soon surpass San Francisco as California's largest city. The world's largest urban mass transit system connects the new downtown, already gridlocked with automobiles, to a sea of sprawling suburbs occupied primarily by Iowans, Nebraskans, and other "middle-class, middle-brow, mid-westerners," as the Anglo-Americanization of Los Angeles is nearly complete. The total population of Mexicans, Blacks and Asians (now mainly Japanese) is probably close to the smallest proportion it has ever been – or will ever be – in the whitened Protestant city and county of Los Angeles.

The sprawl is built in fields of citrus, grapevines, and vegetables (with L.A. County now the richest agricultural area in the country); on scattered oil deposits (with the county's "black gold" suburbs also leading the country in petroleum production); to serve the growing motion picture industry (Hollywood is annexed in 1910 and Charlie Chaplin appears in his first Mack Sennett comedy in 1914); and to enjoy the beach (although Abbott Kinney's dream of creating an Italianate city of culture at Venice-by-the-Pacific had died by 1914 and most of its canals would soon afterward be filled in). The Los Angeles Aqueduct had been opened the year before to William Mulholland's stirring words: "There it is! Take it!" and a budding aircraft industry would soon boom during the war years.

Back in El Pueblo, there were other stirrings. After public oratory had been banned throughout Los Angeles several years earlier, two rostrums "for free speech" are built in 1914, grudgingly acknowledging and commemorating the Plaza's significant role as a political rallying-point for more than a century. In many ways, this marks the end of the Progressive Era in Los Angeles, a decade and a half in which the Plaza and the area around it had been one of the most lively centers of radical democratic activity in North America. The new CITADEL-LA just to the south played a key role in the governmental reform movements that brought California the initiative, the referendum, and the recall. Using the Plaza as its meeting place and revolutionary staging ground, the exiled Partido Liberal Mexicano (PLM) published the newspaper *Regeneracion* and waged its struggle against the Mexican dictator, Porfirio Diaz. Two of its leaders, the Magnon brothers, were tried and convicted of breaking US neutrality laws in the huge Court House nearby, as thousands of Mexicans flocked to the trial from the crowded Sonoratown and Chavez Ravine barrios to the north of El Pueblo. And then there was the labor movement.

In addition to the 1910–12 trial of the Magnon brothers (who were helped by the International Workers of the World and the local Socialist Party), another major trial received widespread attention during the same period. It involved the alleged "anarchic unionists" who, it was claimed, were responsible for the explosion of the downtown offices of the *Los Angeles Times*, which killed 20 people. Milling around the Court House during the 1911 trial, called by labor leaders the new "Battle for Los Angeles," were such national luminaries as Clarence Darrow (who agreed to defend the accused), Lincoln Steffens, Louis Brandeis, and Samuel Gompers.

At issue was more than the guilt or innocence of the accused McNamara brothers. For the local Socialist Party and its national supporters, at a time when the party seemed almost sure of winning the next mayoral election, it was the culmination of a long struggle against the "open shop" and particularly against its most powerful supporters, General Otis and Harry Chandler of the *Times* newspaper empire and the equally powerful Merchants and Manufacturers Association. The political battle was led by the local labor leader Job Harriman, who was also the Socialist Party candidate for mayor. It was felt by many that the entire socialist movement in America would hinge around what happened in Los Angeles in 1911. What happened was the utter defeat of the Socialist Party and the struggle against the open shop.

By 1914 and the onset of the war, Los Angeles had begun fading into the periphery of American struggles for radical democracy. Job Harriman moved in 1914 into the high desert well to the north of the CITADEL-LA to help found a small socialist community called Llano del Rio, later to be called by Mike Davis in his *City of Quartz*, "Open Shop Los Angeles's utopian antipode." But on the horizon was a new centering on Los Angeles, the beginning of the creation here of the world's most powerful war machine. It would trigger the greatest industrial and population boom of any city in the 20th century.

1940: **Paris** World War II, Germans enter Paris in June, Pétain signs armistice, de Gaulle leads Free French resistance, Lascaux caves are discovered (also: Trotsky assassinated in Mexico, Richard Wright's *Native Son* and Raymond Chandler's *Farewell, My Lovely* are published, US population reaches 132 million after slowest decade of growth in the history of the census).

El Pueblo The first freeway opens, connecting downtown to Pasadena. The census reports a City population of more than 1.5 million and a County population of 2.8 million, after Los Angeles countercyclically expands during the Great Depression. And El Pueblo takes on a radically different look.

A filthy alley running with sewage in the late 1920s, Olvera Street has been "theme-parked" as a "picturesque Mexican market place" and tourist attraction, a product of the work, a decade earlier, of Christine Sterling, Plaza de Los Angeles, Inc., the *Times* (of course) and a gang of prison laborers provided by the city. These public-private Anglofications recreate El Pueblo as an Americanized monument to the founding of Los Angeles. The principle of historic preservation is established and at least the shells of many of the old buildings are en route to being preserved: Avila Adobe (built in 1818), Pelanconi House (1855), Sepulveda House (1887), Masonic Temple (1858), Merced Theatre (1870), Pico House (1869), the Old Firehouse (1884), the Garnier Block (1890). The opening of Union Station in 1939, however, forces the relocation of Chinatown to a new site just north of El Pueblo.

The only building continuing to function in its original capacity is the Plaza Church–La Placita, Church of Our Lady the Queen of Angels. It will continue to do so up to the present. The free speech rostrums seem to have disappeared, but the Plaza still is an important gathering place for Mexicans trying, against all odds, to maintain their cultural heritage. This makes the Plaza a prime target for hispanophobia, such as the "repatriation movement" that flourished during the Great Depression and aimed at deporting everyone of Mexican ancestry. Another round of nativist hysteria and violence would occur during the war years against the *pachuco* gangs, leading up to the Zoot Suit riots of 1943. And in 1941, Executive Order 9066 begins the relocation of Japanese-Americans to concentration camps and the legal looting of Japanese assets, especially in the neighborhoods adjacent to El Pueblo, a more sophisticated and nationally supported version of the Chinese Massacre of 1871. During the war, a USO canteen is established near the Plaza and becomes a haven for US servicemen, who play a key role in the racial rioting bubbling out from El Pueblo.

1968: Paris The streets and universities explode in student riots and demonstrations aimed at attaining *le droit à la ville* – the right to the city, unrest brings back memories of the Paris Commune, Lefebvre and the situationists achieve widespread popularity, de Gaulle demands vote of confidence (also: the invasion of Czechoslovakia, the assassinations of Martin Luther King and, in Los Angeles, Robert Kennedy, riots in Chicago, Mexico City hosts Olympic Games, Nixon elected President).

El Pueblo The streets of Los Angeles have already exploded in the most violent urban uprising in 20th-century American history. The Watts rebellion and riots of 1965 result in 35 deaths, 4,000 arrests, $40 million in property damage, and the renaming of Central Avenue south of downtown "Charcoal Alley." By 1968, the Black population of Los Angeles had grown to more than 600,000, eight times the 75,000 who were living here in 1940, a product of one of the largest city-focused migrations of a single racial-ethnic group ever to occur anywhere. In 1970, Los Angeles would explode again in the anti-Vietnam War Chicano Moratorium, the biggest Latino demonstration in US history. Los Angeles was deeply unsettled in 1968, as were most American cities.

By this time, El Pueblo had become a State Historic Park. The Music Center complex had just been completed across the 101 Freeway, which had effectively cut off the Plaza from the new Civic Center. Boxed in by the freeway on the south, Union Station on the east, Chinatown on the north, and Fort Moore Hill on the west, the shrunken El Pueblo had become reduced to the Olvera Street tourist strip, a cluster of old buildings being slowly restored, a Plaza used primarily for tourist services and occasional Mexican-American celebrations (such as Cinco de Mayo and the blessing of the animals at Easter), and the still functioning La Placita church. The heterotopological qualities of the site, however, maintained its sustaining significance as the symbolic center of Latino life and culture in Los Angeles, almost entirely unseen as such by the Anglo majority.

1989:   The time line runs out into the present, with an arrow pointing to the citadel wall and a faint picture of the old Avila Adobe floating above and almost out of sight. The year of the exhibition is an active one, however, as control of El Pueblo Historic Park shifts from a complex city, county, and state administration to the newly renamed (putting *recreation* first) Recreation and Parks Department of the City of Los Angeles, celebrating its own centennial during this French bicentennial year. As reported in the local press, performances are staged in the Plaza by the Asian-American Ballet, the Xipe Totec Aztec Indian dancers, and Somebody Special, Inc., "a drill team of teen-age girls who boogie to Motown sounds." And a time capsule is buried under Olvera Street containing representative ethnic arts and crafts, feathers from the endangered California condor, a preserved grunion, and park-sponsored T-shirts.

In the years following these celebrations, El Pueblo has entered still another round of possible deconstruction and reconstitution. For some, it remains in the way of downtown expansion, an anomalous and anachronistic piece of eminently developable land in the heart of a now "world class" city. For others, it is the signifying site of the contemporary region's spectacular multiculturalism. For many, it is the irrevocable heart and soul of Mexican-Chicano/Chicana Los Angeles or, at least, an important Latino commercial and cultural stronghold worth preserving and enhancing. This renewed struggle

Pueblo--People--Polis

... 1789 - 1830 - 1848 - 1871 - 1889 - 1914 - 1940 - 1968 - 1989 ...

Power--Police--Palimpsest

over time and space will certainly be worth watching in the future.

Looking back, El Pueblo has been the primordial PALIMPSEST of the City of Angels, prepared from its origins to be written upon and erased over and over again in the evolution of urban consciousness and civic imagination. In Foucauldian terms, it has also become a "new kind of temporal heterotopia," one which combines the fleeting time of the festival site or vacation spot with the indefinitely accumulating time of the museum or library. As such, like its disneyfied descendants, it simultaneously serves to abolish histories and cultures and to discover them anew in "other spaces."

Panopticon

Our final stop in the exhibition need not take long, for its rudiments have already been sighted, sited, and cited. It represents an attempt to evoke in a more direct way the *carceral city* that underlies all urban histories and geographies, that everywhere projects the citadel's powers of surveillance and adherence while reflecting also the powers of resistance and defiance. We enter this space through a narrow passageway at the end of the time line wall as it perpendicularly approaches the eastern end of the plotted and mapped CITADEL-LA. At this juncture, the wall map extends inward from roughly the site of El Pueblo to display a sector of enclosures opening eastward from the citadel. Contained here is the largest urban prison population in North America in sites which violently intrude on the unbarred *barrio* of East Los Angeles, where another form of spatial enclosure in practiced.

A takeaway fact sheet provides you with some information: 18,000 inmates fill four county jails, including Men's Central and Sybil Brand, the nation's largest women's prison. In 1988, another 23,000 "non-threatening" prisoners were released early to free up space, provoking then Governor Deukmejian, in his determination to build still more prisons in this downtown wedge, to proclaim that criminals are Los Angeles's "principal export." A newspaper headline tells of neo-Nazi and Ku Klux Klan terrorism within Men's Central aimed at intimidating Black deputies and inmates alike with cross-burnings, gang-baiting taunts, and Nazi memorabilia. The message is clear: one is never free from the legacy of racism in Los Angeles, even when behind bars.

By now, you have entered a dark, cell-like room lit by a bare red bulb. The chicken wire cloud that floats Gehryshly over the rest of the gallery spills into the enclosure and tightly meshes an interior wall (the backside of the time line). On all sides are ominous photographs of the newest addition to the carceral city of Los Angeles, the Metropolitan Detention Center, a federal "administrative" facility squeezed into the eastern edge of the CITADEL-LA near the old Federal Building and the US Courthouse, just across the freeway from El Pueblo.

At about this time, a contest was held to design a symbolic "gateway" extravaganza for Los Angeles to commemorate its surpassing role as an immigration entrepot: the Statue of Liberty and Ellis Island rolled into one. The winning plan proposed a "steel cloud" grill-work surmounting the freeway at this very spot, filled with restaurants, museums, gift shops, and aquaria (with live sharks) floating above the traffic below and between the prison and El Pueblo. It remains unbuilt. Across the way to the south of the federal prison sits a more complementary landmark: the LAPD headquarters in Parker Center.

Dominating the interior space at the gallery is a giant, eerily beautiful picture of the east face of the Federal Detention Center photographed at night by Robert Morrow. The same shot would later appear on the cover of Mike Davis's *City of Quartz* (1990). (Mike also helped with the exhibition.) The evening shades of this Bastille-like frontage are broken only by a bright red light staring from atop the roof (and mimicked by the bare bulb hanging in the room). This jewel of a jail looks like a modernized castle keep, with slitted openings that allow the inmates the mock freedom to reach one arm through and wave and shout at passersby. The only visible bars are hung on the outside of the building in a perfect dissimulation of the carceral enclosure, feigning not to have what it most certainly has.

In my first visit to the actual site, I waved back to the incarcerated prisoners, after entering a fern-filled atrium just to the right of the main entrance. Out of nowhere, armed guards appeared to shoo me away from such open communication with the one-armed inhabitants. The building, it seems, is not meant to be seen. Despite its location and its involvement in such sensitive local issues as immigration control, insider dealing, and financial fraud (I think Michael Milken, of junk-bond fame, spent some time here), the Metropolitan Detention Center was built with remarkably little local awareness. Even today, few law-abiding downtown workers or visitors know of its existence.

A more recent visit to the new federal complex that has grown between the old Federal Building and the Metropolitan Detention Center introduces another dimension to our journey in the CITADEL-LA. It is well worth a little contemporary detour. To guide you, I extract from a photographic essay by Dora Epstein, my Research Associate in the Department of Urban Planning at UCLA, prepared in early 1994 after her first visit to the peculiar sites of Roybal Plaza.

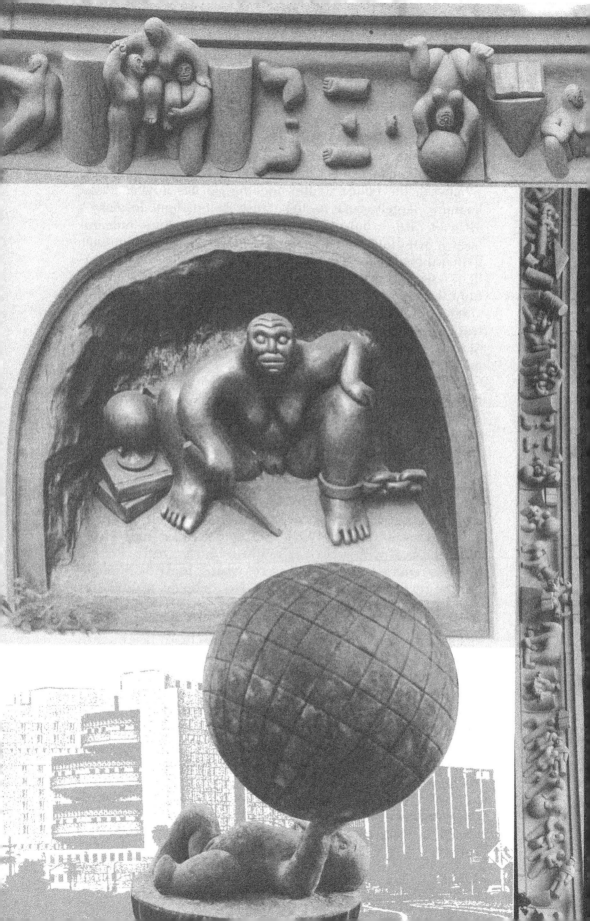

I needed directions to the Roybal Building, so I called a building maintenance manager listed in the phonebook. She told me
to park in an empty lot on Temple Street, and walk over to the
plaza. She told me to keep walking past the police station, past
the "drive-by," and toward the jail [the Metropolitan Detention
Center]. I thought, is this LA, or is this war?

Between the police station and the jail lies Roybal Plaza.
Looming largely next to the street is Jonathon Borofsky's
Molecule Men, a burnished steel sculpture of four interconnected two-dimensional male (?) figures. Throughout their
flattened bodies is a Swiss cheese of holes, so that the LA sunlight reflects harshly off their substance, while passing peace-
fully through their lack. They stand about 30 feet high, and
although it may represent the particles of our being, the violence of the image is clear. They look as if they've been shot.
Molecule Men was, indeed, the "drive-by" the manager had
mentioned.

I looked at the *Molecule Men* from one angle, and I saw
through their bodies the street where police car after police car
cruised past. I looked again from another angle, and this time it
was the jail, concrete and fortress-like. For a second, I was
caught, feeling as if I was the one who had done the shooting. I
was presumed guilty, trapped by an environment of surveillance, discipline, and punish.

I ascended the plaza stairs to the courtyard. Rimming the
courtyard, just at the boundary of the jail, is the pleasant neo-
Classical frieze, *The New World*, by Tom Otterness. In the shape
of a semi-oval with a circular fountain in the center, *The New
World* is reminiscent of a Roman colonnade, imitated throughout the centuries in stately gardens and monuments. Like its
predecessors, *The New World* tells its "story" in the elaborate
cornices above the columns. On opposite ends of the frieze,
small dough-like figures "help" each other "climb" the columns
from the level of the plaza to the cornice. The women are on one
side of the frieze, the men on the other; all are laden with geo-
metric objects. Beginning at the outermost cornice from the center of the frieze, the figures seem to march towards the center.
The women come from the sea, and carry globes and books as
they march. The men elect a king, and escort him in a car. At
the center, chaos reigns, as the king is overthrown. The "palace"
is toppled, the car is overturned, and the king is taken hostage,
shackled, crushed by an elephant, and finally, dismembered.
[ES: The king's remains can be seen floating separately near the
center of the frieze: a crown, two arms, two legs, a heart, and a
penis.]

In the middle column, at viewer's eye level, an angry bronze female squats in her cave. Knife in hand but shackled as well, she viciously guards "the new world" in the fountain of mist before her. In the middle of the fountain is a dazed nude baby in bronze lying on her back, staring up at the globe in her hand.

When Otterness' *The New World* was installed in the Roybal Plaza, it came as part and parcel to the unveiling of the slick new federal building and jail. Obviously, it had passed the GSA (federal Government Services Administration) stipulation that public art should "reflect the dignity, enterprise, vigor, and stability of the American National Government." In fact, it had been unanimously approved by the GSA Art-in-Architecture panel. However, when a judge from the federal court building next door declared that the works were, due to the two nude bronze figures, obscene, the GSA quickly removed the statues from the plaza. Why any of the hundreds of nude figures in the frieze were not considered obscene escapes me, but what seemed to bother this judge, and later Roybal himself [Roybal was the first Mexican American from Los Angeles to serve in the US Congress], was the genitalia of the baby. I imagine that these two men were, like many people, torn between the titillation that women's genitalia often cause, and the revulsion that they felt by being titillated. Certainly, female genitalia, unlike symbolic and real representations of male genitalia, are unusual, if not hidden, to the Western gaze. To see them then so prominently "displayed" caused many of these men to match the rise in their pants with a rise in their anger. After all, if they felt so base, surely a "real" pervert would threaten to deface their beautiful monument to "democracy." The two bronze nudes were removed until a compromise was reached to partly obscure the figures – the woman would squat behind a rosebush, the mist around the baby would be raised.

What was missed, in this brief attention, was the fact that the artwork, in its entirety, was designed to obscure something much more obscene: the boundary of the jail. If one looks carefully between the columns, razor-like barbed wire can be seen. If one steps away from the frieze and dares look up, the concrete mask of the jail merges in the eye with the whiteness of the columns. There is a purposeful architectural fit between the two disparate objects. We accept the pleasance of the colonnade, because it allows us to disregard its imposing mate.

As spectators and employees on lunch go gawking by, they take delight in the whimsy of Otterness' work. The dough-like figures are playful, rounded, human Ewoks, like the

Pillsbury dough boy or roly-poly Gumbys. Except for the bronzes (and this may be another reason why the bronzes were rejected), they smile in a goofy "Have a nice day!" way. Often described in newspapers as a battle between the sexes, I cannot help but feel that the whimsy of the figures made the media miss the point. The "sexes" do not battle. They merely come from opposite sides of the frieze. Although it may seem that women have "won" out in the end (the bronze figures again; maybe that also offended the eminent judges), both sexes seem to actively topple the king. *The New World*, then, is not about a gender-driven madness, but rather a happy revolution. The king's demise is the center of the story, the new generation after the king its culmination. Thus, while we are goaded into the delight of the figures and the dismissal of the site, we are also being called upon to question the authority, to sum up the power and the resources to come together to create a new world – between the police station, the courts, and the jails....

I remembered the [second, federally charged] Rodney King beating trial and the news conferences which had taken place in that plaza, but I could not recall ever seeing the frieze before. I asked a security officer at the metal detector inside the Roybal Building where the news conferences had taken place. He told me that they had moved to the left of the fountain, on the steps. Why? The mist arising from the fountain kept getting on the camera lenses.

Back in the room, another placard is spotlighted and written in bold-faced lettering. It is adapted from Gwendolyn Wright and Paul Rabinow, "Spatialization of Power: A Discussion of the Work of Michel Foucault," published in *Skyline* (1982). It helps us understand where we are.

For Foucault, SPACE is where the discourses about POWER and KNOWLEDGE are transformed into actual relations of power. Here, the knowledge in the forefront is that of aesthetics, of an architectural profession, of a science of planning. But these "disciplines" never constitute an isolated field. They are of interest only when one looks to see how they mesh with economics, politics, or institutions. Then both architecture and urban planning offer privileged instances for understanding how power operates.

The PANOPTICON, Jeremy Bentham's proposal for radially planned institutional buildings, is by now the most famous instance of a concretization of power applied through

architecture. Foucault came upon Bentham's 1787 plan while studying reforms in eighteenth-century hospital and prison architecture, and took it as the paradigmatic example of the interworkings of SPACE, POWER and KNOWLEDGE in disciplinary society. The PANOPTICON is the "diagram of a mechanism reduced to its ideal form," a combination of abstract schematization and very concrete applications.

Think of the bold-faced implications. Amidst its multiplicity of expressions, every city is to some degree a panopticon, a collection of surveillant nodes designed to impose and maintain a particular model of conduct and disciplined adherence on its inhabitants. Those who do not, or actively threaten not to, adhere are in one way or another incarcerated in specialized enclosures, be they asylums and prisons or ghettoes and Skid Rows. This agglomeration of the spatial means of social control, the geographical centers of power, lies at the very origins of urbanization, at the generative source both of the most ancient hydraulic civilizations as well as the institutional regulation that made possible the proletarian cities of industrial capitalism. The spatial concentration of power for social production and reproduction was what the first cities were primarily for. It has remained central to urban life ever since, even as new functions were added. Such commanding geographical presence differentiates the urban from the rural, adherents from the not yet adherent, *polites* (citizens) from *idiotes* (hicks, rubes, the uncivilized).

These origins should never be forgotten. But this does not mean that the carceral city operates simply and directly along the extended visual lines of Bentham's original Panopticon or that its disciplinary technologies and heterotopologies remain constant over time. Too much happens in the city for this to be true. But then again, the major transformations of modern society, especially the growing role of the centralized state in maintaining surveillance and adherence via national citizenship and institutionalized identity politics, should not obscure the continuing power of the city and citadel.

City and State participate together in the more invisible processes of "normalization" that pervade and sustain patriotic allegiance and representative (as opposed to participatory) democracy. The power of the citadel is not just expressed in face-to-face sociality but also in the more abstract, less visible, psychogeographical realm of the authoritative social "system." I emphasize this point because modern political theory and critical social science too quickly abandoned their roots in the polis and in what might be called the specificities of the urban. In the modernist mainstreams, the city was virtually discarded as a central object of knowledge and theory-building for such abstract (and usually de-spatialized) notions as the role of the state, social relations of production and class struggle, nationalism, market competition, the social system, the philosophy of history. Discourses on the sociality and historicality of human life continued to acknowledge that social and political processes took place *in* cities, but this was seen as an unavoidable coincidence rather than a consequential relation or determination. There was relatively little that mattered, at least in the new age of 20th-century-modernity, that was specifically *of* cities, directly imbricated in the real or imagined realms of the urban. For at least the past century, the urban has become epiphenomenal to mainstream theorization of the constitution of modern societies and the making of modern histories.

Today, with the reassertion of spatiality in critical thought and theory, the specificity of the urban is receiving greater attention than it has been given for a century and a half. A new field of *critical urban studies* is taking shape around the trialectical nexus of space, knowledge, and power and the interpretation of urban sites and spaces as simultaneously perceived, conceived, and lived. More is being discovered in the urban than even the staunchest urbanists of the past ever dreamed was there.

Back to the Beginning

After this long tour of the CITADEL-LA seen through the "eye of power," we are back to the beginning, to the postmodern reassertion of space (and the city) in critical social theory, to the spatial critique of historicism and despatialized social science, to the pathbreaking voyages of Lefebvre, Foucault, and hooks, to the cultural critiques of the silencing of gender and race in modernist epistemologies, to a new ontology of spatiality, to the radical openness of Thirdspace, and to new kinds of remembrances.

8

Inside Exopolis: Everyday Life in the Postmodern World

The postmodern confusion of time and space, in which temporal continuity collapses into extension and spatial dimension is lost to duplication, transforms urban culture into a gigantic hologram capable of producing any image within an apparent void. In this process, time and space are transformed into icons of themselves and consequently rendered into scenarios.

(Celeste Olalquiaga in *Megalopolis*, 1992, 19)

It's a theme park – a seven hundred and eighty-six square mile theme park – and the theme is "you can have anything you want."

It's the most California-looking of all the Californias: the most like the movies, the most like the stories, the most like the dream.

Orange County is Tomorrowland and Frontierland, merged and inseparable. 18th century mission. 1930s art colony. 1980s corporate headquarters.

There's history everywhere: navigators, conquistadors, padres, rancheros, prospectors, wildcatters. But there's so much Now, the Then is hard to find. The houses are new. The cars are new. The stores, the streets, the schools, the city halls – even the land and the ocean themselves look new.

The temperature today will be in the low 80's. There's a slight offshore breeze. Another just-like-yesterday day in paradise.

Come to Orange County. It's no place like home.

(From an advertisement for *The Californias*
California Office of Tourism)

"Toto, I've Got a Feeling We're Not in Kansas Any More"

As the spin-doctoring wizards of the California Office of Tourism proclaim, you can have *anything* you want in iconic Orange County, where every day seems just like yesterday but where the ever-present *Now*-ness of tomorrowland makes the *Then* hard to find; where every place is off-center, breathlessly on the edge, but always right in the middle of things, smack on the frontier, nowhere yet now/here like home. To its avid promoters, Orange County is a theme park-themed paradise where the American Dream is repetitively renewed and made infinitely available: as much like the movies as real-reel life can get. This resplendent bazaar of repackaged times and spaces allows all that is contemporary (including histories and geographies) to be encountered and consumed with an almost Edenic simultaneity.

Orange County represents itself as a foretaste of the future, a genuine phenomenological *recreation* of everyday life in a brilliantly recombinant postmodern world, beyond Oz, beyond even the utopic late-modernisms of Disney. The most "California-looking" of all the different Californias, Orange County leads the way in the very contemporary competition to identify the Happiest Place on Earth. If any other place is still in the running, it is purely through faithful simulations of the original. And every day more simulations of Orange County are springing up – around Boston, New York, San Francisco, Chicago, Washington, Dallas-Fort Worth, Miami, Atlanta – propelling what is fast becoming the most spectacular transformation of urban landscapes, and of the language we use to describe them, since the industrial city first took shape in the 19th century.

It is almost as if the urban is being reinvented in places like Orange County to celebrate the end of the millennium. And *you* are *there* whether you like it or not, looking at the coming attractions being screened well beyond modernity's urban fringe. Suspend disbelief for the moment, control the cynical sneers, and enjoy the strangely familiar ride into paradigmatic postmodernity. It will soon be coming to your neighborhood.

Introducing Exopolis

Some have called these amorphous implosions of archaic suburbia "Outer Cities" or "Edge Cities"; others dub them "Technopoles," "Technoburbs," "Silicon Landscapes," "Postsuburbia," "Metroplex." I will name them, collectively, *Exopolis*, the city without, to stress their oxymoronic ambiguity, their city-full non-cityness. These are not only exo-cities, orbiting outside; they are ex-cities as well, no longer what the city used to be. Ex-centrically perched beyond the vortex of the old agglomerative nodes, the Exopolis spins new whorls of its own, turning the city inside-out and outside-in at the same time, unraveling in its paths the memories of more familiar urban fabrics, even where such older fabrics never existed in the first place.

The metropolitan forms we have become used to – with dominating downtowns and elevated Central Business Districts, definable zones of residential and other land uses concentrically and sectorally radiating outward from the tightly packed inner city to sleepy dormitory suburbs, population density gradients neatly declining from core to periphery – are now seemingly in the midst of a profound socio-spatial deconstruction and reconstitution. The new metropolis is exploding and coalescing in multitudes of EPCOTs (experimental prototypical communities of tomorrow), in improbable cities where centrality is virtually ubiquitous and the solid familiarity of what we once knew as urban melts into air.

We're certainly not in Kansas any more, but neither are we in old New York or Chicago – or even modern Los Angeles, the centrifugal ur-Exopolis that is now being left behind in the wake of its endlessly repetitive contemporary simulations. We are now inside Exopolis, and as we tour its postmodern geographies I will lead you along with some old (ca. 1984–90) and a few more recent news-clippings from my files, to add a little sense of history. But before we begin, I will also introduce another guide to our Thirdspace entourage, a late addition who carries with him his own distinctive maps and clippings of the territory being explored.

Re-imaged-in-LA: A Little Bit of Baudrillard

Jean Baudrillard has already made his presence felt in earlier chapters. His work, however, takes on particular significance inside Exopolis, for he, more than anyone else I can think of, has effectively mapped out one of its essential qualities, the infusion and diffusion of *hyperreality* into everyday life, an interpellation that Baudrillard describes as emanating from "the precession of simulacra." Hyperreality and the precession of simulacra – exact copies of originals that no longer exist or perhaps never really existed in the first place – represent contemporary elaborations upon what I have described as the simultaneously "real-and-imagined." Instead of the combinatory hyphens, Baudrillard collapses the terms on to each other, suggesting a growing incapacity to distinguish between them: real-imagined as re-imaged-in-LA.

As our ability to tell the difference between what is real and what is imagined weakens, another kind of reality – a hyperreality – flourishes and increasingly flows into everyday life. It is not the hard empirical reality of materialist science or Marxism; nor is it quite the simultaneously real-and-imagined. It is something different and, if not new, then certainly more prominent and affective in the contemporary world than ever before. Baudrillard has not been alone in drawing our attention to the fact that reality is no longer what it used to be. Here, too, there has been a flurry of neologistic invention aimed at capturing the meaning and growing significance of hyperreality, especially in connection with the electronic media that so effectively carry its messages. The two terms which come most quickly to mind are "virtual reality" and "cyberspace" (with the latter spawning such specialized variations as "cyburbia" and "cyberia").

I am also reminded of Celeste Olalquiaga's descriptions of an endemic urban psychasthenia, "a disturbance in the relation between self and surrounding territory." One's own body/self is "lost in the immense area that circumscribes it," abandoning its own identity "to embrace the space beyond ... camouflaging itself into the milieu," into "represented space." Bodies become like cities in what she calls a "poetic condensation." History is replaced by geography, stories by maps, memories by scenarios, with everything connected to "the topography of computer screens and video monitors" (see chapter 7).

These variations and extensions can be wrapped into the summative concept of hyperreality and used to open up new opportunities (and new dangers) in the challenging task of making theoretical and practical sense of the contemporary world, of Thirdspace, and, more particularly, of everyday life in Exopolis, for it is inside the outer core regions of Exopolis that hyperreality and its accompanying psychasthenias have become most heavily condensed. Other travels in hyperreality will eventually be noted, but it is Baudrillard who will, initially at least, lead the way.

In following Baudrillard, however, we must be careful to avoid the dangers of "disappearing into the desert" with him, for he has pushed his contemporary critique to ridiculous extremes of political immobility. As a caveat against the numbing politics of Baudrillard's political projections from the precession of simulacra and, peculiarly enough, as a useful clarification of his central arguments, I extract from a critical review of Baudrillard's travels in the hyperreality of *America* (1988), written by Zygmunt Bauman, the leading moralist of postmodernism.[1]

The world in which we once confidently spoke of change and renewal, of trends and directions, was a solid one in which we

[1] Zygmunt Bauman, "Disappearing into the Desert," *Times Literary Supplement*, December 16–22, 1988; and Jean Baudrillard, *America*, London and New York: Verso, 1988. See also Bauman's *Postmodern Ethics*, Oxford, UK and Cambridge, USA: Blackwell, 1993.

could tell the difference between an idea and its referent, a representation and what is represented, the image and the reality. But now the two things are hopelessly confused, says Baudrillard.... Take the most important of his concepts: that of simulation ("feigning to have what one hasn't"). Simulation, you might think, consists in pretending that something is not what it in fact is; this does not alarm us because we feel that we know how to tell the pretence from reality. Baudrillard's simulation is not like that, however; it effaces the very difference between the categories true and false, real and imaginary. We no longer have any means of testing pretence against reality, or know which is which. Nor is there any exit from this quandary. To report the change involved, we must say that "from now on" the "relationship has been reversed", that the map, as it were, precedes the territory, or the sign the thing. Yet such talk is itself illegitimate, for with simulation rampant, even the words that we use "feign to have what they haven't", ie, meanings or referents. In fact we do not know the difference between the map and the territory, and would not know it even if we had our noses pressed up against the thing itself.

Forewarned to avoid the no-exit extremes of contemporary Baudrillardism, we can begin to unpack and repackage Baudrillard's earlier journeys to add to our understanding of Thirdspace and to assist in our excursions through the looking glass of the Orange County Exopolis.

There are many ways to connect Baudrillard to *Thirdspace*. He, more than anyone else, has taken an intellectual path that creatively interweaves Lefebvre and Foucault with strands of Nietzsche, surrealism, and situationism. Although he never explicitly chose space as his revealing medium and repeatedly disavowed any lasting theoretical influence from either Lefebvre or Foucault, Baudrillard can nevertheless be seen as a heterotopological metaphilosopher of everyday life, driven by questions of material production and the thickening veils of *la conscience mystifiée*, the mystification of collective consciousness. He too has been a nomadic metaphilosopher, erasing many of his own constructs as he moves on to different horizons of inquiry and critique.

What I want to capture here, however, and bring with us as we enter the Exopolis in Orange County are some gleanings from "The Precession of Simulacra," the first half of a small book called *Simulations*, published in 1983.[2] In his rambling prose, Baudrillard

[2] Jean Baudrillard, "The Precession of Simulacra," in *Simulations*, tr. P. Foss, P. Patton, and P. Beitchman, New York: Semiotext(e), 1983, 1–80.

presents a remarkably clear picture of the *epistemology* of hyperreality amidst a text that quickly becomes filled with a "Moebius-Spiralling Negativity" that threatens to devour his own intentions (1983: 30). The text begins with a passage from *Ecclesiastes*, reminding us of the ancient roots of the term simulacrum: "The simulacrum is never that which conceals the truth – it is the truth which conceals that there is none. The simulacrum is true." Baudrillard turns, appropriately enough, next to Borges.

> If we are able to take as the finest allegory of simulation the Borges tale where the cartographers of the Empire draw up a map so detailed that it ends up exactly covering the territory (but where the decline of the Empire sees this map become frayed and finally ruined, a few shreds still discernible in the deserts ... [a] ruined abstraction ... rotting like a carcass, returning to the substance of the soil, rather as an aging double ends up being confused with the real thing) – then this fable has come full circle for us, and now has nothing but the discrete charm of second order simulacra.
>
> Abstraction today is no longer that of the map, the double, the mirror or the concept. Simulation is no longer that of a territory, a referential being or a substance. It is the generation by models of a real without origin or reality: a hyperreal. The territory no longer precedes the map, nor survives it. Henceforth it is the map that precedes the territory – **PRECESSION OF SIMULACRA** – it is the map that engenders the territory and if we were to revive the fable today, it would be the territory whose shreds are slowly rotting across the map, whose vestiges subsist here and there, in the deserts which are no longer those of the Empire, but our own. *The desert of the real itself.*   (1983: 1–2)

As always, Baudrillard flamboyantly overstates his point to drive home the importance of his underlying argument, that something unusual is happening today in the relation between the real and the imaginary, reality and its representations. How this has come about is eventually condensed in what he calls "the successive phases of the image," an epistemological timeline that I consider one of Baudrillard's most significant achievements. These phases can be interpreted, for our purposes, as a sequence of critical *epistemes* in the development of post-Enlightenment Western modernity, or more simply, different models for making practical sense of the world.

The first critical episteme is captured in the metaphor of the *mirror*, with the image being "the reflection of a basic reality." Practical knowledge is derived from our ability to comprehend in rational thought the sensible "reflections" from the real empirical world, sorting out the accurate, good, useful information from the accompanying noise and distortion. This is essentially the epistemology of modern science and the scientific method. It continues in force to this day in the human, biological, and physical sciences as the dominant epistemology, and is still recognizable, despite its detractors, as an important foundation for critical, emancipatory thought and practice, that is, making practical sense of the world in order to change it for the better.

A second critical episteme was developed most systematically in the 19th century, although, as for the first, ancestors can be found much earlier. Its metaphorical embodiment was not the mirror but the *mask*, a belief that the "good" reflections potentially receivable from the empirical world of the real are blocked by a deceptive shroud of false or counterfeit appearances. In Baudrillard's words, the directly mirrored image "masks and perverts a basic reality" (1983: 11). Practical knowledge and critical understanding thus require an unmasking, a demystification of surface appearances, a digging for insight beneath the empirical world of directly measurable reflections. The systematic exposition of this mode of critical discourse is closely related to the development of various forms of structuralism, from Marx, Freud, and de Saussure, to more contemporary cultural criticism in art, literature, and esthetics (where some might say that an untheorized form of exploring what is hidden behind appearances has always existed). This alternative episteme significantly shaped the discourse on modernity in the *fin de siècle* and has probably been the dominant counter-epistemology for explicitly critical theory and practice throughout the 20th century.

For Baudrillard, a third episteme emerges alongside the others in the late 20th century, ushering in a new critical – and poststructuralist – epistemology metaphorized around the *simulacrum*. Here the image begins to mask "the *absence* of a basic reality," indicative of a transition from "dissimulation" (feigning not to have what one really has, the "evil" lies derived from the appearance of things) to the increasing "liquidation of all referentials," the substitution of signs or representations of the real for the real itself; in other words, to "simulations" (feigning to have what one hasn't). The precession of simulacra threatens the very existence of a difference (and hence of our ability to differentiate) between the true and the false, the real and the imaginary, the signifier and the signified. Like the Exopolis, things seem to be turning inside-out and outside-in at the same time, erasing or at least blurring all familiar categorical boundaries.

The increasing empowerment of simulations and simulacra in the contemporary world has generated many of the epistemological stances associated with the so-called "postmodern condition": the deconstructive attack on all inherited modernist epistemologies and the rise of a more flexible, recombinant, and broadly poststructuralist critical theory; the resort to relativism, radical pluralism, eclecticism, and pastiche in an effort to avoid totalizing or essentialist metanarratives; the recognition of an invasive hyperreality that dislocates and decenters the subject and its (especially his) referents; the crisis of representation and the radical negation of the sign as value; the opening up to and disordering of difference and otherness; the substantive attention to the media and to popular culture as affective and revealing locations for the production of hyperreality; the search for tactical and strategic niches of resistance rather than universal programs for emancipatory social action. All spring, in one way or another, from the realization that postmodern reality is no longer what it used to be.

Baudrillard, however, goes one step further, arguing that we have already moved into a fourth phase, one in which the image "bears no relation to any reality whatever: it is its own pure simulacrum." As he adds, "the whole system becomes weightless, it is no longer anything but a gigantic simulacrum – not unreal, but a simulacrum, never again exchanging for what is real, but exchanging in itself, in an uninterrupted circuit without reference or circumference (1983: 10–11).

At this point, we must suspend our acceptance of the Baudrillardian progression, for it leads us to little else but the nihilistic "ecstasy" or bovine immobility that he seems to recommend as a response to the total conquest of hypersimulation. But whether Baudrillard himself believes we have irresistibly entered this holistic new age or is instead strategically challenging us with exaggerated signals of its impending total absorption is still open to question. What is clearer, however, from this little bit of Baudrillard, is the wholesale intrusion of everyday life by an effulgent hyperreality that is redefining what we mean by the real-and-imagined. With this awareness in mind, we can now enter the Exopolis of Orange County and more effectively explore its postmodern scenarios.

Scenes from Orange County and a Little Beyond

The following scenes from the Orange County Exopolis are wrapped around newspaper clippings and a few more academic references from other travelers in and through its hyperrealities. To guide you, here is a map of the territory.

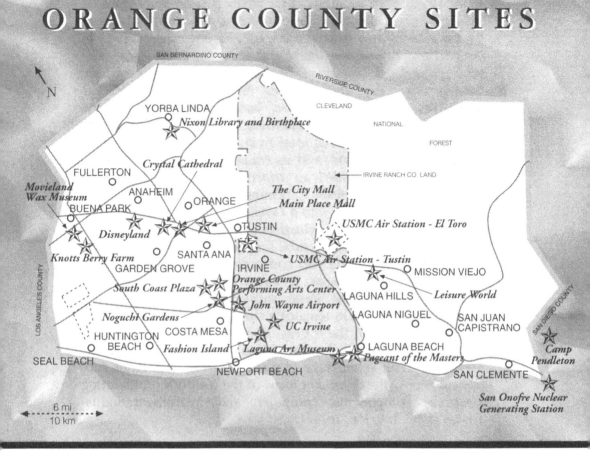

Scene 1: A Mythology of Origins

We begin with the words of the economic geographer, Allen Scott, who has pioneered the exploration of Orange County's Firstspace.[3]

> We may take Orange County as a paradigmatic example of the new patterns of industrialization and urbanization that are now being laid down on the American economic landscape. In particular, Orange County epitomizes with great clarity the basic geographical character of new growth centers in the Sunbelt, with their transaction-intensive economies, their deeply segmented local labor markets and regressive labor relations, and their high-technology production systems geared intimately to federal defense and space expenditures. . . .

[3] Allen J. Scott, "New Frontiers of Industrial-Urban Development: The Rise of the Orange County High Technology Complex, 1955–1984," chapter 9 in *Metropolis: From the Division of Labor to Urban Form*, Berkeley and Los Angeles: University of California Press, 1988.

In the mid-1950s, Orange County was in essence a quiet backwater given over to agricultural pursuits with some modest industrial production.... The population was small, and residential activity was mainly confined to a few communities in the northern half of the county and along the coast. Some suburban tract development was beginning to make itself evident, however, as the Los Angeles built-up area expanded southward.... The uncluttered landscape of the county, together with its abundant recreational facilities and varied natural environments, offered ideal living conditions for its rising middle-class population, and its conservative political inclinations (then as now) made it an attractive place for business....

Sometime in the 1960s ... Orange County manufacturers began to draw together into a *complex* in the true sense, i.e., a congeries of interlinked industries sharing a common pool of labor and various infrastructural services.... By the early 1970s, the high-technology complex had become as tightly organized in geographical space as it apparently was in economic space.... [Today,] the loose subsystem of plants around Anaheim and Fullerton remains a strong element of the overall industrial pattern of the county. In addition, the subsystem in and around Irvine has developed into an extraordinarily dense and tightly knit assemblage of manufacturers. This latter subsystem is now the dominant focus of the county's proliferating electronic components, computer, and instrument industries.   (1988: 160, 166, 167–71)

The exopolis here is depicted first of all as industrial and industrious, a transactional tapestry efficiently knotted into a series of flexible manufacturing and service complexes, great swarms of businesses tied up in hive-like clusters to capture the new "scope" economies of postfordist technology. No longer bound by the rigid hierarchical demands of mass production and assembly lines, a new kind of industrialization was begetting a new kind of "peripheral" urbanization, an offset urban form. This tightly manufactured landscape of flexible economic specialization backed by hyperconservative local politics is the seedbed of Exopolis, not only in Orange County but around nearly every other major American metropolis.

Scott's mapping shows clusters of symbols representing high-technology plants multiplying like weeds over the siliconed landscape of this latest reflowering of the garden city. The densest and busiest of these industrial parklands is the Irvine "subsystem," the master-planned field of operations of the Irvine Company, owner of one-sixth of the land of Orange County and packager of the largest New

Towns in America. To the north, Anaheim and Fullerton stake out another, older complex, while several other hardworking clusters of industrial activity lie scattered about, each a pointillist focus in the urbanizing landscape.

To call this densely knit assemblage of manufacturers and their subcontracted servants simply "postindustrial" is surely to miss the point. And just as surely the area is no longer "sub" urban. Orange County may have no dominant city in the traditional sense, no easily identifiable center or skylined downtown to signal the Modernist citadel and CBD, but it is a metropolis none the less, a Standard Metropolitan Statistical Area of around 2.6 million inhabitants, an industrial capitalist city of a new kind, a city whose "flexible accumulation" of wealth and power is indicative of a restructured political economy. Industrialization thus concretely sets the basic reality of the exopolitan scene.

Scott, attached to the pre-postmodern vision of urban dynamics that has dominated the (First)spatial analysis of industrial restructuring, searches for a rational order in the bubbling postmodern complexity. He calculates the geographical "center of gravity" for the whole Orange County industrial flywheel and imposes hypothetical concentric patternings around the weighty center of industrial production and employment in the hope of discovering more of those distance-frictioned regularities on which indigenous spatial science so heavily depends. Some fuzzy sets of statistical concentricities emerge to satisfy the scientific search, but what is most revealing about this exercise is what it inadvertently shows about the inner circle itself. The center of gravity of the Orange County industrial complex is itself almost empty of high-technology industry, forming a curious doughnut-hole of maximum accessibility that pivots the propulsive subsystems around an apparently hollow core. How exquisitely exopolitan for the middle to be missing, for the center to be gravely defined by the weight of its productive periphery.

But there is more to this seemingly empty inner circle than first meets the eye. Contained within it are some of the peak population densities of Orange County and a large chunk of the Latino barrio of Santa Ana, the county's choice pool of cheap labor and a prime stopover site for thousands of undocumented immigrants who cross over the Mexican border just one county away. There is also, close to the intricate interchange of three freeways and the concrete cut of the Santa Ana River, a large shopping mall semaphorically named "The City." A small stuccoed prison and a vast new hospital complex line such streets as City Drive and Metropolitan Avenue, further signaling this citadellian ambition. And then, rearing up in all its translucent-pomposity, is the Crystal Cathedral of Garden Grove, the Philip

Johnson-designed white-steel-and-silver-glass extravaganza of tele-visual prayer, with its spiky tower soaring to 124 feet, to compete with nearby Disneyland's Matterhorn for peak visibility on the lowflung Orange County skyline. Perhaps this is the center of centers after all. . . .

Extraordinary sites, capitals of fiction become reality. Sublime, transpolitical sites of extraterritoriality, combining as they do the earth's undamaged geological grandeur with a sophisticated, nuclear, orbital, computer technology.

(Baudrillard, 1988: 4)

All the signifiers are here in this inner circle to begin a different mapping of the origins of Exopolis. From its vantage point, we begin to see beyond the industrial base of Orange County and recognize other spaces to explore. We should never lose sight of that industrial basic reality, for it concretely undergirds every Exopolis. But we must move on to the softer cities of illusion and aspiration.

Scene 2: Iconic Emplacements

Spiraling outward from the industrial nucleations that mark the exopolitan frontier is an extraordinary galaxy of locales. Place by place, they too provide paradigmatic reflections of the orbital Exopolis. Umberto Eco introduces our immanent tour of these spaces of representation in his own *Travels in Hyperreality*.[4]

At Buena Park, California, in the Movieland Wax Museum, Jean Harlow is lying on a divan; on the table there are copies of magazines of the period. On the walls of the room inhabited by Charlie Chaplin there are turn-of-the-century posters. The scenes unfold in a full continuum, in total darkness, so there are no gaps between the niches occupied by the waxworks, but rather a kind of connective decor that enhances the sensation. As a rule there are mirrors ... duplicated by an astute play of corners, curves, and perspective, until it is hard to decide which side is reality and which illusion. Sometimes you approach an especially seductive scene, a shadowy character is outlined against the background of an old cemetery, then you discover that this character is you. . . .

But when it comes to spiritual emotions nothing can equal what you feel at the Palace of Living Arts in Buena Park. . . . It is next to the Movieland Wax Museum and is in the form of a Chinese pagoda. . . . The Palace of Living Arts is different, because it doesn't confine itself – except for some statues – to presenting reasonably faithful copies. The Palace reproduces in wax, in three dimensions, life-size and, obviously, in full color, the great masterpieces of painting of all time. Over there you see Leonardo, painting the portrait of a lady seated facing him:

[4] Umberto Eco, *Travels in Hyperreality*, tr. W. Weaver, San Diego: Harcourt, 1986.

She is Mona Lisa, complete with chair, feet, and back. Leonardo
has an easel beside him, and on the easel there is a two-dimensional copy of La Gioconda: What else did you expect?

(1986: 13, 18)

I must interrupt Eco's travels to point out that the Palace of Living
Arts (today no longer in existence) was itself an imitation of a far
more impressive and technologically advanced display of lively simulations, the Pageant of the Masters, held each summer at Laguna
Beach, on the other side of Orange County. There the masterpieces
are painted – with acrylics to deflect the stage lighting and with
shadows and form-fitting indentations – so that live models can slip
in to pose immobile for the required one and a half minutes in "magical tableaux" that still attract over 100,000 visitors every year. The
1990 pageant, for example, featured 23 "live recreations" of paint-
ings (culminating, by tradition, with *The Last Supper*) and various
"living versions" of such objects as a gold ornament, a carved
wooden altar piece, a butterfly brooch, Japanese dolls, and Jules
Charet posters. Here "still life" is radically redefined.

Let us now return to Eco.

The theme of our trip . . . is the Absolute Fake; and therefore we
are interested only in absolutely fake cities . . . with cities that
imitate a city, just as wax museums imitate painting. . . . There is
an embarrassment of riches to choose from. . . . But we'll stick to
the Western myth and take as a sample city the Knott's Berry
Farm of Buena Park. . . . Here the whole trick seems to be
exposed; the surrounding city context and the iron fencing (as
well as the admission ticket) warn us that we are entering not a
real city but a toy city. But as we begin walking down the first
streets, the studied illusion takes over . . . the realism of the
reconstruction . . . life size and executed with absolute fidelity
. . . the customer finds himself participating in the fantasy
because of his own authenticity as a consumer.   (1986: 40–1)

Knott's Berry Farm bills itself as the oldest themed amusement park
in the world, an entertainment enclosure that celebrates the "wholesome aspects of an idealized and simpler America." Its founding
father's conservative family values set the fundamentalist tone for
"old" Orange County's fervently right-wing political traditions and
the entrepreneurial dreams that fed not only the theme-parking of
Exopolis but also its postfordist industrialization. Today, the family
values that control the park descend not from Knott but from Robert
Welch, the founder of the John Birch Society.

Although it is still popular, Knott's simpler America was eclipsed by its more subtly modern corporate neighbor, Disneyland. Our semiological journeyman describes this extraordinary site, this brightly coded *semeion* (Greek for both sign and specific location in space, a coincidence, by the way, that makes all semiotics inherently spatial):

> a "degenerate utopia" ... an ideology realized in the form of a myth ... presented as at once absolutely realistic and absolutely fantastic ... a disguised supermarket, where you buy obsessively, believing that you are still playing ... more hyperrealistic than the wax museum, precisely because the latter still tries to make us believe that what we are seeing reproduces reality absolutely, whereas Disneyland makes it clear that within its magic enclosure it is fantasy that is being absolutely reproduced.... But once the "total fake" is admitted, in order to be enjoyed it must seem totally real ... Disneyland tells us that faked nature corresponds much more to our daydream demands ... that technology can give us more reality than nature can ... here we not only enjoy a perfect imitation, we also enjoy the conviction that imitation has reached its apex and afterwards reality will always be inferior to it.   (1986: 43–6)
>
> Thus, on entering his cathedrals of iconic reassurance, the visitor will remain uncertain whether his final destiny is hell or heaven, and so will consume new promises.   (1986: 58)

Eco's tour unveils the symbolic birthplaces of first-wave hyperreality and opens up these primordial sites to Secondspatial interpretation as representations and re-presentations of reality. But too often the semiotic observer and other postmodern passers-by stop here, in the county's cannily crystalized cathedrals of iconic reassurance, going no further than these first-wave reference points and their screaming symbols. This is a mistake, for there has been a second wave that has carried hyperreality out of its tightly localized manufactory enclosures in the old theme parks and into the geographies and biographies of everyday life, into the very fabric and fabrication of the lived spaces in Exopolis.

Today the artful simulations of Disneyland seem almost folkloric, crusty residuals of a passing era. Now you do not just choose to visit these hyperreality factories at your leisure, *hyperreality visits you* every day wherever you choose to be. The original simulacra are being simulated again, to ever higher powers and lengthening chains, all over the map. And the new map that is appearing is one that resists conventional modes of cognitive understanding. It resembles those maps in shopping malls that tell you exactly where you are with little directive arrows. Only now the map is filled with an infinite number of arrows, each telling the onlooker YOU ARE HERE. Every place, every where, is being affected by the transcending diffusion of hyperreality.

Scene 3: The Diversions of Yorba Linda

There are engines and anchors in every country, and they change over time. When I was born in Yorba Linda, the area was an anchor – an agricultural region and something of a playland. It is now the engine of progress in America, an area where entrepreneurs are gathering to drive the American dream forward. Look at the educational infrastructure, the corporate infrastructure, the political leadership, and you see America's future. It is a dynamic, forward-looking place. Its peoples and products are changing America.... [Its political leaders] are *responsible* for the peaceful revolution of the East Bloc. Some of my old friends ... will remain stalwarts of freedom. We owe them a great deal of thanks for their patient support of a sound defense that helped bring about global change. Others are new, shining stars ... among the most talented additions to the United States Congress in the last twenty-five years. In time, with seniority's assist, they will become superstars.... Far from being out of step, all these leaders and others are actually playing the tune.

Thus spake Richard Milhous Nixon.[5] The same tune continued to be played at the grand opening of the Nixon Library and Birthplace in the first of his hometowns, Yorba Linda, once a Quaker settlement and now a large-lot, zoned-for-horses, almost lily-white municipality that advertises itself as the "Land of Gracious Living" tucked away in the northern reaches of Orange County. Public television had a sniggering segment covering the celebration. On the show, bicoastal Buck Henry hosted an essentially East Coast view of the spectacle, with its thousands of red, white, and blue balloons; Tom Brokaw interviewing Richard Milhous himself, and all kinds of Republican presidents present for a make-believe summit. Only Jimmy Carter, who was supposed to be in Ethiopia, was not there among the allegedly living American presidents.

From the peaceful garden court, filled with (transplanted) indige-

---

[5] Interview with Richard Milhous Nixon, *Orange Coast*, July 1990: 77. *Orange Coast* is one of those boosterish urban glossies that have proliferated in recent years. The same issue contained the information mentioned earlier about that year's Pageant of the Masters at Laguna Beach.

nous trees and bushes, Buck pointed to the difficult-to-find subterranean library, which had earlier stirred some controversy over whether it would be opened to anyone who might find anything in it critical of Nixon. It was later decided that this would be impossible. We were also shown around the Birthplace, a $400,000 simulation of the original home built by Nixon's father from a $300 Sears Roebuck kit. Buck smiled as Nixon's deep voice piped in to reminisce about his first nine years in Yorba Linda amidst all kinds of reproduced memorabilia, including the humble family piano that Dick's mother often whipped him into playing, and the mnemonic sound of a passing train to recreate a sense of history. Could it all be real? You bet!

The biggest enclosure was the audiovisual museum sitting atop the Library, tracing some chosen episodes of the Nixon career. From the grand entranceway, lined with *Time* magazine covers, one could march through his packaged political biography in a series of magical tableaux as real as those of the Pageant of the Masters. Buck took us quickly past the early years, through Mrs Nixon's Passage (filled mainly with old dresses) to the dimly lit Watergate Room. There we listened to snippets from the (remaining) famous tapes and simultaneously read the matching transcripts, as if the correspondence of sight and sound would somehow add authenticity. After viewing a photomontage of Nixon's last day in the White House and a large picture of his emotional two-fingers-raised helicopter farewell, we entered the Presidential Forum. Here the spectator can have what the information sheets call a "conversation" or a "dramatic interaction" with the 37th president via a "touchscreen" video system. Buck buttoned in one of the more than 400 canned questions and we watched another representation of a representation of a representation, like a set of infinite mirrors.

By far the most fascinating site was the World Leaders Sculpture Room, where Nixon, Reagan, Ford, and Bush arrayed themselves amid "bronze-tone" life-size figures of what were described as "ten of the century's greatest political leaders," from Churchill to Mao, posed in "characteristic wardrobe and stances." All the figures looked so real, but you are warned not to touch them for fear that they would topple over.

The television segment ended with appropriate Eastern smugness, showing the Park Avenue building that rejected Nixon's attempt to find a new home in New York City. The intention was clearly to distance New York not only from Nixon but also from the hyperreality of Southern California, the incomprehensibility of it all. The videosimulation of the event, filled with snide detachment, seemed at the end to close off much more than it illuminated. I wondered what Baudrillard would make of it.

Watergate. Same scenario as Disneyland (an imaginary effect concealing that reality no more exists outside than inside the bounds of the artificial perimeter) ... concealing that there is no difference between the facts and their denunciation.

(1983: 26–7)

Another TV show completes our visit to Yorba Linda. Many broadcasts of it were seen all over the world in April, 1994, shortly after Nixon's death and the apartheid-ending elections in South Africa, nearly two years to the day after the incendiary explosions of the Los Angeles riots and rebellion. The simultaneities and juxtapositionings were amazing, as Nixon was re-simulated by the media and a veritable army of spin-doctors, with TV maps showing the world how to get to Yorba Linda. Everyone offered their everlasting sound-bites and sites: Ben Stein, the inventive speechwriter, plugging his book on the "real" Richard Nixon; the still-not-a-crook Spiro Agnew looking longingly at four ex-presidents standing Mt Rushmore-like in chiseled profile by the Sears catalogue boyhood house; Colson, Liddy, and the rest of the Watergate crew revising history again, blaming Stockman for it all; Julie Eisenhower being given a furled flag while the background sound piped in Taps, prerecorded earlier. Interviewers stopped passers-by, including the pilots of the fly-by planes, to collect one-phrase summaries of the essential Nixon legacy. It was, in its terminating way, a surpassing orgy of recreative hypersimulation conducted by some of the world's leading experts. Every other memory seemed to disappear.

Scene 4: UCI – A Campus by Design

Looking out over the empty hills of the Irvine Ranch almost 30 years ago, planner William Pereira searched his mind for a powerful metaphor to match the UC campus he envisioned.

Pereira's aim ... was "to establish a heart and a sense of place" that would offer the first students a feeling for "the destiny of the campus."

The plan became a strong symbolic presence for Irvine Ranch, then a dry rolling wilderness under hot white skies. At the heart of the campus was a series of concentric rings – the innermost containing undergraduate facilities, the outer one housing graduate and research buildings. This ring-within-a-ring metaphor was intended to express a student's progress, from the self-absorbed concentration of the first years of study to the wider circle of the world beyond the campus.

While the late architect's master plan was bold, the buildings he fleshed it out with in the 1960s and 1970s were, in the view of many observers, overscaled and boringly detailed. Campus wags dubbed the modernist concrete boxes that enclose UCI's inner ring mall "a bunch of giant cheese graters."

Now all this is changing....[6]

Prehistory all over again: the ring-within-a-ring representing progress and modernity, disenchantment flourishing when the outer limits are reached, the urban memories of old progressive dreams falling apart in the orange sunlight. Now modernity itself is being displaced by deliberately postmodern architectonics, by a new kind of *campus* (field, plain, level space) designed amidst the empty hills, leveling off, evening out, offering transplanted heartlands, a new and different destiny, an enveloping hypersimulation of unlimited expansion, reconcentration, recentering. A new urban space takes shape as the inner rings are left behind as stale plots from a departed past.

The story begins with the Irvine Ranch-cum-Company, landed benefactor to the University of California branch-ranch that bears its name. Exemplary promoter of the manufactured landscapes of Exopolis, the Irvine Company is hard at work filling in the missing middles, creating its own ex-centric bundles of urban identity.

Under the supervision of campus architect David Neuman, UCI has inaugurated a $350-million expansion that, by 1992, will feature more than 20 major new complexes designed by some of the best U.S. and international architects. "UCI is growing up," Neuman explained. "Its character is changing from a suburban college into an urbane campus with an ambition to be academically and architecturally first-rate.

Orange County's sleepy second-rate sub-urban conglomerations are growing up into a whirled series of dreamily dispersed semi-urbane

[6] This and subsequent quotations are from Leon Whiteson, "Campus by Design – UC Irvine Hopes to Avoid Boring, Boxy Buildings and Add a Degree of Sophistication As It Expands," *Los Angeles Times*, December 12, 1988.

gatherings.[7] The county abounds with these synthetic identities, with what Baudrillard called "artificial paradises" in which "space lends a sense of grandeur even to the insipidity of the suburbs ... a miracle of total availability, of the transparency of all functions" (Whiteson, 1988: 8). The transformation is nowhere better signaled than in microcosmic UCI, which some have translated as "Under Construction Indefinitely," and which has become a virtual architectural theme park, with a building each from a covey of the world's trendiest architects.

On the eastern quadrant, Whiteson writes, "Charles Moore's Italianate Alumni House and Extension classroom complex – described by one critic as a stage set for an opera by Puccini – plays off a colonnaded Graduate School of Management by New York-based Venturi, Rauch, Scott-Brown." Actually, the playfulness of Moore's Extension Building, located where UCI's specialized spoke for the social sciences "extends" to the outside world, is less operatic than it is televisual. Moore himself saw it as the piazza of some imagined Italo-Spanish-Californian town, bounded by three Baroque church-fronts and a rancho style verandah, into which, on some dark night, a sworded Zorro might ride and slash his Z in the dust. Sure enough, during the dedication ceremonies, a masked man dressed in black swooped out of the shadows, presented a plaque to establish authenticity, bowed with a smile beneath his penciled mustache, and cut the air with three swipes of his sword. This instant memory, fantastically faked, made all the necessary reel-world connections.

---

[7] I can't resist thinking of Sartre, who had something (or Other) to say about such terms as "series," "gatherings," "identity," and "field":

> Through this opposition between the Other and the same in the milieu of the Other, alterity becomes this paradoxical structure: the identity of everyone as everyone's action of serial interiority on the Other. In the same way, *identity* (as the sheer absurdity of meaningless dispersal) becomes synthetic: everyone is identical with the Other in so far as the others make him an Other acting on the Others; the formal, universal structures of alterity produce the *formula of the series (la raison de la série)*.

He goes on (and on), in his *Critique of Dialectical Reason* (the Verso paperback, 1982: 264) to add:

> In the formal, strictly *practical*, and limited case that we have been examining, the adoption of the serial mode remains a mere convenience, with no special influence on the individuals. But this simple example has the advantage of showing the emergence of new pratico-inert [*sic* – it should be "practico-inert" of course] characteristics: it reveals two characteristics *of the inactive human gathering*. The visible unity, in this case, in the time of the gathering ... is only an *appearance*; its origin for every observer to whom this totality is revealed, [however,] is integral to *praxis* in so far as it is a perpetual organisation of *its* own dialectical field [Latin = *campus*].

Elsewhere, more playfulness abounds. To the south, quirky, raw-looking fabrications by Frank Gehry and Eric Owen Moss present images of children toying with giant building blocks. On the west, a "Food Satellite Center" designed by Morphosis to serve the Humanities Department "mocks solemnity with a row of free-standing columns that will have the air of an instant architectural ruin." Across the mall to the west, a pop-postmodern science library by James Stirling is banded in multi-colored stucco. To the north of the library, a folksy, red-tiled, Fine Arts Village by Robert Stern; and to the south, a horizontally striped green glass and plastic Biological Sciences Unit by Arthur Erickson, commentator on the centeredness of the CITADEL-LA. Whiteson goes on:

> In the view of national magazines such as Progressive Architecture, several lesser-known designers, many from California, have contributed some of their best work to UCI's expansion.
>
> Among them is L.A.-based Siegel, Sklarek, Diamond's boldly Cubist Student Services Building near the administration center. Project architect Kate Diamond said designing for the campus presents a special challenge. "At present (UCI) tends to lack identity, so it needs buildings that are very sculptural and strongly modeled to provide a real sense of place."
>
> In Stern's view, the "basic blandness" of the early Modernist buildings can work in the designer's favor, providing "a kind of blank canvas on which the architect can paint a more artful and resonant picture."

Reactions to this orbital and exceptionally contemporary *ringstrasse*, and to UCI's attendant search for identity (or what Sartre footnotedly defined earlier (note 7) as "the sheer absurdity of meaningless dispersal") have been mixed. From a student-user of Gehry's new facility: "Looks like a hardware store." Gehry's deconstructive response: "The engineers who occupy my building are interested in how things are put together. So I gave them an architectural metaphor that takes its clues from the assembly of components you might find in a machine. You can see how the whole thing functions because it is broken down into its major pieces, then reassembled as a working complex." As he would say later, after winning the international competition to design Walt Disney Concert Hall atop Bunker Hill in downtown Los Angeles, "I feel you work with what you have and like it. You take your clues from your surroundings, and rejoice."

The UCI Chancellor responded more coyly: "I don't have to like it, but it draws attention, and it's important that people come to see us." Conservative members of the County Board of Supervisors worry, however, that the new architecture might upstage the "unremarkable buildings" that dominate the neighboring City of Irvine and create too much traffic, with all those people coming to see UCI, Orange County's dizzy-kneed University of Californialand. Charles Moore (tongue in cheek?) concludes Whiteson's review with a question: "Who would imagine that a campus seemingly stamped forever with the curse of architectural mediocrity would raise a phoenix of fine design from the ashes of dullness?"

Scene 5: Spotting the Spotless in the City of Irvine

Lying just outside, in the exocampus of UCI, the Irvine Company is busy manufacturing other centers and nodes, ceaselessly creating more absolutely real fakes in order to simulate the appearance of urbanity. Here we enter another but related scene.

> Throw out those visions of pool tables and dart boards. Forget about pickled eggs and older waitresses who call you "hon." The city of Irvine [now with around 100,000 inhabitants] just got its very first bar, and none of the above are anywhere to be seen.
>
> For the Trocadero – not surprisingly – is Irvine incarnate, a so-Southern California watering hole located across the street from UC Irvine and characterized by its owners as "an upscale, traditional Jamaican plantation."
>
> ... as the very first real bar in Irvine history, the Trocadero is as much a symbol as it is a saloon. The Trocadero's owners and site were both hand-picked by the Irvine Co., which controls 50% of the city's retail space and has spent decades carefully molding the retail mix in this spotless [the perfect spatial adjective] suburb. The scrupulous planning has been so successful that enterprises such as dive bars and massage parlors can only be found on the wrong side of the city's boundaries....
>
> ... As owners Mark and Cindy Holechek [pronounced whole-check?] say of their latest endeavor, [it is] a bar where patrons can graze on appetizers including fresh oysters injected with Stolichnaya and topped with orange hollandaise [of course]....
>
> About a year and a half ago, the development company approached Mark Holechek to design and run this bar-to-be and its very trendy kitchen. Holechek, at the time, was co-owner, along with brother-in-law Chuck Norris of action-film fame, of a successful Newport Beach bar called Woody's Wharf.
>
> Holechek was also engaged to the former Cindy Kerby, a

modeling school owner who just happened to be Miss California/USA 1981, third runner-up for Miss USA in the same year, and voted by her cohorts as Miss Congeniality and Most Photogenic.

What's a barkeep to do when faced with such an opportunity? Holechek sold the Wharf, married Kerby and went on an extended honeymoon in the Caribbean ... collecting ideas for the proposed pub.... The product of all that honeymoon research was ... a Honduran mahogany bar and back bar to suggest "manliness," Holechek said, marble-topped tables, ceiling fans, palm trees and primary colors for a "feminine" touch.[8]

Why did it take so long for Irvine to get a bar like the Trocadero, the story asks? The answers are illuminating. They speak of the grand existential dilemmas facing the makers of Exopolis. "First, there's the history problem. Irvine doesn't have one." This "history problem" exists everywhere in Orange County, where, we have been told, even the land and the ocean look new. As the mayor said, "One thing I've learned is that you cannot telescope the evolution of an urban community into a matter of years. These things take time. You talk about Venice or Los Angeles, it's 100 years of history [at least!]. Here, history in a municipal sense is 20 years old."

Then, there is the "Irvine Company problem." Company representatives explained the delay as "a question of place and time: 1988 and Campus Drive are the right ones; any earlier and anywhere else are the wrong ones." The mayor added: "when you own most of the developable real estate, you can pretty much proceed at your own pace and discretion." Obviously, the place and time were right. "When we went before the Planning Commission to tell them our idea," Cindy Holechek said, "they gave us a standing ovation, they were so pleased to finally have a bar here."

Scene 6: Roots and Wings

Swinging over to the west bank of the UCI campus, we find another gathering of masterplanned spaces of a different sort cloistered in a corridor running along the Newport Freeway from Santa Ana and Tustin to Costa Mesa and Newport Beach. This is the Grand Axis-Mundi of exopolitan Orange County, a true *Champs Élysées* of commercial development. An appropriate tour guide is provided by an old copy of the *Airport Business Journal*, a thick monthly serving the

[8] Maria L. La Ganga, "Mixing It Up In Irvine – City's First Bar Carries the Official Stamp of Approval for Style, Location, Appeal," *Los Angeles Times*, November 5, 1988.

huddled masses next to John Wayne International Airport.[9]

> It's been said before, but we'll say it again – the MacArthur
> Corridor is, development-wise, booming – at a mind-boggling
> rate. We all get tired of the millennia of urbanized cities and
> communities staking their claim that they are the place to be.
> Developers along the MacArthur Corridor need only to sit back
> and let their projects speak for themselves.   (1984: 6)

You can add your own comments on this wondrously phenomenological introduction to the making of histories and geographies. The
hype goes on:

> Not since 1849 have Californians witnessed anything quite like
> it: a massive stampede of fortune seekers eagerly laying claim
> to any piece of land they can lay their hands on, their eyes fired
> with a burning hunger, crawling all over each other for a
> chance to cash in on untold riches.
>   Only this time it isn't gold they're breaking ground for – it's
> office buildings. MacArthur Boulevard, once a two-lane asphalt
> path running through orange groves and tomato patches, has
> widened like a flooded river, its waters rich in development
> dollars and its banks giving root to towering office complexes.
>   If you think Orange County has grown in the last 20 years,
> wait until you see what's going to happen in the next ten, for all
> signs point to this section of the County ... becoming *the* major
> financial center in the county, and perhaps California, and per-
> haps the United States.   (1984: 18)

The showpieces of many of the mixed-use projects taking root in the
rich floodwaters seeping outward from the MacArthur Corridor are
fake lakes. A project manager bubbles over on these veritable pools
of urbanity.

> People are attracted to water. Orange County's becoming very
> urban – it's evolved from a rural area very quickly. The people
> of Orange County like a sense of openness and they like the
> romance and the ambiance that water brings. We felt it was a
> good investment to have that water. The community certainly
> likes it, we do have a residential population here. On any given
> evening there are joggers around the lake, people walking their
> dogs – we've managed to create a park-like setting that is some-

[9] "Boom Town" (Editorial) and "The Sky's the Limit! Steel and Concrete Fly as
Developments Boom," *Airport Business Journal*, September 1984, 18.

thing more than just a large office complex with a tunnel effect and a concrete jungle. (1984: 20)

The article goes on to identify "The Creme" of MacArthur Corridor's business parks. The billion dollar Koll Centers (one North and another South), the Jamboree Center and MacArthur Court (both developed by the Irvine Company) anchor the still growing *Irvine Business Complex* just east of John Wayne Airport. The complex is confined on its other side by UCI and its vast Wildlife Preserve, forcing growth to slash diagonally and vertically from the Tustin Marine Corps Helicopter Station (with its huge blimp hangar) into Newport Beach.

Just south of the airport cluster, almost overlooking Newport Harbor, the Civic Plaza, Pacific Mutual Plaza, an art museum and a country club occupy the shores of the massive Fashion Island shopping mall. All together, they comprise the *Newport Center* business complex, another circular site masterplanned by William Pereira and the Irvine Company, and now being constantly reconstructed. Together, the overlapping Newport and Irvine complexes, should they ever reach their maximum planned development, promise to contain upwards of 100 million square feet of office space.

This loaded zone is fantastically reproduced in miniature in the Irvine Exhibit, formerly housed at the Jamboree Center. I vividly remember my first and only visit, accompanied by Allen Scott and a group of Scandanavian geographers and planners from the Nordic Institute. To get to the exhibit you must pass through a colonnade of transplanted palm trees and a set of revolving doors to an imposing security desk, where you are asked to leave your cameras behind (for iconic reassurance?). Disarmed, you are led to the plush seats of a small theater that vibrates with anticipatory technologies.

The whole front wall is a split-screen panorama upon which is soon projected a dazzling array of scenes – of birds and babies, sunsets and shorelines, family outings and businessmen's lunches, clouds and lakes (always lakes) and cuddly animals – all set to stereophonic music and soothing voices announcing the supraliminal messages that appear, in resolutely capitalized words, on the brochure clutched in your hands:

THERE ARE ONLY TWO LASTING THINGS WE CAN GIVE OUR CHILDREN. ONE IS ROOTS. THE OTHER IS WINGS.... ROOTS AND WINGS.... BOTH IN THE COMMUNITY AND THE NATURAL ENVIRONMENT, A BALANCE MUST BE ESTABLISHED IF THE INTEGRITY OF THE SYSTEM IS TO REMAIN SECURE.... A CONSTELLATION OF TECHNOLOGY AND BUSINESS.... ADVENTURES FOR THE MIND....

A GIFT OF TIME. A GIFT OF FAMILY.... THE IRVINE
EXHIBIT.... WE HAVE THE DREAM. WE HAVE THE PLACE
WHERE WE CAN PUT DOWN ROOTS. WHERE OUR LIVES
CAN TAKE WING.

The air is almost drugged with an effort to make you believe, to
make you want to consume new promises. But suddenly the flashing
pictures stop and the screen-wall becomes transparent, a shimmer-
ing gossamer film behind which a secret room appears. Still in your
seat, the music still throbbing in your ears, you realize that the whole
floor of the secret room is moving, tilting up before your eyes, coming at you slowly to fill up the wall with a portentous overview of
Irvine Earth, an exact model of the real world of ROOTS AND
WINGS. You gather your belongings and move toward the alluring
model as it slowly, seductively, tilts back to receive you, to embrace
you in person. It is a fascinating site, detailed down to lane markers
on the freeways and the loose dust where new homes and offices are
being built.

But this totalizing "area model" is not enough. After a brief lecture, a guide takes you through marbled halls and up steel capsule
lifts to another floor, where the model is itself reproduced in pro-
gressively larger-scaled closeups. You are moved, room by room,
closer and closer to the ultimate one-to-one correspondence between
representation and reality, the map and the territory. The final stop
is a space almost entirely filled with a giant structure very much like
the building you are in, exact in nearly every detail, offices lit and
filled with miniature accouterments, including little people and tiny
framed pictures on the walls (made by computers, you are proudly
told). You feel like peeking into the second floor to see if you too can
be seen there, peeking into the second floor....

The experience is finally capped when the beaming guide pushes
a button and an apparently solid outside wall disappears, revealing
a huge window on to the palm-colonnaded entranceway and the
surrounding buildings and grounds of the Irvine Business Complex.
It is a beautiful sight, so much bigger than its replica behind you, so
much more beautiful than your actual memories of having seen it on
the ground just an hour ago. You thank the guides, walk back to the
security desk, retrieve your camera, and exit, noting how disappoint-
ing and dull the real columns of palm trees look in comparison to
their artful imitations.

I could not help but think of Jean Baudrillard's reflections on
California, when he too stopped off at Irvine.

Still, there is a violent contrast here ... between the growing
abstractness of a nuclear universe and a primary, visceral,

unbounded vitality, springing not from rootedness, but from the lack of roots, a metabolic vitality ... in work and in buying and selling. Deep down, the US, with its space, its technological refinement, its bluff good conscience, even in those spaces which it opens up for simulation, is the *only remaining primitive society*. The fascinating thing is to travel through it as though it were the primitive society of the future, a society of complexity, hybridity, and the greatest intermingling, of a ritualism that is ferocious but whose superficial diversity lends it beauty ... whose immanence is breathtaking, yet lacking a past through which to reflect on this.   (1988: 7)

Scene 7: It's a Mall World After All

On the other side of John Wayne Airport, mainly in the city of Costa Mesa, the Irvine empire has its chief competitor, an offside super-center with ferocious intimations of becoming the true upscale downtown simulacrum of Orange County. Here one finds the rest of the business parks: South Coast Metro Center, Center Tower, Home Ranch, Town Center – the names virtually reek with anticipatory pretensions. And there is more here – in what is collectively called *"South Coast Metro ... the shape of the future"* – than merely intimations of urbanity. Its immanence, too, is breathtaking.

When America gets around to culture, the pioneers used to say, America will make culture hum. Except for places like Texas, there's nowhere the frontier spirit hums better than in affluent Orange County, which finally has symphony, opera, ballet, Broadway musicals, you name it, in a $73-million Orange County Performing Arts Center, known by its awful acronym OCPAC. Victory over any barbarian past is signified by a mighty triumphal arch.

But this isn't imperial Rome. It's the nebulous, non-urban realm of Orange County. The arch doesn't command intense life at the Forum, but at South Coast Plaza, the vast shopping mall and high-rise office development owned by Henry T. Segerstrom and his family, along the San Diego Freeway at Costa Mesa.

Never mind that the arch is a structural fake. Its reddish granite cladding is pure veneer, covering a trussed inner frame, all angles and squares, that has nothing to do with a rounded form. The great forward wall is nothing more than a free-stand-ing *screen*, an enormous advertisement, cut open in the shape of an arch.

Yet the superficial effect is grand ... the great symbolic portal

– which turns out to be not a real entrance at all – swells majestically across the front of Segerstrom Hall, the 3,000-seat auditorium that is OCPAC's pride and joy. At night, when the building is lit, the arch acts as a monumental proscenium for the social drama attending the performance, revealing open terraces that are crowded on warm evenings, glittering and mirrored spaces within – spectacularly walled in glass – through which a colossal "Firebird" sculpture by Richard Lippold crashes outward into the void, flashing brightly colored metal plumage.

There could be no better emblem for Orange County, crashing through provincialism to the big-time world of music and art.... Despite many architectural flaws, Segerstrom Hall is, functionally, the finest multipurpose facility of its kind in the country.[10]

So too, one might say, is all of Orange County: a structural fake, an enormous advertisement, yet functionally the finest multipurpose facility of its kind in the country, the "primitive society" of our future. How did it grow? The Swedish Segerstroms came to Orange County nearly a century ago to farm, and they claim to still be the world's largest lima bean producers. But it is culture without the agri that the sons of Segerstrom dig today. Henry T., for one, is now "building a city" where anything you want can be found along the orange brick road that winds its way through the "phalanx of showy department stores" and "many-arched portals" of the "curiously insubstantial" *South Coast Plaza*, California's largest and most profitable shopping mall, with nearly 3 million square feet of space and almost 10,000 parking places, Nordstroms, Mayco, Sears, Bullocks, Saks Fifth Avenue, Robinsons, the Broadway, over 200 other "specialty" stores and boutiques, and more than a half billion dollars of taxable retail sales. Around and inside the C. J. Segerstrom and Sons office buildings and the tall South Coast Plaza Westin Hotel are sculptures by Henry Moore, Alexander Calder, Joan Miró. Nearly at the intersection of Town Center and Park Center Drives and not far from the Center Tower and Town Center office complexes (centers growing everywhere), just next to "the charming little South Coast Repertory Theatre [that] perks up to the right" looms, "like the ascent to a shrine," the ceremonial ramp of OCPAC, rising toward the brightly lit auto pavilion drop-off point and "lofty, glass-enclosed lobby where the 'Firebird' flies outward above the broad terrace" of Segerstrom Hall itself. There is a superb view of the

---

[10] Allen Temko, "The Fine Sound and Flawed Design of a Grand Orange County OCPAC," *Los Angeles Times*, December 20, 1987.

plazas and mesas of the coastal plain in the distance from this extraordinary site.

Just around the corner, quietly hidden, stands "the remarkable glass and steel gates by Los Angeles artist Claire Falkenstein that lead to Segerstrom's most exalted public gift to the environment," Isamu Noguchi's "California Scenario," a spectacularly calm stone and water garden nestled nonchalantly in an eerie scene that both mirrors and hides from the buildings around it.

> Noguchi set himself the super-human task of fusing time and space into an art that makes Stonehenge modern and subsumes the idea of modernism to man's most primitive impulses, delivering him back – truly civilized – to the endless cycled rhythms of nature.[11]

Noguchi's California Scenario is a piece-full urban oasis where civilized nature is preserved in fabricated stone and water. I was once shown around it by a younger son of Segerstrom, whose primary, visceral, unbounded vitality was indeed breathtaking. He had worked personally with Noguchi to shape the scenario, to seek that balance between the "human community" and the "natural environment" that secures "the integrity of the system" and makes it soar to new heights – using much the same words as the ROOTS AND WINGS-themed Irvine Exhibit. Dressed tightly in Italian silk, he coolly spoke of family and farming in the blistering heat as he explained to me the symbolism of the various sculpted forms. As we approached one of my favorite spots, a pile of large fairly regularly-shaped grey-green stones turned orange by the late afternoon sun, he recounted how each stone was carefully cut and shaped in a particular Japanese village that specialized in making natural-looking objects from Nature so that no one can tell the difference. He watched Noguchi assemble these stony simulacra – perfect copies of non-existent originals – into a stimulating pile that would eventually be named in his family's honor: Ode to the Lima Bean.[12]

"I want to excentre myself," says Baudrillard in *America* (1988: 28), "to become eccentric, but I want to do so in a place that is the centre of the world." Perhaps here, in the illusive calm of the California Scenario, one comes as close as possible to the center of the contemporary world, the "highest astral point," the "finest orbital space."

[11] William Wilson, "Trail of Ageless Art is Noguchi's Legacy," *Los Angeles Times*, January 2, 1989, written soon after Noguchi's death.
[12] Sitting at the base of Ode to the Lima Bean, I introduced Orange County to a wider audience in a film produced by Vic Lockwood and the British Broadcasting Corporation (BBC) for the Open University series, "Understanding Modern Societies". See, in the most visual sense, "L.A.: City of the Future," London: BBC (1992).

Scene 8: Cities That Are Doubles of Themselves

Thus far we have concentrated on the breathtaking industrial-cum-commercial-cum-cultural core landscapes of Exopolis, only hinting at the existence of residential populations. It is time now for another spin outward, to the sleeping margins of Exopolis, the super-dormitories of the southern half of Orange County, another vast zone of re-centering.

> Mission Viejo – swim capital of the world, mecca for medalists, home of the perfect 10 high dive, three competition swimming pools but only one public library – is nestled alongside a freeway in the rolling hills of south Orange County.
>
> Billed by its developer as "The California Promise," it has emerged as the epitome of the American Dream.
>
> During the [1984] Olympic Games, 200 of the world's best bicycle racers pedaled for the gold here past neat lawns, $1-million lake-front homes, a private gated community and 200,000 cheering fans.
>
> Swimmers and divers trained here before reaping a harvest of Olympic medals, nine gold, two silver and one bronze – more medals than were won by France or Britain or, for that matter, 133 of the 140 nations taking part in the Games....
>
> The world-famous Nadadores swim and diving teams train here, and are subsidized by the Mission Viejo Co., the developer. But there also are three wading pools, four hydrotherapy pools, a 25-meter Olympic diving pool, 19 lighted tennis courts, 12 handball and racquetball courts, five volleyball courts, two outdoor basketball courts, men's and women's saunas, two weight rooms, four outdoor playgrounds, a multipurpose gym, 19 improved parks, four recreation centers, a 125-acre man-made lake, two golf courses, and three competition pools (one a 50-meter Olympic pool), and more – all built or donated, some still owned and operated, by the company.[13]

A local real-estate saleswoman and Municipal Advisory Council member summed it up: "It's a community that offers a great life style – a house in the suburbs, and your children kept busy. I think this is an area that people will flock to." Another member of the Council asks, "How can anybody from the East have anything but desire to move out here with us? I guess you'll just have to excuse us gloating about it. I couldn't speak too highly of the community – I'm in love with it." Some, however, are less sanguine. A forty-year-old housewife feels "out of step."

> "It's just a status thing here," she explained. "You must be happy, you must be well-rounded and must have children who do a lot of things. If you don't jog or walk or bike, people

---

[13] Mark Landesbaum and Heidi Evans, "Mission Viejo: Winning is the Only Game in Town," *Los Angeles Times*, August 22, 1984.

wonder if you have diabetes or some other disabling disease."

"I think it puts a tremendous amount of pressure on the children," she said. "We need a little more Huck Finn around here ... more time to kick tin cans down the street."

She later asked a reporter not to identify her because her comments could create friction for her husband with his business and golfing friends, who, she said, "are *very Mission Viejo.*"

Mission Viejo speaks for itself through the ventriloquy of the Mission Viejo Company, since the late 1970s part of the gargantuan empire of Philip Morris Inc., but with deeper roots going back to the large landholding families that continue to dominate the present-day political economy, even under new corporate umbrellas. The crafty corporate connections, however, are worth contemplating, for they too deal in illusion.

With its purchases first of General Foods and then Kraft Foods some years ago, the Philip Morris company has become one of the largest corporate conglomerates in America. Mission Viejo's staplemates now include not only the coughless Marlboro man and the tastes-good/less-filling Miller Lite beer drinker but also the good-to-the-last-drop coffee spouses of Maxwell House, the wish-I-were-an Oscar Meyer wiener, the Velveeta and Miracle Whip team of breadspreads, and the Kool-Aided and Tang-flaked babies – a cradle-to-grave conglomeration if there ever was one. And General Foods bolsters the list with even more mimetic products: Pop-Rocks, Shake'n Bake, Dream Whip, Stove Top stuffing, Jell-O, Cool Whip, and an array of brands that make it the largest coffee seller in the world: Maxwell House, Yuban, Sanka, Brim, Maxim, Mellow Roast (a "grain-extended product"), and the sugar-laden International Coffees.

Kraft's "foods" are still more specialized and further removed from the real things. Kraft has always concentrated on manufacturing ersatz dairy products. Some years ago, they marketed an "engineered" cheese made with vegetable oil instead of butterfat. This cheese "analog," as it was officially classified – it was to be called Golden Image – caused some problems with the National Cheese Institute. Since processed cheese is an imitation cheese to begin with, what do you call an imitation of an imitation? The National Cheese Institute suggested that perhaps the category should be identified as "Golana," because "it is pleasant sounding ... and is analog spelled backwards."[14] No one, it seems, suggested "Murcalumis."

---

[14] This "cheese" story has been adapted from M. Moskowitz, M. Katz, and R. Lovering, eds, *Everybody's Business – The Irreverent Guide to Corporate America*, New York: Harper and Row, 1980: 52.

Imitations and analogs of the corporate New Town of Mission Viejo (itself an imitation of the corporate New Town of Irvine) are filling up the frontierlands of south Orange County, lining the Saddleback Valley and other areas with a sprawl of coalescing urblets. Like the originals, they reach out for specialized residential niche markets and tightly package the local environment and life style, to the point of prescribing the colors you may paint your house, what may be done to the provided decor, whether you may hang an American (or other) flag outside your front door, and how best to keep up with the residential theme (Spanish Colonial, Greek Island, Cervantes' Spain, Capri Villa, Uniquely American). Everything is spelled out in lengthy contracts with the developer, which some claim venture too close to private sector socialism for their tastes. Housing struggles focus around permissions and exceptions. May I construct a basketball hoop? May I line my pool with black tile? Do I dare to paint it peach?

Mission Viejo is now cloning itself toward the beach in the new development of Aliso Viejo, while just inland there is an even bigger, 5,000-acre New Town Urban Village in the works, billing itself as Rancho Santa Margarita, "... where the west begins. Again." "The First Frontier of the 21st Century." Just across the way, where the 1984 Olympic Games Pentathlon was held, Arvida Disney (along with Chevron and City Federal Savings and Loan) initiated the development of the upscale "resort and residential community" of Coto de Caza, building upon its pre-existing facilities for riders of horses and hunters of pheasant, partridge, dove, quail, and clay pigeons. It continues to grow today under new corporate roots and wings.

Along the coast at Monarch Beach, Tokyo-based Nippon Shinpan Company, which owns Japan's largest credit-card service, purchased some years ago from an Australian subsidiary firm a 231-acre bite out of the last large stretch of undeveloped coastal property in Southern California. A new planned development was designed around a "world class" golf course not far from the very world class Ritz-Carlton Hotel, a reminder (if any is needed) of the growing internationalization of Exopolis. The local is becoming global with blinding speed.

Also an integral part of the residential panorama are two Elderly New Towns, aging rather nicely in Orange County. Leisure World, the gated retirement community in the Laguna Hills, is considered the largest retirement community in America. Around it has developed another specialized orbital landscape.

> The one-time bean fields outside Leisure World ... have sprouted at least nine securities brokerage houses, five banks, 12 savings institutions and numerous other money handlers.

The institutions have turned a five-block area outside Leisure World's main gate off El Toro Road and Paseo de Valencia into a supermarket of financial services. And many more brokers, bankers and lenders are a short distance away.

Retirement communities attracting brokers is not unusual. For instance, [the parent company of one local securities franchise is] based in the nation's premier retirement city, St. Petersburg, Fla. But Leisure World, which opened in September, 1964, is different. The development has an intense concentration of money and is located in a growing part of an affluent county. . . .

"It's the largest growth of upper wealth in the county". . . . About 21,000 mostly retired people (the average age is 76 years old) live in 12,736 units on 2,095 acres. Their homes, from apartment cooperatives to single-family dwellings, range in price from $40,000 to $400,000 – and the price does not include the land, which is owned by three Leisure World housing corporations. . . .

The five bank branches in the mini-financial center outside the main gate of Leisure World [where one of the first major buildings was called Taj Mahal] collectively reported total deposits of more than $343.1 million [in 1985]. . . .

The residents often were captains of industry – retired corporate executives, bank executives, publishing executives and successful doctors, dentists, and lawyers. At least three retired Army generals and two Navy admirals, along with a retired German U-boat captain from World War I, live in Leisure World.[15]

The generals and admirals that live in this targeted El(derly) Dorado of Leisure World bring to mind another fitful presence in the Exopolis, a more secretive series of theme parks being exposed with the in-fill of the outer spaces.

Scene 9: On the Little Tactics of the Habitat

Marine Lt. Col. William J. Fox was still angry about the Japanese attack on Pearl Harbor as he flew over the wide open farmlands of Orange County. It was 1942 and Fox was searching for "just the right place" for a mainland airfield where Marine Corps aviators could be trained for the campaign to regain the Pacific.

---

[15] James S. Granelli, "Brokerages Find a Gold Mine in Leisure World," *Los Angeles Times*, February 2, 1986.

As he swept over a tiny railroad whistle-stop called El Toro, Fox spotted a sprawling plot of land covered with bean fields and orange groves.

It was perfect: few and far-away neighbors; close to the ocean so pilots could practice carrier landings; within range of desert bombing ranges, and near Camp Pendleton, the Corps' then-new 125,000-acre troop-training base in northern San Diego County.

"Orange County was an ideal place for military bases," Fox, now 92 and a retired brigadier general, recalled.... "It was all open country.... There was hardly anyone living there."

Today, that Orange County airfield – the El Toro Marine Corps Air Station – is under a siege Fox could not have imagined 46 years ago. Tightly packed housing tracts have brought tens of thousands of neighbors creeping closer and closer to the base fences. High-rise buildings, shopping centers and industrial parks are popping up around the airfield. A stone's throw from the base, long lines of automobiles, motorcycles and trucks stream through the "El Toro Y" – the southern intersection of two of Southern California's busiest freeways, Interstates 5 and 405.

With the advance of urban development have come the volleys of complaints about the thunderous screams of low flying Marine Corps jets.

The peacetime assault on El Toro is not unique. Base commanders from Boston to San Diego and Seattle to Jacksonville Fla., are defending their ground against well-organized community groups, environmental activists, land-hungry developers and demanding local political leaders.... The Navy now reports that base neighbors have affected some kind of training or maneuvers at every one of its 65 air stations in the United States....

Although the military could not provide exact figures on how much it spends annually in its battle against what it terms encroachment by civilians, the amount runs well into millions of dollars. The Air Force, for example, ... has more than 100 people nationwide who deal with the problem daily. In addition, the Air Force has spent $65 million for land alone to increase buffer areas around air bases.[16]

The battlelines drawn here identify most of the collective actors struggling for territorial security and recentered identity in the outer

---

[16] George Frank, "Urban Sprawl: A New Foe Surrounds the Military," *Los Angeles Times*, December 24, 1988.

core regions of the Exopolis. Together these actors and their actions provide an apt introduction to the peculiar postmodern politics of the exopolitan borderlands. In no particular order they include: (1) platoons of besieged military commanders and weapons-testers "defending their ground" against the peacetime assault on their once pristine fortresses, deeply disturbed at being discovered but willing to enter into preliminary negotiations to maintain their footholds; (2) "well organized community groups" and homeowners' associations fighting, privately and publicly, for their property rights and values against all outside encroachments upon the premises and promises they have so faithfully purchased; (3) "land-hungry developers" and self-proclaimed "community builders" hunting for new room to accumulate and construct their spectacular capitals of fiction at all costs, a few under the guise of public-spiritedness and family values; (4) small clusters of "environmental activists" desperately searching for sanctuaries to protect the many endangered species of the Exopolis; (5) "demanding local political leaders," seeming to be stunned by it all, attempting impartially to turn every which way to profitably serve their fractious constituencies and maintain their own positions; and (6) the consultant media and imagineering specialists who selectively inform while performing as geopolitical spin-doctors to it all, shaping the vantage points and defining the battlelines of nearly all the little exopolitan wars.

This hexagonal contest for socio-spatial control (with its notable absence of alien ethnicities and the voices of the well-barricaded residential poor) effectively describes the recentered local politics of Orange County and other Outer Cities. Everything here seems to revolve around emplacement and positioning, or what Michel Foucault once described as "the little tactics of the habitat" and the "micro-technologies of power." In the peculiar political geometry of Orange County, where every point in space lays special claim to being the central place, these six-sided local wars over habitat and turf – or what might be called the "habitactics" of the Exopolis – occasionally assume a grander, more global, scope. A few more newspaper clippings arising specifically from the military presence in the late 1980s illustrate this globalization. Let me suggest for each an appropriate musical accompaniment, just to keep you humming.

"Up a Laser River"
A flash fire at a test facility near San Juan Capistrano will delay indefinitely the final testing of the missile-killing Alpha laser [capable also of obliterating whole cities if necessary], a key component of President Reagan's "Star Wars" defense initiative, the Pentagon disclosed.... The fire broke out ... at the sprawling 2,700 acre TRW Inc. plant in southern Orange

County when a worker opened a valve at the wrong time, officials said.... With the vacuum chamber contaminated by smoke and debris, officials said it is impossible to conduct experiments in which the laser beam would be produced and tested in space-like conditions.[17]

"SONGS of Silence"
In a special way, the people of San Clemente are a microcosm of today's American psyche, reflecting the complacency and occasional concern about the nation's problems.... Three nuclear reactors are now being used to generate ... electricity at the largest and, potentially, the most dangerous nuclear power plant west of the Mississippi.... The plant represents one of the most terrifying threats imaginable, short of full-scale war, yet people seldom speak of that. They live in the "Basic Emergency Planning Zone," with its questionable promise that, in the event of something untoward, there will be a sure and hasty evacuation of everyone. The people hereabouts depend on that, even as they depend on assurances that it won't ever be needed.... What will it take to arouse them? ... I have told some people that a single reactor meltdown at [SONGS – the San Onofre Nuclear Generating Station] could cause 130,000 early deaths, 300,000 latent cancers, and the evacuation of 10 million people.... They listen, but they cannot allow themselves to consciously accept such grim processes. They choose not to live in fear.[18]

"Tern, Tern, Tern"
[Camp] Pendleton, located almost midway between the fast growing areas of Orange and San Diego counties, has ... become the last sanctuary for many endangered species. Pendleton troops engaged in field warfare training must be careful to avoid the nesting areas of the lightfooted clapper rail, the Belding's Savannah sparrow or the California least tern.... At Camp Pendleton, a chain reaction wreck ... on busy Interstate 5 occurred after a cloud of dust from maneuvering tanks blanketed the highway, cutting visibility to near zero. Just to the north, Irvine community leaders became alarmed in 1986 over safety when a CH-53E Super Stallion transport helicopter based in Tustin came down for an emergency landing near a

---

[17] Richard Beene and John Broder, "Fire at TRW Delays Final Testing of Alpha Laser, Key 'Star Wars' Weapon," *Los Angeles Times*, January 29, 1988.
[18] Fred Grumm, "Make-Believe Sunshine in the Shadows of San Onofre," *Los Angeles Times*, October 11, 1987.

residential area. And earlier this year, an out-of-control Navy F-14 returning to Miramar Naval Air Station in San Diego crashed at a suburban airport near the base, killing one and injuring four others seriously.... "The West Coast is right now in the lead as far as these encroachment problems go," [a Marine Colonel] said.[19]

It was recently announced that the El Toro Marine Corps Air Station, as part of the military cutbacks of the 1990s, would be shut down completely in the near future, setting off a frenzy of competitive ideas for its re-use. Still scheduled to survive, however, is the huge Camp Pendleton, which has continued to stir up new habitactical wars, including one over refuge-seeking undocumented migrants squatting on Camp grounds to evade the most authoritative checkpoint along the California coast, symbolized in freeway signs with a silhouette of a Latino mother and child rushing across to avoid the murderous traffic.

Scene 10: Scamscapes: The Capitals of Fiction Become Reality

Every day, life in Orange County seems to move one step beyond Eco's early vision of Disneyland as "a degenerate utopia ... an ideology realized in the form of a myth ... presented as at once absolutely realistic and absolutely fantastic," and deeper into Baudrillard's world of hypersimulation, where "it is no longer a question of a false representation (ideology), but of concealing the fact that the real is no longer real." Perhaps more than any other place, more than the desert, more than either New York or Los Angeles, Orange County verifies Baudrillard's iconoclastic vision of the anesthetic enchantments of hyperreality. It is the generative utopia, a make-believe paradise that successfully makes you believe in make-believing, the most irresistably California-looking of all the Californias, the most like the movies and TV "situation comedies," the most like the promised American Dream.

Under these transcendental conditions, it is no surprise that image and reality become spectacularly confused, that the difference between true and false, fact and fiction, not only disappears but becomes totally and preternaturally irrelevant. In this final scene from the inventive Exopolis of Orange County, another postmodern geography is explored and brought up to date. It represents the landscape of hyperreality as a fraudulent or better, metafraudulent *scamscape*, an ecstatic playground for the habitactics of make-believe.

---

[19] George Frank, "Urban Sprawl: A New Foe Surrounds the Military," see footnote 16, p. 271.

We begin, appropriately enough, in the heterotopia of the *boiler room*.

> Orange County holds the dubious title of "the fraud capital of
> the world," according to U.S. Postal Service inspectors. Five
> inspectors working out of the Santa Ana post office will handle
> mail-fraud complaints involving as many as 10,000 victims this
> year, said ... the inspector who heads the team. [He] estimated
> that Orange County suffers losses of $250 million a year in
> fraud.... Orange County's affluence and the large number of
> retired people living here combine to make the area a favorite
> of con men....
>
> Postal inspectors say the hottest current schemes involve pre-
> cious-metal futures. Underground boiler room operations typi-
> cally convince investors they can reap huge profits and then
> spend their money on parties, drugs, and cars.... The operators
> usually disappear about the time the investors become suspi-
> cious.
>
> Another popular scam, envelope-stuffing, is difficult to trace
> because victims who send in money to learn how to participate
> are usually too embarrassed to admit they've been had.[20]

More than 200 boiler rooms – named for the intensity of activity in the
barest of spaces – now punctuate the scamscape of Orange County,
reportedly generating activities that gross over $1 billion a year. The
typical boiler room is, at its base, a telemarketing business serving to
collect money for charities, public television stations, credit card and
loan applications, as well as for promoting investment schemes
promising quick and easy profits. The basic work is done by "wholesome youth" on telephones with high-tech "confidencers" used to sift
out all background noise. Often starting out their employment careers,
the youthful telemarketers are trained by experts who instill confidence and entrepreneurship with promises of advancement to higher
levels of take-home pay (much of it tax-free, it is hinted). The densest
agglomeration of boiler rooms is probably in and around Newport
Beach, one fraud inspector said, "simply because it sounds classier
over the phone than, say, Pomona."

Competition in this telephonic cyberspace is intense and profitable, especially when dealing with "investment opportunities" that
are promoted by young telemarketers reading from prepared scripts
infused with the habitactics of make-believe. According to authorities, the average boiler room fraud victim loses between $40,000 and
$50,000. One was reported to have invested $400,000 based on a single telephone call, and another, a 90-year-old widow in Nebraska,

[20] Untitled article in the Los Angeles *Daily News*, March 24, 1987.

sent off $750,000 to a young man who told her he was "a native Nebraska boy brought up with high morals" and was working his way through college. It is not uncommon for some of the boiler rooms to have a gross take of $3 million a month, certainly competitive with narcotics and, it would seem, infinitely more wholesome. During one police raid, a placard was found on a salesman's desk. Effectively capturing the sincerely duplicitous honesties of the boiler room, another of the magical enclosures of the Exopolis, it proudly proclaimed: WE CHEAT THE OTHER GUY AND PASS THE SAVINGS ON TO YOU.

But the boiler room is only one of the many indications that the "seven hundred and eighty-six square mile theme park," the place where "you can have anything you want," has become the most active scamscape on earth. The Defense Criminal Investigative Service office in Laguna Niguel, investigating fraud in the defense industry, is the largest in the country, befitting Southern California's world leadership in the devastating production of offensive weaponry. In the late 1980s, before the defense industry burn-down in Southern California, the DCIS filed nearly 100 indictments and recovered more than $50 million from local cases involving much more imaginative practices than the famous examples of the $640 toilet-seat and the even more costly screwdriver. In highly specialized Orange County, the key frauds involved product substitution and the falsification of test results. Its most controversial case forced the bankruptcy and closure of a local firm that produced "fuzes" for the warheads of Phoenix air-to-air missiles – the weapons of choice for "Top Gun" Navy jet pilots. The fuze's frightening function is appropriately ambivalent: it both detonates the warhead and prevents it from detonating prematurely. Local workers confidently boasted of their prowess on embossed metal signs that said: THE BEST DAMNED FUZES IN THE WORLD ARE MANUFACTURED BEHIND THIS DOOR and hinted that this made thorough testing unnecessary and superfluous. Pentagon procurers and DCIS officials, however, worried that the difference between the two functions of detonation and antidetonation was being insufficiently attended to on the factory floor. To this day, no one really knows which viewpoint was correct.[21]

---

[21] Such duplicitous honesties continue today throughout Southern California, still "the nation's capital for defense fraud," even with the significant reductions in federal expenditures. It was reported in 1995 that 75 percent of all criminal fines collected across the country in the previous four years came from Southern California, the most serious case being the bypassing of required inspections and the falsification of test records in the Lucas Aerospace plant in the City of Industry. It was alleged that a faulty gearbox produced at the plant was responsible for the crash of a $40 million F-18 jet fighter during the 1991 Gulf War. See Ralph Vartabedian, "Cases of Defense Fraud Boom Amid Cutbacks," *Los Angeles Times*, March 26, 1995.

Other variations on the recurrent theme of fraud involve bankruptcy and foreclosure scams, stockswindling (junk bonds were invented over the county border in Beverly Hills), computer crimes, environmental regulation crimes, real estate cons, many types of insurance fraud (including a particularly serious problem, fed by a small army of what are called "capper lawyers," who stage automobile accidents and occupational injuries to obtain their percentages), Medicare and Medicaid fraud, and an amazing variety of other scams and cons that densely layer everyday life with the ideologic of hyperreality. But these are small change compared to two explosions that have hit the scamscape in the past decade. Some sort of mythic peak of financial fraud was reached in the late 1980s in the Savings and Loan industry debacle. Symbolizing this meltdown was a "thrift" headquartered in Irvine but connected all over the US, and especially to the junk-bond empires that continue to rise and fall in Southern California. So costly were the unregulated scams of Charles H. Keating's unthrifty Lincoln Savings and Loan company that its sincerely duplicitous honesties and the larger riot of S&L failures it connected with will probably cost taxpayers as much as $500 billion to cover up and repair. Keating and most of the other S&L criminals insisted that they did nothing wrong, they played the deregulated financial games the way the law said they should be played. In this financial crisis, the scamscape surpassed itself, transforming mere fraud into *metafraud*, and "false representation" into a ceremonial concealment of the fact, as Baudrillard says, "that the real is no longer real."

In December, 1994, the scamscape exploded again as the entire Orange County government declared bankruptcy after it was found that the public sector ran its finances very much like the bankrupt Savings and Loan companies, with a touch of the boiler room added for good measure. Just as Keating and his companies had come to symbolize a new entrepreneurial model of people-oriented investment in the mid-1980s, the Orange County investment fund, run by a folksy tax collector with the almost dementedly appropriate name of Citron, had become a nationally known model of the "new fiscal populism," artfully playing the profitable financial games of the 1990s. Constrained by the tax-revolt metafraud of Proposition 13, passed in 1978 as a deceptive simulation of popular democracy,[22] all local governments in California were forced to be fiscally innovative to make up for the downsizing of property tax revenues. Citron-led

---

[22] For an excellent analysis of the California tax revolt and the passage of Proposition 13, with its massive shift of resources from the public to the private sector, see Clarence Y. H. Lo, *Small Property versus Big Government: Social Origins of the Property Tax Revolt*, Berkeley and Los Angeles: University of California Press, 1990 (paperback with new preface and epilogue, 1995).

Orange County, with almost no supervision from the elected Board of Supervisors (as usual a *faux* governmental body filled with retired colonels and other amateur political simulacra), leapt head on into the magical realism of the new bond market, with its lush derivatives, reverse repos, and inverse floaters. Using local government's main comparative advantage – its (significantly reduced post-Prop 13) ability to raise tax revenues and issue municipal and "general obligation" bonds – the Orange County Investment Fund gambled the county's well-being on the bond market and on the continuing drop in interest rates. By early 1994, the fund had reached at least $7 billion; by the end of the year it was $1.7 billion shorter and what has been described as America's richest and most highly educated (and Republican) county was forced into the largest municipal bankruptcy in US history.

Finale: The Precession of Exopolis

It is not just in Orange County where everything is possible and nothing is real. Creatively erosive postmodern urban geographies are being invented at a furious pace in every region of the country. The simpler worlds of the artificial theme parks are no longer the only places where the disappearance of the real is revealingly concealed. This ecstatic disappearance permeates everyday life, enabling the hyperreal to increasingly influence not only what we wear and what we eat and how we choose to entertain ourselves; but also where and how we choose to live, who and what we vote for, how government is run, and also how we might be agitated to take more direct political action not just against the precession of simulacra but *within* it as well.

Simulations increasingly dominate contemporary politics at the local, state, national, and international levels. Reality is spin-doctored for us by a new caste of specialists who determine exactly what will be revealed and what will be concealed. Pollsters take the temperature of this fickle body politic as sim-citizens march together to the fine-tuning of such metafrauds as "the taxpayers revolt," "small government is good government," "the magic of the market," "telephonic democracy," "the end of history," and "the triumph of capitalism." A new mode of regulation seems to be emerging spontaneously from this diffusion of hyperreality, plugging us into the new economic machinery of virtual reality and cyberspace, protected by elaborately carceral systems of social control and leading us to the promised lands of postmodern re-enchantment, where tax cuts for the rich magically benefit the poor and social spending for the poor is seen as hurting those that receive it.

The Exopolis stretches our imaginations and critical sensibilities in much the same way it has stretched the tissues of the modern metropolis: beyond the older tolerances, past the point of being able to spring back to its earlier shape. For those who choose to struggle against its alluring and illusive embrace, stubbornly modernist modes of resistance are certainly not enough. New tactics and strategies must come into being which draw upon an assertively radical and postmodern subjectivity to take us beyond the Baudrillardian message and messenger to a new cultural politics of identity/difference/resistance, to a radical and creatively postmodern spatial praxis that recognizes the hyperreality of everyday life in the intertwinings of the Inner and Outer City, and seeks to transform the real-and-imagined by forcefully spin-doctoring alternative images, strategic simulacra that subvert from within and work toward transforming our lived spaces of representation.

Unmasking the corruption, deceit, greed, emptiness, the tinseled superficialities and exploitative social relations of our contemporary lifeworlds, is still useful. Indeed, much of this chapter is just such a critical unmasking of what seems to lie behind the radiant surface of the Exopolis. What has been revealed, however, is that practically everything discoverable beneath the surface is also another mask. To dig even deeper will bring us *back to the surface*, to an Orange County that continues to function symbolically as an exceedingly attractive lived space, undemystifiable because its (hyper)reality *is* mystification itself. What the journey through the Exopolis of Orange County begins to suggest to those seeking an end to exploitation, domination, and subjection is that we need not only to declaim and demystify but also, from a critical space of radical openness, to *re*claim and *re*mystify hyperreality in a determined continuation of progressive political projects. To do otherwise verges on both anachronism and anachorism (inappropriate location in space).

9

The Stimulus of a Little Confusion: A Contemporary Comparison of Amsterdam and Los Angeles[1]

All these elements of the general spectacle in this entertaining country at least give one's regular habits of thought the stimulus of a little confusion and make one feel that one is dealing with an original genius.

(Henry James, experiencing the Netherlands in
Transatlantic Sketches, 1875: 384)

What, then, is the Dutch culture offered here? An allegiance that was fashioned as the consequence, not the cause, of freedom, and that was defined by common habits rather than legislated by institutions. It was a manner of sharing a peculiar – very peculiar – space at a particular time ... the product of the encounter between fresh historical experience and the constraints of geography.

(Simon Schama, *The Embarrassment of Riches: An Interpretation
of Dutch Culture in the Golden Age*, 1988: xi)

[1] The original version of this chapter was published by the Centrum voor Grootstedelijk Onderzoek (Center for Metropolitan Research) of the University of Amsterdam in 1991 as part of its series of "Texts of a Special Lecture." Sponsored by the city of Amsterdam, the CGO supports a program of visiting professors on urban research. I was a visiting professor in the Spring of 1990 and presented a lecture on "The Changing Relations of City and Suburb in Los Angeles and Amsterdam," from which the first publication derived. The essay was subsequently reprinted (without its footnotes) in M. P. Smith, ed., *After Modernism: Global Restructuring and the Changing Boundaries of City Life*, New Brunswick, NJ and London, UK: Transaction Publishers, 1992: 17–38; and, in full, in Leon Deben, Willem Heinemeijer and Dick van der Vaart, eds, *Understanding Amsterdam*, Amsterdam: Het Spinhuis, 1993. I have reworked the essay to fit this chapter of *Thirdspace*.

On Spuistraat

In Amsterdam in 1990, I dwelt for a time on Spuistraat, a border street on the western flank of the oldest part of the Inner City. Squeezed in between the busy Nieuwezijds Voorburgwal (literally, on the "new side" of the original settlement, in front of the old city wall) and the Singel (or "girdle," the first protective canal moat built just beyond the wall), Spuistraat runs roughly North–South starting near the old port and the teeming Stationsplein, where the Central Railway Station sits blocking the seaview, pumping thousands of visitors daily into the historic urban core. At half its length, Spuistraat is cut by Raadhuisstraat, the start of the main western boulevard axis branching off from the nearby Royal Palace (once the Town Hall or *raadhuis*) and tourist-crammed Dam Square, where the city was born more than 700 years ago in a portentous act of regulatory tolerance (granting the local settlers toll-free use of the new dam across the Amstel River, Amstelledamme becoming Amsterdam). Recalling my stay in Amsterdam provides another opportunity to explore other urban spaces of representation.

After passing the old canal house where I lived, the street ends in what is simply called Spui, or "sluice," once a control channel connecting the Amstel and the older inner city canal system with the great bib of concentric canals that ring the outer crescent of the Inner City, or as it is popularly called, the Centrum. The Spui (pronounced somewhere in between "spay" and "spy") is now a short broad boulevard lined with bookstores, cafes, a university building, the occasional open-air art fair, and the entranceways to several popular tourist attractions, ranging from the banal (Madame Tussaud's) to the enchanting (the Begijnhof and just beyond, the Amsterdam Historical Museum). The "city museum" offers the most organized introduction to the historical geography of Amsterdam, with roomsful of splendid imagery bringing to life what you see first upon entering, a panoramic model that sequentially lights up the city's territorial expansion in stages from 1275 to the present. Just as effective, however, as a starting-point for an interpretive stroll through Amsterdam's Centrum is the Begijnhof, or Beguine Court.

The Begijnhof is a small window on to the urban imaginary of contemporary Amsterdam. It is filled with that bewildering Dutch mix of the familiar and the incomprehensible that so attracted Henry James in 1874 and later inspired Simon Schama's brilliant interpretation of the "moralizing geography" of Dutch culture in its 17th-century Golden Age, *The Embarrassment of Riches*.[2] One enters the

---

[2] Simon Schama, *The Embarrassment of Riches: An Interpretation of Dutch Culture in the Golden Age*, Berkeley and Los Angeles: University of California Press, 1988.

Begijnhof through an arched oak door off Spui square, an innocently unmarked opening to an enticing microcosm of civic refuge and peaceful respite in a cosmopolitan Dutch world of ever-so-slightly repressive tolerance. Before you is a neat quadrangle of lawn surrounded by beautifully preserved and reconstructed 17th- and 18th-century alms-houses, nearly every one fronted with flower-filled gardens. A restored wooden house dates back to the 15th century, one of two survivors of the many fires which burned down the old city (and the original Begijnhof) before the more substantial Golden Age. The other survivor is located in a different kind of refuge zone along the Zeedijk, today known as the "boulevard of junkies," the tolerated and planned resting place for the city's residential corps of aging hard-drug addicts.

There are also two small churches in the Begijnhof, one dating back to 1392 but built again in 1607 and known since as either English Reformed or Scottish Presbyterian. Here the fleeing English Pilgrim Fathers prayed before setting sail on the *Mayflower*, comfortable in their temporary but dependable Dutch haven. On one of my visits, the church was filled with the concerted voices of the Loyola College choir from New Orleans, singing American spirituals to the passers-by. The other church, a clandestine construction in 1655, was originally a refuge for Catholic sisters escaping post-Reformation Calvinist religious purges. One of its stained-glass windows commemorates the epochal "wafer miracle" of 1345, an event that boosted Amsterdam into becoming a major medieval pilgrimage center and began its still continuing and far reaching internationalization.[3]

The Beguine Court was originally founded one year after the miracle as a sanctuary for the Beguines, a Dutch lay sisterhood that sought a convent-like life but with the freedom to leave and marry if they wished, an early marker of the many Dutch experiments with what might be called engagingly flexible inflexibility. Today, the Begijnhof continues to be home to *ongehuwde dames* (unmarried ladies) who pay a nominal rent to live very comfortably around the lawned quadrangle and proffer their services to the ill and the elderly. Despite the flocking tourists, it remains a remarkably peaceful spot, a reflective urban retreat that succeeds in being both open and closed at the same time, just like so many other paradoxical spaces and places in the refugee-filled Amsterdam Centrum.

I lived just around the corner in another of these artfully preserved places and spaces, a relatively modest variant of the more than 6,000

---

[3] The "miracle" apparently occurred when a sick man, unable to swallow the communion bread, spit it into a fireplace where it remained whole and unburnt. The site of this holy event is commemorated under a small glass window embedded into the sidewalk near where the 14th-century Chapel of the Holy City formerly stood.

"monuments" to the Golden Age that are packed into the sustaining Centrum, the largest and most successfully reproduced historic inner city in Europe. With a frontage that seemed no wider than my driveway back home in Los Angeles, the building, like nearly all the others in the Centrum, rose four storeys to a gabled peak embedded with a startling metal hook designed for moving furniture and bulky items by ropes in through the wide windows. Given my sizeable bulk (I stand nearly two meters high and weigh more than an eighth of a ton), I had visions of having to be hoisted up and in myself when I first saw the steep narrow stairwell (*trappenhuis*) to the first floor. But I quickly learned to bow my head and sidle in the doorway.

Golden Age taxation systems encouraged physical narrowness and relatively uniform building facades up front, squeezing living space (and stimulating expansive creativity in interior design) upward and inward from the tiny street- or canal-side openings. The patient preservation yet modernization of these monuments reflects that "original genius" of the Dutch to make the most of little spaces, to literally produce an enriching and communal urban spatiality through aggressive social intervention and imaginative grass-roots planning. In many ways, the preservative modernization of the cityscape of the Centrum has been an adaptive feat on a par with the Dutch conquest of the sea.

Simon Schama roots Dutch culture in this moral geography of adaptation, an uncanny skill in working against the prevailing tides and times to create spaces and places that reinforce collective self-recognition, identity, and freedom. "Dutchness," he writes, "was often equated with the transformation, under divine guidance, of catastrophe into good fortune, infirmity into strength, water into dry land, mud into gold" (1988: 25). In Amsterdam, perhaps more so than in any other Dutch city, these earthy efforts to "moralize materialism" moved out from the polderlands to become evocatively urban, not through divine guidance so much as through secularized spatial planning, enlightened scientific management, and an extraordinarily committed civic consciousness that persists to the present. The canal house simulates this rootedness, enabling one to experience within it the very essence of a liveable city, the agglomeration of individuals into socially constructed lifespaces that are always open to new possibilities even as they tightly enclose and constrain. The lived spaces of the Centrum are popularly designed to make density beautiful as well as accommodating, to flexibly enculturate and socialize without imprisoning, to make the strange familiar, and to add somehow to one's regular habits of thought that entertaining stimulus of a little confusion.

To live in a canal house is to encounter Amsterdam immediately

SPUISTRAAT 230

and precipitously, as my kind hosts from the University of Amsterdam's Center for Metropolitan Research knew well in finding me such strategically located lodgings.[4] The past is omnipresent in its narrow nooks and odd-angled passageways, its flower-potted corners and unscreened windows that both open and close to the views outside. Everyday life inside becomes a crowded reminder of at least four rich centuries of urban geohistory being preserved on a scale and contemporary intensity that is unique to Amsterdam. At home, one is invited daily into the creative spatiality of the city's social life and culture, an invitation that is at the same time embracingly tolerant and carefully guarded. Not everyone can become an Amsterdammer, but everyone must at least be given the chance to try.

The prevailing atmosphere is not that of a museum, however, a fixed and dead immortalization of the city's culturally built environment. The history and geography are remarkably alive and filled with the urban entertainment that makes Amsterdam so attractively familiar and yet so peculiarly incomprehensible, neat and clean and regular but curiously tilted, puzzling, an island of mud not quite entirely turned into gold but transformed enough to make one believe in the creative alchemy of Amsterdam's modestly democratic citybuilders. From my vantage point on Spuistraat a moving picture of contemporary life in the vital center of Amsterdam visually unfolded, opening my eyes to much more than I ever expected to see.

The view from my front windows affirmed for me what I continue to believe is the most extraordinary quality of this city, its relatively (the Dutch constitutionally refuse all absolutes) successful achievement of highly regulated urban anarchism, another of the creative paradoxes (along with the aforementioned and closely related "repressive tolerance" and "flexible inflexibility") that two-sidedly filter through the city's historical geography in ways that defy comparison with almost any other *polis*, past or present. This deep and enduring commitment to libertarian socialist values and participatory spatial democracy is apparent throughout the urban built environment and in the social practices of urban planning, law enforcement, popular culture, and everyday life. One senses that Amsterdam is not just preserving its own Golden Age but is actively keeping alive the very possibility of a socially just and humanely scaled urbanism. Still far from perfection itself, as the Dutch never

---

[4] My thanks to Leon Deben, Dick van der Vaart, and Jacques van de Ven, for sponsoring my stay in Amsterdam, along with the Department of Social Geography of the University of Amsterdam. Very special thanks also to Pieter Terhorst of the Department of Social Geography, who more than anyone else warmly and informatively shaped my understanding of Amsterdam.

cease telling you, Amsterdam is none the less packed with conspicuously anomalous achievements. There is little or no boosterism, no effort to proclaim the achievements or to present them as a model for others to follow. Instead, there is again, *pace* Schama, an unadvertised "embarrassment of riches," modestly reproduced as in the past on the "moral ambiguity of good fortune."

There are many ways to illustrate this peculiar urban genius. For now, the view through my Spuistraat window will do for a start. Immediately opposite, in a building very much like mine, each floor is a separate flat and each storey tells a vertical story of subtle and creative citybuilding processes. It was almost surely a squatter-occupied house in the past and is probably one now, for Spuistraat has long been an active scene of the squatter movement. On the ground floor is an extension of the garage offices next door. There is a small "No Parking" sign on the window but nearly always a car or two is parked in front. Our ground floor, in contrast, is a used book shop, one of the many dozens densely packed in this most literate of Centrums, the place where enlightened scholars from Descartes to Voltaire, Montesquieu, and Rousseau first found the freedom to have their enlightened works published and publicized without censorship.

One cannot avoid noticing that the automobile is an intruder in the Centrum. Spuistraat, like so many others, is a street designed and redesigned primarily for pedestrians and cyclists. Alongside the busy bike path there is a narrow one-way car lane and some newly indented parking spaces, but this accommodation to the automobile is tension-filled and wittily punctuated. The police are always ready to arrive with those great metal wheel clamps and the spectacle of their attachment usually draws appreciative, occasionally cheering and laughing, crowds of onlookers. Traffic is nearly always jammed, yet (most of the time) the Dutch drivers wait patiently, almost meekly, for they know they are guilty of intrusion and wish to avoid the steel jaws of public disapprobation. I was told that the city planners have accepted the need to construct several large underground parking garages in the gridlocked Centrum, but only with the provision that for every space constructed below ground, one space above is taken away.

On the first floor of the house across the way were the most obviously elegant living quarters, occupied by a woman who had probably squatted there as a student but had by now comfortably entered the job market. She spent a great deal of time in the front room, frequently had guests in for candlelit dinners, and would occasionally wave to us across the street, for my wife, Maureen, and I too had our most comfortable living space just by the front windows. On the floor above there was a young couple. They were probably still students and still poor, although the young man may have been working at least part-time for he was rarely seen, except in the morning and late at night. The woman was obviously pregnant and spent most of her time at home. Except when the sun was bright and warm, they tended to remain away from the front window and never acknowledged anyone outside, for their orientation was decidedly inward. The small top floor, little more than an attic, still had plastic sheeting covering the roof. A single male student lived there and nearly always ate his lunch leaning out the front window alone. His space made one wonder whether the whole building was still a "squat" for if he was paying a nominal rent, one would have expected the roof to have been fixed, in keeping with the negotiated compromises that have marked what some would call the social absorption of the squatter movement in the 1980s. Civic authorities have actually issued pamphlets on "How to be a Squatter" in Amsterdam, still another example of creatively regulated tolerance.

This vertical transect through the current status of the squatter movement was matched by an even more dramatic horizontal panorama along the east side of Spuistraat, from Paleisstraat (Palace Street) to the Spui. To the north (my left, looking out the front window) was an informative sequence of symbolic structures, beginning with a comfortable corner house on Paleisstraat that was recently rehabilitated with neat squatter rentals (another contradiction in terms?) above; and below, a series of shops also run by the same group of rehabilitated and socially absorbed squatter-renters: a well-stocked fruit and vegetable market-grocery selling basic staples at excellent prices, a small beer-tasting store stocked with dozens of imported (mainly Belgian) brews and their distinctively matching drinking glasses and mugs, a tiny bookstore and gift shop specializing in primarily black gay and lesbian literature, a used household furnishings shop with dozens of chairs and tables set out on the front sidewalk, and finally, closest to my view, a small hand-crafted woman's cloth hat shop.

This remarkably successful example of gentrification by the youthful poor is just a stone's throw away from the Royal Palace on the Dam, the focal point for the most demonstrative peaking of the radical squatter movement that blossomed city-wide in conjunction with the coronation of Queen Beatrix in 1980. A more immediate explanation of origins, however, is found just next door on Spuistraat, where a new office-construction site has replaced former squatter dwellings in an accomplished give and take trade-off with the urban authorities. And just next door to this site, even closer to my window, was still another paradoxical juxtapositioning, one which signaled the continued life of the radical squatter movement in its old "anarchic colors."

A privately owned building had been recently occupied by con-temporary squatters and its facade was brightly repainted, graffi-toed, and festooned with political banners and symbolic bric-à-brac announcing the particular form, function, and focus of the occupa-tion. The absentee owner was caricatured as a fat tourist obviously beached somewhere with sunglasses and tropical drink in hand, while a white-sheet headline banner bridged the road to connect with a similar squat on my side of the street, also bedecked with startling colors and slogans and blaring with music from an estab-lished squatter pub. I was told early in my stay that this was the most provocative ongoing squatter settlement in the Centrum. It was scheduled to be recaptured by the authorities several days after my arrival, but when I left the situation was unchanged, at least on the surface.

The view south, to my right, on Spuistraat presented another urban trajectory dominated by much more traditional forms of gen-trification. Some splendid conversion, using fancy wooden shutters, modernized gables (no hook here), and vaulted interior designs was transforming an old structure for its new inhabitants, who were much more likely to visit the boutiques and gourmet restaurants in the vicinity than the shops up the road. The transition quickened in a little restaurant row that ranged from what was reputed to be the best seafood place in Amsterdam and one of the grandest traditional centers of Dutch cuisine (called the Five Flies and fed daily by bus-loads of mainly Japanese and German package-tourists), to a variety of smaller cafes, Indonesian restaurants (considered part of Dutch cuisine), and fast-food emporia selling tasty bags of frites.

By the time you reach the Spui on foot, the street scene is awash again with activity and variety. A large bookstore shares one corner with an international news-center, spilling over on to the sidewalk with newspapers, magazines, and academic journals from around the world as well as pamphlets, flyers, and broadsheets announcing more local events. There are beer pubs nearby, as well as an American-style cocktail bar and several representatives of the aston-ishing variety of specialized Amsterdam cafes. One guidebook lists the following cafe types, each with its own internal variations: white, brown, and neo-brown; cocktail bars, gay bars, beer cafes, student cafes (differentiated by dress codes and academic disciplines), liter-ary cafes, chess cafes, ping-pong cafes, theatrical cafes, high-tech cafes, 8–2 cafes, discotheques, and night pubs.

One of the Centrum's best known "white cafes" (just drinks) is located where Spuistraat meets the Spui. It is beginning to lose its yuppie edge, however, to the stand-up, quick service, "old-style" cafe next door, much better able to quench the growing thirst for nostalgia. Nearly adjacent but stoically distanced is a famous radical cafe, where an older clientele sits and glares at the sipping elites across the way. The dense territorialities here are invisible to the casual visitor and they may be blurring even for the Dutch, as the cosmopolitan mixture of Amsterdam takes over, globalizing the local street scene.

Just around the corner are a few of the Centrum's hash coffee shops, perhaps the best known of Amsterdam's almost infinite variety of meeting-places. Their heady smoke flows out to fill the nostrils of passers-by. While living on Spuistraat, I had the good fortune to be taken on an amazing tour of the entire inner Centrum by Adrian Jansen, author of a fascinating analysis of the geography of hash coffee shops in Amsterdam.[5] Jansen is the Baudelaire of the Centrum, a geographer-*flâneur* of the inner spaces of the city. He reads the Centrum on foot – he reputedly does not even own a bicycle – and writes on such topics as "Funshopping" and the Dutch taste for Belgian beer and beerhalls. In his treatise on cannabis in Amsterdam, which ranges from Felliniesque poetics to stodgy classical location theories (e.g. Hotelling's famous "ice-cream vendor" model), Jansen describes life in Siberia, the name of a hashish coffee shop he took me to on our foot-tour.

> Compared to the "Tweede Kamer," "Siberia" is a much larger coffee shop. Not only does it offer a large variety of softdrinks, but they also serve excellent coffee. The espresso machine guarantees high quality, as it does in most coffee shops. The table football game is in constant use. Some people come in to play cards. Others play a kind of home-made skill game, in which a bicycle bell sounds if someone is not doing too well. Not every visitor buys hashish or marijuana, and not everyone pays. A man, clearly on his way home from the beach, orders yoghurt and disappears soon after eating it. Two men from Surinam enter. One has a story about his jacket being stolen. He asks the coffee shop owner for a loan, since he and his friend want a smoke. He gets it, but the shopkeeper makes him promise to return the money tomorrow. (I have reason to believe that the two use hard drugs as well. They are often to be seen at the "Bridge of Pills," a spot near my home where hard drugs change hands. Their hollow cheeked faces show small inflammations.) Hashish and marijuana are offered here in prepacked quantities; in small bags worth ten or 25 guilders. The ten guilder bags appear to be the most popular. The shop owner turns to me and says, "Hey, tell me, what do you think is the

---

[5] A. C. M. Jansen, *Cannabis in Amsterdam: A Geography of Hashish and Marijuana*, Muiderberg: Coutinho BV, 1991.

best coffee shop around?" A difficult question.   (Jansen, 1991:
14)

My journey with Jansen opened up many spaces I would never have
seen on my own, or with any other tour-guide. Thank you, Adrian.
Here's to your health.

Back on Spuistraat, the panorama being explored seems to concentrate and distill the spectrum of forces that have creatively rejuve-
nated the residential life of the Centrum and preserved its
anxiety-inducing *overvloed* (superabundance, literally overflood) of
urban riches. At the center of this rejuvenation has been the squatter
movement, which has probably etched itself more deeply into the
urban built environment of Amsterdam than in any other inner city
in the world. To many of its most radical leaders, the movement
today seems to be in retreat, deflected if not co-opted entirely by an
embracing civic tolerance. But it has been this slightly repressive tolerance that has kept open the competitive channels for alternative
housing and counter-cultural lifestyles not only for the student
population of today but for other age groups as well. It has also
shaped, in distinctive ways, the more "acceptable" gentrification
process and helped to make it contribute to the diversity of the
Centrum rather than to its homogenization, although this struggle is
clearly not yet over.

This contemporary residential rejuvenation of Amsterdam
requires some geohistorical explanation. Decentralization in the
1930s began emptying the inner city of offices and manufacturing
employment, and postwar suburbanization continued the process in
a heightened flow of residential out-migration not just to the poly-
centered urban fringe but beyond, to such Christallerian new towns
as Almere and Lelystad, planned and plotted in hexagonal lattices
on the reclaimed polders of isotropic Flevoland.[6] As has happened in
every century after the Golden Age, the continued life and liveliness
of the Centrum was threatened by exogenous forces of modernization. A turning-point, however, was reached in the 1960s, as cities
exploded all over the world in often violent announcements that the
postwar boom's excesses were no longer tolerable to the under-
classes of urban society. A contrapuntal process of urban restructuring was initiated almost everywhere in an effort to control the
spreading unrest and to shift economic gears in an attempt to
recover the expansionary capitalist momentum. A comparison of the

[6] The descriptive terms here – Christallerian, hexagonal, isotropic – refer to the
"central place theory" that became so widely espoused by quantitative geographers
in the 1970s and just as widely applied by urban and regional planners all over the
world. The Dutch experiment with such imaginative spatial planning was particularly interesting.

urban restructuring of Amsterdam and Los Angeles over the past 25 years becomes appealing at this point, for each has in its own way been paradigmatic. But I do not wish to leave my Spuistraat vantage-point too quickly, for it continues to be revealing.

The contemporary residential renaissance of Amsterdam's Centrum, more effectively than any other place I know, illustrates the power of popular control over the social production of urban space. It has been perhaps the most successful enactment of the anarcho-socialist-environmentalist intentions that inspired the urban social movements of the 1960s to recover their "rights to the city," *le droit à la ville*, as it was termed by Henri Lefebvre, who visited Amsterdam many times and whose earlier work on everyday life inspired the Amsterdam movements. As mentioned in chapter 1, Lefebvre was particularly influential in the COBRA (Copenhagen–Brussels–Amsterdam) movement that formed in 1949 to reject the arrogantly rational modernization of state planning in the immediate postwar period and to release the pleasure of art in popular culture and everyday life. COBRA disbanded in 1951, but its inspiration and distinctive brand of situationism continued to live on, especially in Amsterdam.

More familiar contemporary paths of urban restructuring can be found in and around Amsterdam, but the Centrum's experience verges on the unique. Uncovering this uniqueness is difficult, for it has been covered over by more conventional wisdoms, right and left, that see today only either a continuation of "creatively destructive" decentralization emptying the urban core of its no longer needed economic base (and hence necessitating more drastic forms of urban renewal to fit its new role); or the defeat and co-optation of the most radical urban social movements by the governing powers (leading too easily to a sense of popular despair over what now is to be done in these once radically open but now closing spaces of resistance). Both views can be argued with abundant statistics and effective polemics, but when seen from the outside, in a more comparative and global perspective on the past 25 years of urban restructuring, a third view of Amsterdam emerges.

In 1965, while Watts was burning in Los Angeles, a small group of Amsterdammers called the Provos (after their published and pamphleted "provocations") sparked an urban uprising of radical expectations and demands that continues to be played out on Spuistraat and elsewhere in Amsterdam's "magical center" of the world. The Provos were activated in the previous summer and had rallied their famous "happenings" nearly every Saturday evening around *Het Lieverdje*, a bronze statue of a smiling street urchin that still stands in Spui square. At first the movement focused, with conscious irony, on an anti-tobacco campaign (the statue had been donated by a local cigarette manufacturer), but soon the Provos' provocations spread to anti-war, anti-nuclear, and anti-pollution protests.

Their "White Bikes Plan" (whereby publicly provided bicycles would be available for free use throughout the city) symbolized the growing resistance to automobile traffic in the Centrum that would far outlive the plan itself.[7] Today, the network of bicycle paths and the density of cyclists is probably the highest in any major industrial or even post-industrial city; urban planners routinely publicize their distaste for automobile traffic while flexibly accommodating its inevitability; and the people continue to take free public transport by simply not paying on the subway, tram, or bus. If the free-riders are caught (by characteristically soft enforcers, usually unemployed youth hired as fare checkers), they make up names, for the Dutch were unique in pre-1992 Europe in having no official identification cards. Driving licenses, the universal stamp and regulator of personal identity in America, are nearly superfluous in the Netherlands and certainly not open to easy inspection. Integration into the European Community is today forcing the introduction of identity cards, but depend on the Dutch to find ways to keep them out of sight.

The Provos concentrated their eventful happenings in both Dam and Spui squares and managed to win a seat on the City Council, indicative of their arousal of wider public sympathies. Their creative challenges to hierarchy and authority lasted only for a few years, but they set in motion a generational revolution of the "twentysomethings" (my term for the youthful households comprised mainly of students between the ages of 20 and 30 that today make up nearly a quarter of the Centrum's population) that would dominate the renewal of the Centrum over the next two and a half decades. In no other major world city today are young householders, whether students or young professionals, in such command of the city center.

After 1967 the movement was continued by the Kabouters (Sprites or Gnomes), who not only promoted a complete ban on cars in the Centrum but developed plans for a full-scale greening of Amsterdam, with city-based farms, windmill-generated electricity, more open-space green belts, and special ministries for the elderly and the poor. In 1970, the "Orange Free State" was declared as an alternative popular government rallying around the last Provo city councillor, a key figure in the movement who was named ambassador to the "old state" and who today sits again on the Amsterdam

---

[7] A pollution-free electric car was also designed by a Provo engineer as another means of raising public consciousness against that smoke-belching symbol of consumer society. As far as I can tell, the electric vehicle was never built.

City Council as representative of perhaps the most radical anarchist-Green party in Europe. A few years ago, when it came time to assign a council member to oversee the still-being-negotiated plans to construct a luxury office and upscale housing development in the old Oosterdok waterfront – Amsterdam's anticipated version of London's Docklands or New York's Battery Park City – the same radical anarchist environmentalist became the obvious choice. No better symbol can be found of the continuing impact of the twentysomethings: compromised to be sure, far from having any absolute power, but nevertheless aging with significant virtue, commitment, and influence.

The final renewal came with the full-scale squatter or *kraken* movement, beginning in 1976. The squatters launched their famous "No Housing No Coronation" campaign in 1980 and, for a few days, occupied a building near Vondel Park, declaring the site "Vondel Free State." Armed with helmets, iron bars, and stink bombs, the Vondel state squatters were eventually defeated by an army of 1,200 police, six tanks, a helicopter, several armored cars, and a water cannon. After 1980, the movement did not decline so much as become a more generalized radical pressure group protesting against all forms of oppression contained within what might be called the specific geography of capitalism, from the local to the global scales. Squatters, for example, merged into the women's movement, the anti-nuclear and peace movements, the protests against apartheid (a particularly sensitive issue for the Dutch) and environmental degradation (keeping Amsterdam one of the world's major centers for radical Green politics); as well as against urban speculation, gentrification, factory closures, tourism, and the siting of the Olympic games in Amsterdam.

The greatest local success of the radical squatter movement was ironically also the cause of its apparent decline in intensity and radicalness. This was to keep the right to accessible and affordable housing at the top of the urban political agenda by "convincing the local authorities of the urgency of building more housing for young households and of prohibiting the destruction of cheap housing in the central city for economic restructuring, gentrification, or urban renewal."[8] Nowhere else did so much of the spirit of the 1960s penetrate so deeply into the urban planning practices of the 1980s and 1990s.

---

[8] Virginie Mamadouh, "Three Urban Social Movements in Amsterdam: Young Households in the Political Arena Between 1965 and 1985," revised version (September 1989) of a paper presented at the conference on "The Urban Agglomeration as Political Arena," Amsterdam, June 1989, p. 15. My thanks to Ms Mamadouh for giving me a copy of this informative paper and for chatting with me about urban social and spatial movements in Amsterdam.

The population of Amsterdam peaked around 1965 at over 860,000. Twenty years later the total had dropped to a little over 680,000, but the Centrum had already begun to grow again and after 1985 so has the city as a whole. There are many factors that affected this turnaround, but from a comparative perspective none seem more important than that peculiar blend of democratic spatial planning and regenerative social anarchism that has preserved the Centrum as a magical center for youth of all ages, a stimulating possibilities machine that is turned on by active popular participation in the social construction of urban space. As the prospects for urban social justice seem to be dimming almost everywhere else today, there remains in Amsterdam a particularly valuable embarrassment of geohistorical riches.

Off Spuistraat

Anybody who grows up in Amsterdam invariably finds himself (*sic*) in the area of tension between the imprisonment of the ring of canals and the centrifugal escape via the exit roads of the city. In this area of tension one may not even know major parts of the city and still be an Amsterdammer. . . .

From my very first visit to Los Angeles in the early 1970s, I have had the feeling that the major Dutch cities (with Amsterdam in the lead) deny out of sentimental considerations the fact that they are part of a larger whole (an area as large and diffuse as Los Angeles) and as such completely ignore a dimension of an entirely different order from the one which they traditionally know. (Dutch architect Rem Koolhaas, in *Amsterdam: An Architectural Lesson*, 1988: 112)

At first glance, a comparison of Los Angeles and Amsterdam seems as impossible as comparing popcorn and potatoes. These two extraordinary cities virtually beg to be described as unique, incomparable, and of course to a great extent they are. But they are also linkable as opposite and apposite extremes of late 20th-century urbanization, informatively positioned antipodes that are almost Foucauldian inversions of one another yet are united in a common and immediate urban experience. Let me annotate some of the more obvious oppositions.

Los Angeles epitomizes the sprawling, decentered, polymorphic, and centrifugal metropolis, a nebulous galaxy of urbs in search of the urban, a place where history is repeatedly spun off and ephemeralized in aggressively contemporary forms. In contrast,

Amsterdam may be the most self-consciously centered and historically centripetal city in Europe, carefully preserving every one of its
golden ages (and also those of other enlightened urban spaces elsewhere) in a repeatedly modernized Centrum that makes other rem-
nant mercantile capitalist "Old Towns" pale by comparison. Both
have downtowns of roughly comparable area, but only 1 of 150
Angelenos live in the City's center, whereas more than 10 percent of
Amsterdammers are Centrum dwellers.

Many residents of the City of Los Angeles have never been downtown and experience it only vicariously, on television and film. Very
few now visit it to shop (except for excursions to the garment district's discount stores and the teeming Latino mercados along
Broadway); and surprisingly few tourists take in its local attractions,
at least in comparison to more peripheral sites. Amsterdam's
Centrum receives nearly 8 million tourists a year and is packed daily
with many thousands of shoppers. Amsterdammers may not be
aware of the rest of the city, but they certainly know where the center can be found.

It has been claimed that nearly three-quarters of the surface space
of downtown Los Angeles is devoted to the automobile and to the
average Angeleno freedom and freeway are symbolically and often
politically intertwined. Here the opposition to Amsterdam's
Centrum, second only to floating Venice in auto-prohibition, is
almost unparalleled. It is not the car but the bicycle that assumes, for
the Amsterdammer, a similarly obsessive symbolic and political role,
but it is an obsession filled not with individualistic expression and
automaniacal freedom as much as with a collective urban and environmental consciousness and commitment. This makes all the contrasts even more stark.

Amsterdam's center feels like an open public forum, a daily festival of spontaneous political and cultural ideas played at a low key,
but all the more effective for its lack of pretense and frenzy. Its often
erogenously zoned geography is attuned to many different age
groups and civically dedicated to the playful conquest of boredom
and despair in ways that most other cities have forgotten or never
thought possible. Downtown Los Angeles, on the other hand, is
almost pure spectacle, of business and commerce, of extreme wealth
and poverty, of clashing cultures and rigidly contained ethnicities.
Boredom is assuaged by overindulgence and the bombardment of
artificial stimulation; while despair is controlled and contained by
the omnipresence of authority and spatial surveillance, the ultimate
in the substitution of police for *polis*. Young householders are virtually non-existent. In their place are the homeless, who are coming
close to being half the central city's resident population despite vigorous attempts at gentrification and dispersal.

In compact Amsterdam, the whole urban fabric, from center to periphery, is clearly readable and explicit. From its prime axis of the Damrak and Rokin (built above the former route of the Amstel River), the city unfolds in layers like a halved cross-section of an onion, first in the "old side" and "new side" (with an Old Church and New Church appropriately placed and streets named by their position inside or outside the old city walls), then in the neat crescents of the ringing canals from the inner to the outer Singel girdles, and finally in segments and wedges of inner and outer suburbs, many of which are helpfully named Amsterdam North, East, Southeast, South, and West. This morphological regularity binds Amsterdammers to traditional concepts of urban form and function and encourages its urbanists to be somewhat cautious when confronted with new theories of urban transformation.

In comparison, Los Angeles seems to break every rule of urban readability and regularity, challenging all traditional models of what is urban and what is not. One of America's classic old suburbias, the San Fernando Valley, is almost wholly within the jigsawed boundaries of the monstro-City of Los Angeles, while many inner city barrios and ghettoes float outside on unincorporated county land. There is a City of Industry, a City of Commerce, and even a Universal City, but these are not cities at all. Their combined populations would be outnumbered by the weekday shoppers on the Kalverstraat. Moreover, in an era of what many have called post-industrial urbanization, with cities being emptied of their manufacturing employment, the Los Angeles region continued its century-long boom in industrial growth, in both its core and periphery, at least until the past few years. It is no surprise, then, that Southern California has become a center for innovative and non-traditional urban theory, for there seems little from conventional, established schools of urban analysis that any longer makes sense.

And then there is that most basic of urban functions, housing. One of the most interesting features of the success of the squatter movement in Amsterdam has been the absence of a significant housing shortage. Although much of the Centrum is privately owned, the rest of the city is a vast checkerboard of public, or social housing, giving Amsterdam the distinction of having the highest percentage of public housing stock of any major capitalist city, around 80 percent (although here too there has been a slight turnaround in recent years). Even what the Dutch planners consider the worst of these social housing projects, such as the huge Bijlmermeer highrise garden suburb, has served effectively to accommodate the thousands of migrants from Surinam and other former colonies during the 1970s in housing that appears embarrassingly substandard primarily to the Dutch.

As noted earlier, the squatter movement was more than just an occupation of abandoned offices, factories, warehouses, and some residences. It was a fight for the rights to the city itself, and for the right to be different, especially for the young and for the poor. Nowhere has this struggle been more successful than in Amsterdam. Nowhere has it been less successful than in Los Angeles. In the immediate postwar period, Los Angeles was poised to become the largest recipient of public housing investment in the country, with much of this scheduled to be constructed in or around downtown. In no other American city did plans for public housing experience such a resounding defeat by so ferociously anti-socialist campaigners. The explosion of ethnic insurrections in the 1960s and early 1970s cruelly accelerated the commercial renewal of downtown at the expense of its poor residential inhabitants. On the central city's "new side" grew a commercial, financial, governmental, and high cultural fortress, while the "old side," beyond the skyscraper walls, was left to be filled with more residual land uses: the tiny remnants of El Pueblo on the north, the fulsome Skid Row of cardboard tenements and streetscapes of despair in the center, the Dickensian sweatshops and discount marts of the expansive Garment District on the south, where immigrants continue to work even today in near-slave-like conditions.

And finally, there is the issue of racism. Amsterdam has been built on controlled tolerance and openness to many different "others," from the Huguenots and Jews of the Golden Age to the Surinamese, Turks, Sri Lankans, and Moroccans of today. The Indonesians who came in large numbers in the mid-century have been almost seamlessly absorbed into Dutch culture and citizenship. Their numbers are uncountable even if there was a population census (which there is not in the Netherlands, the Dutch being ever vigilant against government intrusion), for there is no "official" way to tell the difference between and among Dutch citizens of whatever color. To be sure, there remain some significant racial problems in Amsterdam today, especially as large numbers of migrants and refugees from all over the world have recognized the city as a particularly tolerant and safe haven. Amsterdam too has become an increasingly Third World city, and this brings with it the multiplication of potential friction points and situations.

The comparison with Los Angeles, however, is startling. Los Angeles was built on a bedrock of racism and radical segregation. Its history is marked by a continuous sequence of public and private sector partnerships which, at least since the Mexican conquest, have directly or indirectly induced the most violent interracial conflagrations that have occurred in any US city. There have also been, in the past and the present, significant examples of racial tolerance and peaceful coexistence. But these have been as much exceptions as they have been the rule in Amsterdam.

The core of my oppositional comparison is thus amply clear. But what of the periphery, or the "larger whole," as Rem Koolhaas called it? Are there comparative dimensions "of an entirely different order" that are missed when we concentrate only on the antipodal centralities of Amsterdam and Los Angeles? For the remainder of this contemporary comparison, I will set the two cities in a larger, more generalizable context that focuses on contemporary processes of *urban restructuring*. Here, the cities follow more similar paths than might initially seem possible. These similarities are not meant to contradict or erase the profound differences that have already been described, but to supplement and expand upon their emphatic and extreme particularity.

An evolving literature on contemporary patterns and processes of urban restructuring has identified a series of intertwined trends that have become increasingly apparent not only in Los Angeles but in most of the world's major urban regions. Each of these trends takes on different intensities and forms in different cities, reflecting both the normality of geographically uneven development (no social process is ever expressed uniformly over space and time) and the social and ecological particularities of place. Every city thus has its own specific trajectory through this contemporary urban restructuring. At the same time, however, there is a correlatable interconnectedness that nearly every city shares, a common or collective impact of restructuring that is defining an emergent new mode of urbanization in the contemporary world. These "new urbanization" trends will be examined in much greater detail in *Postmetropolis*, a companion volume to *Thirdspace*. Here I will introduce them to provide additional insight to the contemporary comparison of Amsterdam and Los Angeles.

It is appropriate to begin with the geographical recomposition of urban form, for this spatial restructuring process deeply affects the way we look at the city and interpret the basic meaning of urbanization. As with the other trends, there is a certain continuity with the past, lending credence to the argument that restructuring is more of an acceleration of existing urban trajectories than a complete break and redirection. The current geographical recomposition, for example, is in large part a continuation on a larger scale of the decentralization and polynucleation of the industrial capitalist city that was begun in the last half of the 19th century and became periodically accelerated as more and more population segments and economic sectors left the central city.

There are, however, several features of the recent round of polynucleated decentralization that suggest a more profound qualitative shift in the patterning of the urban fabric. First, the size and scale of cities, or more appropriately of urban regions, has been reaching unprecedented levels. The older notion of "megalopolis" seems increasingly inadequate to describe a Mexico City of 30 million inhabitants or a "Mega-York" of nearly 25, stretching from Connecticut to Pennsylvania. Never before has the focus on the politically defined "central" city become so insubstantial and misleading. Complicating the older form still further has been the emergence of "Outer Cities", amorphous agglomerations of industrial parks, financial service centers and office buildings, massive new residential developments, giant shopping malls, and spectacular entertainment facilities in what was formerly open farmland or a sprinkling of small dormitory suburbs. Neither city nor suburb, at least in the older senses of the terms, these reconcentrated poles of peripheral urban and (typically "high-tech") industrial and commercial growth have stimulated the new descriptive vocabulary that was explored in chapter 8 inside the "Exopolis" of Orange County.

The urbanization of suburbia and the rural margins of the metropolis is part of a recentralization of urban activities that is taking place at the same time as sprawling decentralization continues. This "concentrated deconcentration," to use a phrase associated with the literature on "growth poles" and regional planning, reflects the changing competitive positions of different areas within the broader metropolitan region (including not only "greenfield" sites but the inner cities as well) within an increasingly globalized economy. Over the past 25 years, the decentralization of manufacturing and related activities from the core of the older industrial capitalist cities has broken out from its national containment that is, from what were once regional or national "hinterlands." Jobs and factories continue to move to suburban sites or non-metropolitan areas within the national economy, but much more than ever before, they also migrate to hitherto non-industrialized regions of the old Third World, creating a new geographical dynamic of growth and decline that has been changing not only the long-established international division of labor but also the spatial division of labor within urban regions.

This geographical recomposition of the modern metropolis is paradigmatically clear in Greater Los Angeles. Within a radius of 60 miles (100 kilometers) from the "Centrum" of the misshapen City of Los Angeles there is a radically restructured regional metropolis of nearly 15 million people with an economic output and population size roughly equivalent to that of the Netherlands. At this scale, a new mode of comparing Amsterdam and Los Angeles emerges. Following Rem Koolhaas's "centrifugal escape" via the exit roads of the city, we can define another 100-kilometer circle from Amsterdam's Centrum that cuts through Zeeland well south of the Hook of Holland, touches the Belgian border near Tilburg, curves past Eindhoven (where Philips is headquartered) to touch the German border not far from Nijmegen, and then arcs through the heart of Friesland to the North Sea. Most of the nearly 15 million Dutch live within this densely urbanized region and its scale and productivity come remarkably close to matching its Southern California counterpart.[9]

The southwest quadrant of this "Greater Amsterdam" coincides rather neatly with the Randstad, which can, with a little stretching, be seen as the Netherlands' largest Exopolis or Outer City. Here, however, the defining central core is neither an old downtown nor the amorphous urban blobs that characterize Orange County. The "center of gravity" of the Randstad, to refer back to Allen Scott's analytics at the beginning of chapter 8, is a determinedly preserved rural and agricultural area known as the "Green Heart" of Holland.[10] Around the Green Heart are clustered the largest cities of the Netherlands: Amsterdam (700,000), Rotterdam (575,000), The Hague (445,000), and Utrecht (230,000), each experiencing a selective redistribution of economic activities between old central city, suburban fringe, and more freestanding peripheral urban centers. Taken as a whole, the Randstad contains the world's largest port (Los Angeles-Long Beach is now probably second), Europe's fourth largest international financial center (after London, Zurich, and Frankfurt) and fourth largest international airport (Schiphol, surpassed in traffic only by Frankfurt, Paris, and London).

Like the Greater Los Angeles regional metropolis, this Greater Amsterdam region has been experiencing a complex and geographically uneven decentralization and recentralization over the past 25

[9] In chapter 9 of *Postmodern Geographies* (1989), I took a bird's-eye tour above the circumference of the 60-mile (100-kilometer) circle around Los Angeles, discovering an amazing coincidence of military installations and federally owned land along the route. I have been told by some Dutch geographers that a similar circumferential geography of military and other government sites can be found 100 kilometers from Amsterdam. What can be made of such coincidences is an open question.

[10] There is another strange coincidence here, however, in that the Green Heart is a primary source of air and water pollution in the Netherlands. The dangerous effluents arise not from factories or automobiles but from the superabundance of livestock.

years.[11] Each of the largest cities are struggling to maintain their traditional centralities, while some favored peripheral sites blossom into new cities of their own and others remain blissfully rural. Many more detailed regional comparisons might be made along these lines, but I must leave this task to others. For present purposes, I wish only to make one final point about the specifically *regional* scale and scope of urban restructuring.

Too often, analyses of urban restructuring, even when attempting to add a significant cultural dimension to their prevailing political economy perspective, are presented almost exclusively as if everything that is significant occurs only in the spaces of the central city. The surrounding metropolitan fringe and the sprawling Outer Cities are rendered nearly invisible. A sort of "Manhattanitis" sets in, confining interpretation and more general conclusions to the center alone. Much can be learned from focusing on the Centrum and the CITADEL-LA, but to quote Koolhaas again, we must not deny "out of sentimental considerations" our positioning within a larger whole and thereby "ignore a dimension of an entirely different order."

Moving on in our comparative regionalism, another set of processes can be identified that is intricately related to the recomposition of urban form. They relate to something already alluded to: the increasing globalization of the regional metropolis and what the literature refers to as "world city formation." Amsterdam in its Golden Age was the prototypical model of an older world city, generated primarily from the global reach of mercantile capitalism. It has subsequently survived various phases of development and redevelopment to remain among the higher ranks of the new, contemporary world cities, whether combined in the Randstad or not. Los Angeles, of course, is representative only of the newer variety.

What distinguishes the global cities of today from those of the past is the *scope* of internationalization, in terms of both capital and labor. To the control of world trade (the primary basis of mercantile world cities) and international financial investment by the national state (the foundation of imperial world cities) has been added the transnational financial management of industrial production and producer services, allowing the contemporary world city to function at a global scale across all circuits of capital. First, Second, and Third World economies have become increasingly integrated into a global system of production, exchange, and consumption that is sustained by an information-intensive and "de-territorialized" hierarchy of

---

[11] See, for example, Marc de Smidt, "In Pursuit of Deconcentration: The Evolution of the Dutch Urban System from an Organizational Perspective," *Geografiska Annaler* 69B (1987), 133–43; and J. H. T. Kramer, "The Airport of Schiphol: Economic and Spatial Impact," *Tijdschrift voor Econ. en Soc. Geografie* 79 (1988), 297–303.

world cities, topped today by the triumvirate of Tokyo, New York, and London.

Los Angeles and Amsterdam are in the second tier of the restructured world city hierarchy. Before its recent recession, Los Angeles had been growing much more rapidly, leading some to predict that it would join the top three by the end of the century. Amsterdam has been more stable, maintaining its specialized position in Europe on the basis of its concentration of Japanese and American banks, the large number of foreign listings on its Stock Exchange (second in the world to London), the strong and long-established export-orientation of Dutch companies, and its control over Dutch pension funds, reputed to be 40 percent of Europe's total.[12] The banking and financial services sector remains a key actor in Amsterdam's Centrum, feeding its upscale gentrification and drawing strength from the information-rich clustering of government offices, university departments, cultural facilities, and specialized activities in advertising and publishing.

A characteristic feature of increasing internationalization everywhere has been an erosion of local control over the planning process, as the powerful exogenous demands of world city formation penetrate deeply into local decision-making. Without a significant tradition of progressive urban planning, Los Angeles has welcomed foreign investment with few constraints. Its downtown "renaissance" was built on foreign capital to such an extent that at one point almost three-quarters of the prime properties in the Central Business District were either foreign-owned or partially controlled by overseas firms. Before the current recession, the large number of Japanese firms in particular routinely contributed to local political campaigns and fund-raising cultural activities, and have even made loans to the City government to maintain its pension fund program. The internationalization of Amsterdam has been more controlled, as one would expect, but the continued expansion of the city as a global financial management center is likely to pose a major threat in the very near future to many of the "magical" qualities of the Centrum.

The other side of globalization has been the attraction of large numbers of foreign workers into almost every segment of the local labor market, but especially at lower wage and skill levels. Los Angeles today has perhaps the largest and most culturally diverse immigrant labor force of any major world city, an enriching resource not only for its corporate entrepreneurs but also for the cultural life of the urban region. Amsterdam too is fast approaching becoming a

---

[12] H. W. Ter Hart and J. Piersma, "Direct Representation in International Financial Markets: The Case of Foreign Banks in Amsterdam," *Tijdschrift voor Econ. en Soc. Geografie* 81 (1990), 82–92.

"majority minority" city, a true cosmopolis of all the world's populations. With its long tradition of effectively absorbing diverse immigrant groups and its contemporary socialist and socializing governance system, Amsterdam appears to have been more successful than Los Angeles in integrating its immigrant populations, socially and spatially, into the urban fabric.

One achievement is certain: the immigrants are better housed in Amsterdam, for Los Angeles is currently experiencing one of the worst housing crises in the developed world. As many as 600,000 people, predominantly the Latino working poor, now live in seriously overcrowded conditions in dilapidated apartments, backyard shacks, tiny hotel and motel rooms, and on the streets. While Dutch sociologists argue about whether there are any truly homeless people in Amsterdam (one count I heard of found 12 in 1990; there are many more now I am told), Los Angeles is often called the homeless capital of America, with as many as 120,000 people living on the streets for at least a few days in the year.

Intertwined with the geographical recomposition and internationalization of Los Angeles and Amsterdam has been a pervasive industrial restructuring that has come to be described as a trend toward a "postfordist regime of flexible accumulation" in cities and regions throughout the world. A complex mix of both deindustrialization (especially the decline of large-scale, vertically integrated, often assembly-line, mass production industries) and reindustrialization (particularly the rise of small and middle-size firms flexibly specializing in craft-based and/or high technology facilitated production of diverse goods and services), this restructuring of the organization of production and the labor process has also been associated with a repatterning of urbanization and a new dynamic of geographically uneven development.

A quick picture of the changing postfordist industrial geography of the contemporary (post)metropolis would consist of several characteristic spaces: older industrial areas either in severe decline or partially revived through adaptation of more flexible production and management techniques; new science-based industrial districts or technopolises typically located in metropolitan peripheries; craft-based manufacturing clusters or networks drawing upon both the formal and informal economies; concentrated and communications-rich producer services districts, especially relating to finance and banking but also extending into the entertainment, fashion, and culture industries; and some residual areas, where little has changed. It would be easy to transpose this typology to Greater Los Angeles, for much of the research behind it has been conducted there. Although postfordist restructuring has not gone nearly as far in Amsterdam, the transposition is also quite revealing.

Over the past 30 years, the Centrum has been almost entirely leached of its older, heavier industries and 25 percent of its former office stock has been lost, primarily to an impressive array of new subcenters to the southeast, south, and west (e.g. Diemen, Amstelveen, and Sloterdijk) and to the growing airport node at Schiphol (15 kilometers away), where more than 35,000 people are now employed. One might argue, as many Dutch observers do, that this dispersal represents a sign of major decline in the inner city, due in part to a shift from a concentric to a more grid-like pattern of office and industrial development.[13] Just as convincing, however, is a restructuring hypothesis that identifies the Centrum as a flexibly specialized services district organized around international finance and banking, university education, and diverse aspects of the culture and entertainment industries (fashion, especially for the twentysomethings, television and film, advertising and publishing, soft drugs and sex, and, of course, tourism, with perhaps the most specialized attractions in the world for the young and the poor visitor). Except for the University of Amsterdam, which is being pressured to reduce its space in the Centrum, each of these other specialty areas has started to reconcentrate in recent years in the increasingly information-intensive inner city.

Another trend needs to be added, however, before one goes too far in tracing the impact of postfordist industrial restructuring. This is the tendency toward increasing social and economic polarization that seems to accompany the new urbanization processes. Recent studies have shown that the economic expansion and restructuring of Los Angeles has dramatically increased poverty levels and hollowed out the middle ranks of the labor market, squeezing some job growth upward, to a growing executive-professional-managerial "technocracy" (stocked by the largest urban concentrations in the world of scientists, engineers, and mathematicians); and downward, in much larger numbers, to an explosive mix of the "working poor" (primarily Latino and other immigrants, and women, giving rise to an increasing "feminization of poverty") and a domestic (white, African-American, and Mexican-American, or Chicano-Chicana) "urban underclass" surviving on public welfare, part-time employment, and the often illegal opportunities provided by the growing informal, or underground economy.

This vertical and sectoral polarization of the division of labor is reflected in an increasing horizontal and spatial polarization in the residential geography of Los Angeles. Old and new wealth is increasingly concentrated in protected communities with armed

---

[13] H. J. Brouwer, "The Spatial Restructuring of the Amsterdam Office Market," *Netherlands Journal of Housing and Environmental Research* 4 (1989), 157–74.

guards, walled boundaries, "neighborhood watches," and explicit signs that announce bluntly: "Trespassers will be shot"; while the old and new poor either crowd into the expanding immigrant enclaves of the Third World City or remain trapped in murderous landscapes of despair. In this bifurcating urban geography, all the edges and turf boundaries become potentially violent battlefronts in the continuing struggle for the rights to the city.

This bleak and foreboding picture of contemporary Los Angeles can be inverted to describe a brighter side to urban restructuring, but such a flip-flopping description is too simple and would only dilute the need to recognize and respond politically to the urgent problems facing this still expansive, and still politically unresponsive, urban region. Here again, the Amsterdam comparison is both informative and ambiguously encouraging, for it too has been experiencing a process of social and economic polarization over the past two decades and yet it has managed to keep the multiplying sources of friction under relatively successful social control.

The Dutch "Job Machine," for example, shows a similar hollowing out of the labor market, with the greatest growth occurring in the low-paid services sector.[14] Official unemployment rates have been much higher than in the US, but this difference is made meaningless by the contrasts in welfare systems and methods of calculating the rate itself. Overall job growth over the past 30 years, as in almost every other OECD member country, has been much lower than in the US and, except for the producer (i.e. mainly financial) services sector, there has been a decline in high-wage employment, thus limiting the size of the executive-professional-managerial "bulge." Increasing flexibility in the labor market, however, is clearly evident in the growth of "temporary" and "part-time" employment, with the Netherlands having the largest proportion of part-time workers in the EC and perhaps the highest rate (more than 50 percent) in the Western World for women.[15]

Terhorst and van de Ven have examined the particular forms taken in Amsterdam by the widespread expansion of low-wage, often temporary and/or part-time services jobs.[16] Focusing on the revival of personal services, they gently provide evidence of social

---

[14] Pieter A. Boot and Maarten Veraart, "The Dutch Labour Market," *Tijdschrift voor Econ. en Soc. Geografie* 78 (1987), 399–403.

[15] Boot and Veraart, "The Dutch Labour Market," see note 14. The labor market participation rate for women in the Netherlands (34.5 percent in 1985) is relatively low for Europe and much below the figure for the US. Rates of growth in participation have been very high during the 1980s, however, as large numbers of women entered the labor market for the first time, especially in part-time jobs.

[16] Pieter Terhorst and Jacques van de Ven, "The Revival of Personal Services in Cities: An Expression of Polarization?" *Netherlands Journal of Housing and Environmental Research* 1 (1986), 115–38.

and economic polarization in Amsterdam and link this to restructuring processes elsewhere in the world. But again, there is a more human face to this polarization in Amsterdam. Personal services have flourished with expanding immigration, the entry of larger numbers of women into the part-time and/or low-wage workforce, the growth of the informal economy, and the gentrification of the Centrum, but the downside of this process has been ameliorated by, and indeed made to contribute to, the very special nature of the inner city area.

With its exceptional concentration of young, educated, often student households, high official levels of unemployment, still solid social security system, and two distinctive waves of gentrification (one fed by the high-wage financial services sector and the other by multi-job households typically comprised of former students and squatters still committed to maintaining the distinctive quality of life in the Centrum), an unusual synergy has developed around the personal services sector and between various age and income groups. As Terhorst and van de Ven point out, income polarization has been associated with a growing complementarity between the higher and lower income groups with respect to the flexible use of time and space, especially in the specialized provision of such personal services as domestic help and babysitting, late-night shopping, entertainment and catering, household maintenance and repair, educational courses and therapies, fitness centers, bodycare activities, etc.

Such activities in Amsterdam take place primarily in the underground economy and are not captured very well in official statistics. They are also not likely to be a major factor in stimulating rapid recovery from economic stagnation and crisis. But they none the less provide a legitimate and socially valuable "survival strategy" for the poor and unemployed that has worked effectively to constrain the extreme effects of social polarization that one finds in Los Angeles or New York City. Moreover, it is a strategy that draws from the peculiar urban genius of Amsterdam, its long tradition of grass roots communalism, its sensitive adaptation to locality, its continuing commitment to libertarian and participatory social and spatial democracy, and its unusual contemporary attention to the needs of the twentysomething generation.

There is, of course, a dark side as well to this revival of personal services in Amsterdam's Centrum and the Dutch analysts are always careful to point it out. But here again, the comparative perspective produces a different picture. Viewed comparatively, the restructuring of the Centrum over the past 25 years has produced two reconfigurations of the urban political economy that distinguish Amsterdam from other major urban regions. The first arises from what may be the most successful urban implementation of the anarchist, environmentalist, and situationist principles that mobilized the student and other social movements in the 1960s in cities all over the world. That this success is far from complete and continues to be constrained and challenged by both internal and external forces takes nothing away from its distinctive achievement relative to other urban regions.

The second distinctive quality of Amsterdam becomes clearer when conjoined with the first and linked to the current literature on flexible specialization and urban-industrial restructuring. Although it would require much more empirical analysis to demonstrate convincingly, there appears to be taking shape in the Centrum of Amsterdam a new kind of specialized agglomeration that is as reflective of current trends in the world economy as the technopolises of California, the craft-based industrial districts of the "Third Italy," or the concentrated nodes of global financial management and control in Lower Manhattan and the City of London. It is primarily a services agglomeration, although small-scale, technologically advanced, design and information intensive industries remain a significant part of the "complex." Producer services, especially in banking and finance, are also very important, for the Centrum remains the focus for a major world city of global capitalism. The most intense and decidedly flexible specialization, however, is locally focused, extraordinarily innovative, and more advanced in Amsterdam than perhaps anywhere else in the developed world. A large part of its innovativeness lies in the simultaneous preservation and modernization of a unique urban heritage and built environment on a lively human scale – no small achievement in this age of the megacity. That it has not yet become a Disneyfied theme park for tourists is another achievement rooted both in the power of geohistorical traditions of participatory democracy and socially responsible planning, and in the contemporary influence of several generations of vigorous twentysomething activism. This has produced a services complex of remarkable diversity and interpersonal sensitivity, in which basic needs take precedence over market demands to a degree difficult to find almost anywhere else. I have no idea what to call this specialized space, but however it might be categorized it is worthy of much greater attention, analysis, and appreciation.

There is still another restructuring trend that must be mentioned, having to do with postmodernization and the recomposing of the cultural logic and the urban imaginary in the socially polarized, postfordist, exopolitan, and increasingly carceral world city. As noted in earlier chapters, a neo-conservative form of postmodernism, in which "image" replaces reality and simulated and "spin-doctored" representations assume increasing political and economic power, has been significantly reshaping popular ideologies and everyday life all over the world and is fast becoming the keystone for a new mode of social regulation designed to sustain the development of (and control the resistance to) the new postfordist and globalized urban economies. After experiencing Amsterdam, where resistance to the imposition of this neo-conservative cultural and ideological restructuring seems exceptionally strong (although I hear it is less so in The Hague), it is tempting just to add another polar opposition to the comparison with Los Angeles, where this process is probably more advanced than almost anywhere else on earth. But I will leave the issue open for further research and reflection, and conclude this stroll Off Spuistraat with the strangely familiar suggestion that perhaps the entire text of this chapter – and all those before it – revolves around the political challenges implied in this inconclusive last paragraph.

Postscript I: On the Views from Above and Below

The journeys On and Off Spuistraat as well as the spiralling tours of the Citadel-LA and the Exopolis of Orange County raise interesting questions about the appropriate scale of critical urban analysis and interpretation. For example, do we learn more about Amsterdam or Los Angeles or any other real-and-imagined cityspace by engaging in microgeographies of everyday life and pursuing the local view from the city streets; or by seeing the city as a whole, conceptualizing the urban condition on a more comprehensive regional or macrospatial scale? By now, I hope the answer to this question is clear. Understanding the city must involve both views, the micro and the macro, with neither inherently privileged, but only with the accompanying recognition that no city – indeed, no lived space – is ever completely knowable no matter what perspective we take, just as no one's life is ever completely knowable no matter how artful or rigorous the biographer. The appropriate response to the micro vs. macro choice is thus an assertive and creative rejection of the either/or for the more open-ended both/and also. . . .

As we approach the conclusion of *Thirdspace*, it is useful to re-emphasize this point through another look back to the writings of Henri Lefebvre, for his view of the micro–macro relationship continues to be pertinent and challenging. I am assisted in this retrospective glance by the recent translation of Lefebvre's brief introduction to *Espace et politique* (1972), a collection of essays that served both as a companion volume to *Le Droit à la ville* (originaly published in 1968) and as a prelude to *La Production de l'espace* (1974).[17] In this

---

[17] See *Writings on Cities/Henri Lefebvre*, selected, translated and introduced by Eleonore Kofman and Elizabeth Lebas, Oxford, UK, and Cambridge, MA: Blackwell Publishers, 1996: 185–202. Kofman and Lebas give the date for *Espace et politique* as 1973, but my copy of the book published by Éditions Anthropos, *Le Droit à la ville/Espace et politique*, is dated 1972.

introduction to space and politics in the city, Lefebvre presents another of his dialectics of triplicity, one that recomposes the micro–macro relationship around a critical third term: what Lefebvre creatively described as *social space* and I have redescribed, via the trialectics of spatiality, as Thirdspace.

As Lefebvre always insisted, his theorization of social space encompassed "on the one hand the critical analysis of urban reality and on the other that of everyday life," with the former defining the urban condition writ large as a concrete abstraction of the macrospatiality of social life and the latter a similarly concrete abstraction for the socialized microspatiality of the urban. Lefebvre saw both everyday life and urban reality as global concepts and knowledges that are "indissolubly linked, at one and the same time products and production." They "occupy a social space generated through them," to which he added "and inversely" to emphasize that both are also shaped by this (socially produced) social space, giving his theorization its characteristic trialectical turn (1996: 185).

Lefebvre also makes it abundantly clear that this trialectic of critical urban analysis (everyday life–urban reality–social space) is all-encompassing and not reducible to a sociology or a history or a spatial science. "The analysis is concerned with *the whole of practico-social activities* [my emphasis] as they are entangled in a complex space, urban and everyday.... The global synthesis is realized through this actual space, its critique and its knowledge." Also clear is Lefebvre's assertive choice of (social) space as his transdisciplinary focal point, an opening window through which one can thread through all the complexities of the modern (and postmodern) world. His urban trialectic provides a way of thinking that explicitly spatializes not just the urban *per se*, but the "whole of practico-social activities" that he envelops under his broadened concept of the urban condition, at every scale from the most local to the most global. The urban thus becomes for Lefebvre something very much like what I have been describing as Thirdspace: a lived space of radical openness and unlimited scope, where all histories and geographies, all times and places, are immanently presented and represented, a strategic space of power and domination, empowerment and resistance.

For those who are still uncomfortable with such encompassing open-ended-ness, I remind you again of how easy it has been to attach a similarly infinite scope to the historical-cum-sociological imagination, to see everything that exists in the present or ever existed in the past as potentially knowable, at least in large part, through its embracing historicality and sociality. We are accustomed to grounding political praxis, the efforts to change the world for the better, in the contexts of making history and transforming society.

That spatiality and the spatial imagination, and how we engage in critical urban analysis and spatial praxis, are similarly scoped is the most important conclusion to be gained from *Thirdspace* and its journeys.

In *Espace et politique*, Lefebvre applied his critique of the polarizations that have developed around micro vs. macrospatial analysis specifically to the citybuilding professions, with the architect fixed primarily at the micro scale and the planner at the macro. He weaves into this critique a trenchant attack on the privileged powers of visualization and the urge to achieve legibility (*lisibilité*, or readability) that are rampant in the citybuilding professions.

> Legibility [which Lefebvre links to a process of coding-decoding] passes for a great quality, which is true, but one forgets that all quality has its counterpart and its faults. Whatever the coding, legibility is bought at a very high price: the loss of part of the message, of information or content. This loss is inherent in the movement which rescues from the chaos of tangible facts, a meaning, a single one. The emergence of this meaning breaks the network, often very fine and richly disorderly from which the elaboration began. It completes its erasure by *making another thing*. The snare of legibility is therefore everywhere. . . .
>
> (1996: 192)

Lefebvre extends his critique of legibility in ways that would be echoed in more recent writings on everyday life in the city.

> Visual legibility is even more treacherous and better ensnared (more precisely, ensnaring) than graphic legibilitiy, that is, writing. Every legibility stems from a paucity: from redundance. The fullness of text and space never go together with legibility. No poetry or art obeys this simple criterion. At best legibility is blank, the poorest of texts! (1996: 193)

As an alternative to these ensnarements, Lefebvre argued for the need to "determine the junction, the articulation of these two levels, of the micro and the macro, the *near* and *far order*, neighboring and communication," for it is at this scale "that nowadays thought can intervene and intervention be situated" (1996: 194).

I resurrect Lefebvre's calls for a trialectical articulation of the micro and the macro in a critical understanding of the fullness of social space because there has begun to develop in the burgeoning new field of contemporary critical urban studies, especially among those who take an explicitly postmodern position, a subtle, and at times not so subtle, privileging of the micro viewpoint. This privileging is expressed both in a disdain for more holistic readings of the urban condition and a tacit, if not exclusive, preference for the "view from below," the intimate ethnography–geography of everyday life exemplified best in the individual voice of the intensely localized *flâneur*. This is not quite the old division Lefebvre referred to between the architect and the planner, but it is having a similar fragmenting and divisive effect. For many cultural critics of urbanism, macro-viewing the city has become tightly associated with a modernist and masculinist hegemony of the visual, an authoritarian viewpoint that allegedly eradicates difference and heterogeneity, silences subaltern voices, obscures the immediacy and sensuality of urban life, and works only to reproduce its own academic and intellectual hegemony.

One of the most influential sources for this privileging of the "view from below" against any attempt to conceptualize the city-as-a-whole comes from the work of Michel de Certeau, especially from certain passages in *The Practice of Everyday Life* (1984). In his now well-known contemplation of New York City from atop the World Trade Center, de Certeau brilliantly excoriated the view from above as an arrogant voyeurism that homogenizes urban life under the "desire for legibility," the quest for an all-comprehending "signifying vision." Like Lefebvre's tracing of the origins of "abstract space," de Certeau looks back to the early historical development of visual perspective as an archetypal view from above. "The desire to see the city," he writes, "preceded the means of satisfying it. Medieval or Renaissance painters represented the city as seen in a perspective that no eye had yet enjoyed. This fiction already made the medieval spectator into a celestial eye." He goes on to emplant this celestial eye/*I* in a more contemporary site/sight.

> [The view from the World Trade Center] continues to construct the fiction that creates readers, makes the complexity of the city readable, and immobilizes its opaque mobility in a transparent text.... To be lifted to the summit of the World Trade Center is to be lifted out of the city's grasp. One's body is no longer clasped by the streets ... nor is it possessed, whether as player or played, by the rumble of so many differences.... The city's agitation is momentarily arrested by vision. The gigantic mass is immobilized before the eyes. It is transformed into a texturology in which extremes coincide.... The ordinary practitioners of the city live "down below," below the thresholds at which visibility begins.... These practitioners make use of spaces that cannot be seen ... elude visibility.
>
> His elevation [for this is inherently a male gaze] transfigures him into a voyeur. It puts him at a distance. It transforms the

bewitching world by which one was "possessed" into a text that lies before one's eyes. It allows one to read it, to be a solar Eye, looking down like a god. The exaltation of a scopic and gnostic drive: the fiction of knowledge is related to this lust to be a viewpoint and nothing more.

Is the immense texturology spread out before one's eyes anything more than a representation, an optical artifact? It is the analogue of the facsimile produced, through a projection that is a way of keeping aloof, by the space planner urbanist, city planner or cartographer. The panorama-city is a "theoretical" (that is, visual) simulacrum, in short a picture, whose condition of possibility is an oblivion and a misunderstanding of practices.[18]

In these passages, de Certeau recasts Lefebvre's critique of everyday life (from which he occasionally draws insight) into an exaltation of the view from below. But he does so at the expense of a more "elevated" perspective on urban reality and the fullness of Lefebvre's interpretations of social space. He may not have intended to so polarize the micro–macro relation, but nevertheless De Certeau's exaltations have deeply infiltrated the contemporary critical cultural discourse on cities, especially among some feminist urban critics who see a recalcitrant and masculinist voyeurism in nearly all macro-scale "big pictures" of urban life, preferring instead to focus exclusively the intimacies of the local, the body, the street, the everyday. These are powerful and revealing positionings, but at the same time they close off too much, unnecessarily reducing the scope and power of the critical spatial imagination.

We must realize that both the views from above and from below can be restrictive and revealing, deceptive and determinitive, indulgent and insightful, necessary but wholly insufficient. Each has its voyeurs and arrogant readers, its illusions of transparency and legibility. To set them up in antagonistic opposition only constrains critical interpretation and severely limits the possibilities for strategic intervention and radical spatial praxis. Such binarizations as micro vs. macro, the view from above vs. the view from below, as Lefebvre insisted, are never enough. *Il y a toujours l'Autre.* There is always an-Other view.

---

[18] The quoted passages are from Michel de Certeau, *The Practice of Everyday Life*, Berkeley and Los Angeles: University of California Press, 1984: 91–3. My discussion of de Certeau is informed significantly by an unpublished paper by Lawrence Barth, "Immemorial Visibilities: Seeing the City's Difference," draft copy, March 1995. Larry also visited me on Spuistraat in Amsterdam during my stay in 1990 and shared his impressions over drinks at a cafe on the Spui. I would also like to thank Olivier Kramsch for his probings into the work of de Certeau and for his more general research assistance for this book.

Postscript II: A Preview of Postmetropolis

Thirdspace was originally intended to continue on with further journeys to Los Angeles and other real-and-imagined places that would not only extend the discussions of the trialectics of everyday life–urban reality–social space but also apply more directly the characteristic approaches of Firstspace, Secondspace, and Thirdspace epistemologies and interpretive perspectives. Three additional chapters were planned, comprising Part III: Exploring the Postmetropolis, my term for the new concatenations of everyday life and urban reality that have been developing in the wake of thirty years of profound urban restructuring and postmodernization.

The first of these chapters, on the "Timelines and Contexts of Urban Restructuring," was to exemplify the richness of a Firstspace (macro)perspective in analyzing the empirical and geohistorical background to the contemporary postmodernization of Los Angeles, building upon the brief glimpses of the geohistory of Paris/El Pueblo contained in chapter 7. The second, "Representations of Los Angeles," was aimed at exploring the prevailing discourses that have been locally developed to make practical and theoretical sense of these urban restructuring processes, an evocation of the conceived Secondspaces of critical urban theory and an elaboration of the discussion of urban restructuring introduced earlier in the Off Spuistraat comparisons of Los Angeles and Amsterdam. The concluding chapter focused on the events of 1992 inside and outside Los Angeles and, infused primarily with a Thirdspace perspective, would selectively encompass the other two spheres of the spatial imagination (Firstspace and Secondspace) in an effort to open up a distinctive new interpretive realm of its own. As Part III expanded almost to the length of the preceding nine chapters of *Thirdspace*, it was decided to publish it as the core of a companion volume, *Postmetropolis*, to appear in early 1997. A brief preview of *Postmetropolis* thus serves both to close *Thirdspace* and open it up again to further elaboration and reinterpretation.

At the core of *Postmetropolis* is an argument that has only been hinted at in *Thirdspace*. It explores the possibility that the contemporary postmetropolis, exemplified primarily in Los Angeles, has become the scene of a dramatic and still unfolding transition from the crisis-generated restructuring that began some time in the late 1960s–early 1970s, to a *restructuring-generated crisis*, a crisis rooted in the new urbanization processes that have been reshaping the spatiality of urban life over the past thirty years. Two explosive moments of urban unrest punctuate this argument and provide both a local and global window through which to explore some of the specific strengths and weaknesses of the trialectics of spatiality defined in

Thirdspace. The first springs from the Watts Riots of 1965, one of the earliest of the urban uprisings that would spread to dozens of cities around the world in the 1960s. The second revolves around the riots and insurrection that occurred in Los Angeles in late May–early June 1992, after the court decision to acquit the policemen who captured and beat Rodney King.

There are many ways in which the events of 1992 can be seen as a continuation of the same forces and conditions that led to the Watts Riots of 1965. I trace these continuities, especially with respect to the long history of racism and racist cultural politics in Los Angeles. But I also explore and emphasize the significant differences between these two events, the one symbolic of the breakdown of the modern metropolis after two decades of post-war boom and the other signifying the beginnings of a breakdown of the postmodern metropolis after a long period of intensive and extensive restructuring. A similar contrast is explored in the postmodern bankruptcy of Orange County and the earlier, more characteristically modernist, fiscal crises that affected New York and many other cities beginning in the 1960s.

While the Second Los Angeles Uprising hovers in the background of all the chapters in *Thirdspace*, it rises to the forefront of *Postmetropolis*, guiding its exploratory journeys into the perceived, conceived, and lived spaces of Los Angeles. Here too we can be inspired by the stimulus of a little confusion to rethink the meaning of what has happened after the so-called Rodney King trial. I thus conclude *Thirdspace* and begin *Postmetropolis* with two prose poems that respond to the need for a global and local reinterpretation of the events of Los Angeles: 1992. The first is from Homi Bhabha.[19] The second is from my original conclusion to the extended manuscript for *Thirdspace*.

> This twilight moment
> is an in-between moment.
> It's the moment of dusk.
> It's the moment of ambivalence
> and ambiguity
> The inclarity,
> the enigma,
> the ambivalences,
> in what happened in the L.A.
> uprisings
> are precisely what we want to get hold of.

[19] Homi K. Bhabha, "Twilight #1," composed from a telephone conversation with documentary theatre artist Anna Deveare Smith, in Smith, *Twilight – Los Angeles, 1992*, New York: Doubleday Anchor Books, 1994: 232–4.

It's exactly the moment
when the L.A. uprisings could be something
else
than it was seen to be,
or maybe something
other than it was seen to be.
I think when we look at it in twilight
we learn
to. . .
we learn three things:
one, we learn that the hard outlines of what we see in daylight
that make it easy for us to order
daylight
disappear.
So we begin to see its boundaries in a much more faded way.
That fuzziness of twilight
allows us to see the intersections
of the event with a number of other things that daylight
obscures for
us,
to use a paradox.
We have to interpret more in
twilight,
we have to make ourselves
part of the act,
we have to interpret,
we have to project more.
But also the thing itself
in twilight
challenges us
to
be aware
of how we are projecting onto the event itself.
We are part of
producing the event,
whereas, to use the daylight
metaphor,
there we somehow think
the event and its clarity
as it is presented to us,
and we have to just react to it.
Not that we're participating in its clarity:
it's more interpretive,
it's more creative.

            *   *   *

To attempt a conclusion
using Iain Chambers words,[20]
LA 1992 was indeed
a "highly charged punctuation of the cosmopolitan script."
Like so many other events that have marked
the geohistory of the world
since 1989
(the fall of the Berlin Wall and the Cold War's disappearance,
the Gulf War, Tienanmen Square, the Salmon Rushdie affair,
the Columbian quincentenary, the rise of the NICs,
the rebalkanization of southeastern Europe,
the fragmented unification of a Europe of the regions,
and, most recently,
the downfall of Orange County
and the terrorism of Oklahoma City),
the Second Los Angeles Uprising
is igniting
a "new mode of thinking"
that is "neither fixed
nor stable,"
a mode of thinking instigated
by "recent apertures in critical thought"
and tensely pressured by
"the return of the repressed,
the subordinate
and the forgotten
in 'Third World' musics,
literatures,
poverties and populations
as they come to occupy the
economies,
cities,
institutions,
media
and leisure time
of the First World."
To quote Chambers yet again,
this new mode of (re)thinking the postmetropolis
is "open to the prospect of a continual return to events,
to their   re-elaboration
and   revision,"

---

[20] The quoted material is from Iain Chambers, *Migrancy, Culture, Identity*, London: Routledge, 1994: 3.

to a   "re-telling,
              re-citing
and   re-siting
of what passes for historical and cultural knowledge,"
to a   "re-calling
and   re-membering
of earlier fragments and traces that flare up and flash
in our present moment of danger,"
to "splinters of light that illuminate our journey
[to real-and-imagined places]
while simultaneously casting questioning shadows along the
path."

For bell hooks and so many
Others
it is an unconventional mode of rethinking race
and representation,
the relations between the colonizer
and the colonized,
and the alternative images that can work to subvert
the status quo
and change our worldviews
and lifeworlds
at the same time
and in the same
lived spaces.
It is also about   real-and-imagined
                                    bodies
                                                    cities
                                    texts
what Barbara Hooper calls   the "order in place"
about consciously   *spatial praxis*
that aims at   disordering
                              deconstructing
                                        reconstituting
but not destroying or eliminating   difference and reason
                                        identity and enlightenment
                                        philosophy and mastery
It is about   border work/la frontera
                                        a new cultural politics
                                        choosing the margin as a space
                                        of radical openness
                                        and hybridity
about finding meeting places
where new and radical happenings

can occur

about a politics of   deterritorialization - and - reconnection
a politics in which arguments over   **SPACE**   its enclosures
exclusions   internments
become subjects for debate and discussion,
and more important, for
resistance
and
transgression

It is about   postmodern culture
postmodern geographies
heterotopologies
the differences that postmodernity makes.

It is about   race
class
gender

and/also...
It is about   material spatial practices
representations of space
lived spaces of representation

and/also...
It is about   Firstspace
Secondspace
Thirdspace

and/also...

Only one ending is possible: **TO BE CONTINUED....**

Select Bibliography

Anzaldúa, Gloria 1987: *Borderlands/La Frontera.* San Francisco: Spinsters/Aunt Lute Press.
—— ed. 1990: *Making Face/Making Soul.* San Francisco: Spinsters/Aunt Lute Press.
Anzaldúa, G. and C. Moraga, eds 1981: *This Bridge Called My Back.* New York: Kitchen Table: Women of Color Press.
Appadurai, Arjun 1990: "Disjuncture and Difference in the Global Cultural Economy." *Public Culture* 2: 1–25.
Baudrillard, Jean 1988: *America*, London and New York: Verso.
—— 1987: *Forget Foucault.* New York: Semiotext(e).
—— 1983: *Simulations.* New York: Semiotext(e).
Bhabha, Homi K. 1994: *The Location of Culture.* New York and London: Routledge.
—— 1990a: "The Third Space." In *Identity, Community, Culture, Difference*, J. Rutherford, ed., 207–21.
—— 1990b: "The Other Question." In *Out There*, R. Ferguson et al., eds, 71–88.
—— ed. 1990c: *Nation and Narration.* New York and London: Routledge.
Borges, Jorge Luis 1971: *The Aleph and Other Stories: 1933–1969.* New York: Bantam Books.
Butler, J. and Scott, J., eds 1992: *Feminists Theorize the Political.* New York and London: Routledge.
Castells, Manuel 1977: *The Urban Question.* London: Edward Arnold.
Chambers, Iain 1994: *Migrancy, Culture, Identity.* London and New York: Routledge.
—— 1990: *Border Dialogues.* London and New York: Routledge.
—— 1986: *Popular Culture.* London and New York: Routledge.

Davis, Mike 1990: *City of Quartz*. London and New York: Verso.

—— 1987: "*Chinatown*, Part Two?" *New Left Review* 164: 65–86.

—— 1985: "Urban Renaissance and the Spirit of Postmodernism." *New Left Review* 151: 53–92.

Debord, Guy 1977: *The Society of the Spectacle*. Detroit: Black and Red.

de Certeau, Michel 1984: *The Practice of Everyday Life*. Berkeley and Los Angeles: University of California Press.

Deleuze, Gilles and Felix Guattari 1987: *A Thousand Plateaus*. Minneapolis: University of Minnesota Press.

—— 1983: *Anti-Oedipus*. Minneapolis: University of Minnesota Press.

Deutsche, Rosalyn 1991: "Boys Town." *Environment and Planning D: Society and Space* 9: 5–30.

—— 1988: "Uneven Development." *October* 47: 3–52.

Dreyfus, Hubert and Paul Rabinow, eds 1982: *Michel Foucault*. Chicago: University of Chicago Press.

Eco, Umberto 1986: *Travels in Hyperreality*. San Diego: Harcourt.

Fanon, Frantz 1967: *Black Skins/White Masks*. New York: Grove Weidenfeld.

Ferguson, R. et al., eds 1990: *Out There*. Cambridge, MA: MIT Press and New York: New Museum of Contemporary Art.

Foucault, Michel 1988: *Politics, Philosophy, Culture*, L. Kritzman, ed. New York and London: Routledge.

—— 1986: "Of Other Spaces." *Diacritics* 16: 22–7.

—— 1980: "Questions on Geography." In *Power/Knowledge*, tr. C. Gordon. New York: Pantheon: 63–77.

—— 1977: *Discipline and Punish*. New York: Pantheon.

Freidberg, Anne 1993: *Window Shopping*. Berkeley and Los Angeles: University of California Press.

Fuss, D., ed. 1991: *Inside/Out*. New York and London: Routledge.

Gilroy, Paul 1993: *Small Acts*. London and New York: Serpent's Tail.

Gómez-Peña, Guillermo 1993: *Warriors for Gringostroika*. St Paul: Graywolf Press.

Gregory, Derek 1995: "Imaginative geographies." *Progress in Human Geography* 19: 447–85.

—— 1994: *Geographical Imaginations*. Oxford, UK, and Cambridge, MA: Blackwell.

—— 1978: *Ideology, Science, and Human Geography*. London: Hutchinson.

Grossberg, L., Nelson, C. and Treichler, P., eds, *Cultural Studies*. New York and London: Routledge.

Halley, Peter 1987: "Notes on Nostalgia." In *Peter Halley: Collected Essays, 1981–1987*, Zurich: Bischofsberger Gallery.

Haraway, Donna 1992: "The Promises of Monsters: A Regenerative Politics for Inappropriate/d Others." In *Cultural Studies*, Grossberg, Nelson, and Treichler, eds, 295–337.

Harlow, Barbara 1992: "The Palestinian Intellectual and the Liberation of the Academy." In *Edward Said: A Critical Reader*, M. Sprinker, ed., 173–93.

Harvey, David 1989: *The Condition of Postmodernity*. Oxford, UK, and Cambridge, MA: Blackwell.

—— 1973: *Social Justice and the City*. Baltimore: Johns Hopkins University Press.

Hayden, Dolores 1995: *The Power of Place*. Cambridge, MA, and London, UK: MIT Press.

—— 1984: *Redesigning the American Dream*. Cambridge, MA: MIT Press.

—— 1982: *The Grand Domestic Revolution*. Cambridge, MA: MIT Press.

—— 1976: *Seven American Utopias*. Cambridge, MA: MIT Press.

Hebdige, Dick 1990: "Subjects in Space." *New Formations* 11: vi–vii.

Hess, Remi 1988: *Henri Lefebvre et l'aventure du siècle*. Paris: A.M. Métailié.

hooks, bell 1994a: *Outlaw Culture*. New York and London: Routledge.

—— 1994b: "Feminism Inside." In *Black Male*, T. Golden, ed. New York: Whitney Museum, of American Art and Harry N. Abrams Inc., 127–40.

—— 1990: *Yearning*. Boston: South End Press.

—— 1984: *Feminist Theory: From Margin to Center*. Boston: South End Press.

hooks, bell and Cornel West 1991: *Breaking Bread*. Boston: South End Press.

Hooper, Barbara 1992: "Split at the Roots." *Frontiers* 13: 45–80.

—— 1994: "Bodies, Cities, Texts: The Case of Citizen Rodney King," unpublished ms.

Jameson, Fredric 1984: "Postmodernism, or, The Cultural Logic of Late Capitalism." *New Left Review* 146: 53–92.

—— 1992: *Postmodernism, or, The Cultural Logic of Late Capitalism*. London and New York: Verso.

Jones, John Paul, Wolfgang Natter, and Theodore Schatzki, eds 1993: *Postmodern Contentions*. New York and London: Guildford Press.

Keith, Michael and Steven Pile, eds 1993: *Place and the Politics of Identity*. London and New York: Routledge.

Kern, Stephen 1983: *The Culture of Time and Space, 1880–1918*. Cambridge, MA: Harvard University Press.

Koolhaas, Rem 1988: "Epilogue." In *Amsterdam: An Architectural Lesson*, M. Kloos, ed. Amsterdam: Thoth Publishing House, 108–19.

Krause, Rosalind 1983: "Sculpture in the Expanded Field." In *The Anti-Aesthetic*, H. Foster, ed. Port Townsend WA: Bay Press.

Lefebvre, Henri 1996: *Writings on Cities/Henri Lefebvre*. Selected, translated and introduced by Eleonore Kofman and Elizabeth Lebas. Oxford, UK and Cambridge, MA: Blackwell.

—— 1992: *Éléments de rythmanalyse: Introduction à la connaissance des rythmes*. Paris: Éditions Syllepse.

—— 1991a: *The Production of Space*. Oxford, UK and Cambridge, MA: Blackwell.

—— 1991b: *Critique of Everyday Life – Volume I: Introduction*, tr. J. Moore, London: Verso.

—— 1980: *La Présence et l'absence*. Paris: Casterman.

—— 1976: *The Survival of Capitalism*, tr. F. Bryant, London: Allison and Busby.

—— 1975: *Le Temps des méprises*. Paris: Stock.

—— 1974: *La production de l'espace*. Paris: Anthropos.

Loukaitou-Sideris, Anastasia and Gail Sansbury 1996: "Lost Streets of Bunker Hill." *California History* 74: 146–62.

Lugones, María 1994: "Purity, Impurity, and Separation." *Signs* 19.

—— 1990: "Playfulness, 'World'-Travelling, and Loving Perception." In *Making Face/Making Soul*, G. Anzaldúa, ed., 390–402.

Massey, Doreen 1994: *Space, Place and Gender*. Oxford: Blackwell and Cambridge: Polity Press.

Minh-ha, Trinh T. 1991: *When the Moon Waxes Red*. New York and London: Routledge.

—— 1989: *Woman, Native, Other*. Bloomington: Indiana University Press.

Morris, Meaghan 1992: "The Man in the Mirror." *Theory, Culture and Society* 9: 253–79.

Olalquiaga, Celeste 1992: *Megalopolis*. Minneapolis: University of Minnesota Press.

Preziosi, Donald 1988: "La Vi(ll)e en Rose." *Strategies* 1: 82–99.

Price-Chalita, Patricia 1994: "Spatial Metaphor and the Politics of Empowerment." *Antipode* 26: 236–54.

Rabinow, P., ed. 1984: *The Foucault Reader*. New York: Pantheon.

Rabinow, Paul 1984: "Space, Knowledge, and Power." In P. Rabinow, ed. *The Foucault Reader*, 239–56.

Rose, Gillian 1993: *Feminism and Geography*. Cambridge: Polity Press.

Ross, Kristin 1988: *The Emergence of Social Space*. Minneapolis: University of Minnesota Press.

Rutherford, J., ed. 1991: *Identity: Community, Culture, Difference*. London: Lawrence and Wishart.

Said, Edward 1979: *Orientalism*. New York: Vintage.

Schama, Simon 1988: *The Embarrassment of Riches*. Berkeley and Los Angeles: University of California Press.

Scott, Allen J. 1988a: *Metropolis*. Berkeley and Los Angeles: University of California Press.

Serres, Michel 1982: *Hermes: Literature, Science, Philosophy*, J. V. Harari and D. F. Bell, eds. Baltimore: Johns Hopkins University Press.

Smith, Anna Deveare 1994: *Twilight – Los Angeles, 1992*. New York: Doubleday Anchor Books.

Soja, Edward W. 1996: "Planning in/for Postmodernity." In *Space and Social Theory*, G. Benko and U. Strohmayer, eds, Oxford, UK, and Cambridge, MA: Blackwell.

—— 1995: "Postmodern Urbanization." In *Postmodern Cities and Spaces*, S. Watson and K. Gibson, eds, 125–37.

—— 1993: "Postmodern Geographies and the Critique of Historicism." In *Postmodern Contentions*, J. P. Jones, W. Natter, and T. Schatzki, eds, 113–36.

—— 1992: "Inside Exopolis." In *Variations on a Theme Park*, M. Sorkin, ed., 94–122.

—— 1991a: "The Stimulus of a Little Confusion." Amsterdam: Center for Metropolitan Studies, Texts of a Special Lecture, 1–37.

—— 1991b: "Henri Lefebvre 1901–1991." *Environment and Planning D: Society and Space* 9: 257–9.

—— 1990: "Heterotopologies." *Strategies* 3: 6–39.

—— 1989: *Postmodern Geographies*. London and New York: Verso.

Soja, Edward W. and Barbara Hooper 1993: "The Spaces That Difference Makes." In *Place and the Politics of Identity*, M. Keith and S. Pile, eds, 183–205.

Sorkin, Michael 1992: *Variations on a Theme Park*. New York: Hill and Wang-Noonday Press.

Spain, Daphne 1992: *Gendered Spaces*. Chapel Hill: University of North Carolina Press.

Spivak, Gayatri Chakravorty 1990: *The Post-Colonial Critic*, S. Harasym, ed. New York and London: Routledge.

—— 1988: *In Other Worlds*. New York and London: Routledge.

Spivak, G. C. and Guha, R., eds 1988: *Selected Subaltern Studies*. New York: Oxford University Press.

Sprinker, M., ed. 1992: *Edward Said: A Critical Reader*. Oxford, UK, and Cambridge, MA: Blackwell.

Watson, Sophie and Kathy Gibson, eds 1995: *Postmodern Cities and Spaces*. Oxford, UK, and Cambridge, MA: Blackwell.

West, Cornel 1990: "The New Cultural Politics of Difference." In *Out There*, R. Ferguson et al., eds.

White, Hayden 1993: Review of *The Production of Space*, by Henri Lefebvre. *Design Book Review*: 29/30 (Summer/Fall 1993), 90–3.

—— 1987: *The Content of the Form*. Baltimore: Johns Hopkins University Press.

—— 1978: *Tropics of Discourse*. Baltimore: Johns Hopkins University Press.

Wilson, Elizabeth 1991: *The Sphinx and the City*. London: Virago Press.

Wolin, Richard 1992: *Walter Benjamin: An Aesthetic of Redemption*. New York: Columbia University Press.

Wright, Gwendolyn and Paul Rabinow 1982: "Spatialization of Power: A Discussion of the Work of Michel Foucault." *Skyline*: 14–15.

Name Index

Subject Index

Printed and bound by CPI Group (UK) Ltd, Croydon, CR0 4YY

27/10/2024

14580392-0001